Materials for Marine Systems and Structures

Treatise on Materials Science and Technology

VOLUME 28

MATERIALS FOR MARINE SYSTEMS AND STRUCTURES

EDITED BY

DENNIS F. HASSON

Department of Mechanical Engineering
U.S. Naval Academy
Annapolis, Maryland

C. R. CROWE

Naval Research Laboratory
Washington, D.C.

TREATISE ON MATERIALS SCIENCE AND TECHNOLOGY

VOLUME 28

ACADEMIC PRESS, INC.
Harcourt Brace Jovanovich, Publishers
Boston San Diego New York
Berkeley London Sydney
Tokyo Toronto

ACADEMIC PRESS, INC.
1250 Sixth Avenue, San Diego, CA 92101

United Kingdom Edition published by
ACADEMIC PRESS INC. (LONDON) LTD.
24-28 Oval Road, London NW1 7DX

Library of Congress Cataloging-in-Publication Data

Materials for marine systems and structures.
(Treatise on materials science and technology; v. 28)
Includes bibliographies and index.
1. Ocean engineering—Materials. I. Hasson, Dennis F. II. Crowe, C. R. (C. Robert) III. Series.
TA403.T74 vol. 28 620.1'1 s [620'.4162] 87-28979
[TC1650]
ISBN 0-12-341828-3

PRINTED IN THE UNITED STATES OF AMERICA

88 89 90 91 9 8 7 6 5 4 3 2 1

*TO OUR WIVES,
JEAN AND PAM*

Contents

4 Marine Corrosion

G. H. Koch, J. A. Beavers, and W. E. Berry

5 Corrosion Control

R. Heidersbach and J. S. Smart

6 Metallic Materials for Marine Structures

J. F. Jenkins

7 Concrete Marine Structures

B. C. Gerwick, Jr.

Contributors

Numbers in parentheses refer to the pages on which the authors' contributions begin.

J. A. BEAVERS (121), Battelle, Columbus Division, 505 King Avenue, Columbus, Ohio 43201

W. E. BERRY (121), Battelle, Columbus Division, 505 King Avenue, Columbus, Ohio 43201

C. R. CROWE (1), Naval Research Laboratory, Washington, D.C. 20375

R. A. DALRYMPLE (35), Ocean Engineering Group, Department of Civil Engineering, University of Delaware, Lewes, Delaware 19716

J. R. DEPALMA (89), U.S. Naval Oceanographic Office, NSTL, Mississippi 39522-5001

S. C. DEXTER (35), College of Marine Studies, University of Delaware, Lewes, Delaware 19958

B. C. GERWICK, JR. (351), Department of Civil Engineering, University of California, Berkeley, California.

D. F. HASSON (1), Department of Mechanical Engineering, U.S. Naval Academy, Annapolis, Maryland 21402

R. H. HEIDERSBACH (245), Department of Metallurgical Engineering, California Polytechnic State University, San Luis Obispo, California 93407

J. F. JENKINS (277), Materials Science Division, Naval Civil Engineering Laboratory, Port Hueneme, California 93043

N. KOBAYASHI (35), Ocean Engineering Group, Department of Civil Engineering, University of Delaware, Newark, Delaware 19716

G. H. KOCH (121), Battelle, Columbus Division, 505 King Avenue, Columbus, Ohio 43201

B. J. LITTLE (89), Naval Ocean Research and Development Activity, NSTL, Mississippi 39529–5004

M. M. SALAMA (389), Petroleum Research and Development Department, Conoco, Inc., Ponca City, Oklahoma

J. S. SMART (215), Amoco Production Company, PO Box 3092, Houston, Texas 77253

J. M. THOMAS (415), Failure Analysis Associates, 2225 E. Bayshore Road, Palo Alto, California 94303

Preface

The oceans cover almost three fourths of the earth's surface. Hidden within and below this important resource are a food-bearing ecosystem, and energy and mineral reserves, respectively. The ecosystem even controls our climate *à la* El Niño. In the forties of this century, it became technologically feasible to extract fossil energy by the construction of offshore drilling rigs. Also, improvements in ship construction by the use of tougher materials and improved welding techniques provided the capability for the safe shipment of oil. In addition, the same developments led to the fabrication of thin walled austenitic stainless steels for the bladders of liquefied natural petroleum gas (LNPG) ship carriers. An understanding of the various aspects of the ocean, such as, tides, salinity, temperature, oxygen concentration, wave characteristics, and wind and water currents, is mandatory to provide for structural reliability in high performance ships and offshore structures. One part of the approach of this treatise is to provide this environmental background, including marine corrosion and biofouling, as an information source. Since most ocean structures and systems are metallic, the other part this treatise emphasizes is the importance of metallurgical factors such as, mill practice, microstructure, thermomechanical treatment, and fabrication procedures (e.g., welding). Also included are corrosion behavior and protection. In addition, information is included on large concrete offshore structures. The overall approach is to emphasize the synergism between environment, materials, and design. The purpose of this treatise is to provide an integrated approach, utilizing the environmental information of the ocean scientists, materials science, and structural integrity principles as they apply to offshore structures and ships.

This volume is aimed at the professional scientist and engineer, as well as graduate students in the fields of ocean and marine engineering and naval architecture and associated fields (e.g., aquaculture and ocean mining). It includes an overview article by the editors on the treatise theme of materials and their performance in marine systems and structures. There are two articles on the marine environment: S. C. Dexter, R. A. Dalrymple, and N. Kobayashi discuss all aspects of the marine environment including winds, waves, ice, and chemical and biological characteristics; and B. J. Little and J. R. DePalma discuss and present examples of micro- and macrobiofouling. Marine corrosion, with the emphasis on metallic materials, is treated extensively by G. H. Koch, J. A. Beavers, and W. E. Berry. R. Heidersbach and J. S. Smart present methods for corrosion protection of marine structures. An overview of metallic materials in marine structures is presented by J. Jenkins and includes methods for materials selection along with their behavior in the surface, near surface, and deep ocean environments. Novel marine structural materials and mooring systems for advanced deep water offshore structures are discussed in the papers by B. C. Gerwick and M. M. Salama, respectively. The volume concludes with J. M. Thomas' article on all the elements, such as material selection, testing, design, construction, operation, inspection, and monitoring, that make up the state-of-the-art fracture control practices to insure the structural integrity necessary to provide reliable marine structures.

The editors express their sincere appreciation to the contributors who have taken valuable time from their demanding schedules and commitments to share their outstanding knowledge, accomplishments, and ideas. Their contributions come at a propitious time when our nation's natural mineral and energy resources need to be efficiently and safely extracted from their hidden source—the ocean.

Dennis F. Hasson
C. Robert Crowe

1

Materials and Their Performance in Marine Systems and Structures

DENNIS F. HASSON

U.S. Naval Academy
Annapolis, Maryland

and

C. ROBERT CROWE

Naval Research Laboratory
Washington, D.C.

I. Introduction

In recent years major changes have been made to the philosophy and practice used to design and fabricate marine structures. These changes have come about in part because marine structures have become larger, more complex in their design and fabrication, and more expensive to build and to maintain. The changes have also occurred because major advances in engineering have provided methodologies to account for complex material and structural behavior.

The transformation of materials into a successful marine structure is a multidisciplinary endeavor (Figure 1) that includes a design stage and a manufacturing stage. In the design stage, the usual engineering goal is to plan a structure which when fabricated will function in the manner intended, be

ISBN 0-12-341828-3

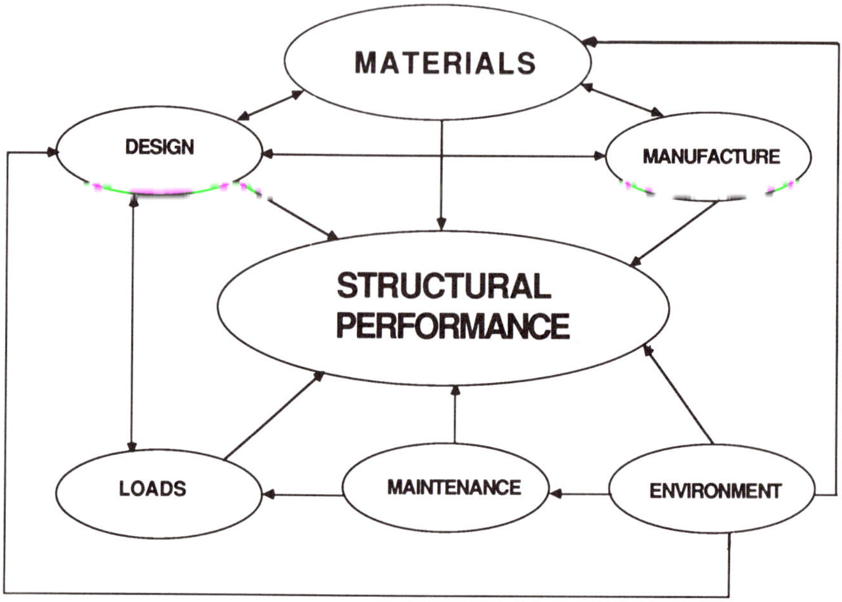

Fig. 1. Schematic showing the interactive nature of the parameters affecting structural performance in the marine environment.

economical to produce and maintain, be durable, and be reliable. The main design functions to assure these characteristics involve determining overall structural configuration, operating stress levels, local stress levels and stress concentrations, material selection, and selection of fabrication processes and finishes. The use of modern numerical methods for calculating static and dynamic stresses and the development of fracture mechanics for subcritical crack growth in fatigue, corrosion fatigue, and stress corrosion cracking (SCC) provide a means to evaluate and design for defect tolerance.

The materials selected for fabrication and the processes used to convert the mill products into the structure during fabrication play a significant role in the success of the structure. This is especially true for marine structures because of the strong effects of the marine environment on the performance of materials. Such factors as chemistry, microstructure, heat treatment, effects of welding, surface finish, and protective finish are important parameters. All of these factors relate in an interactive way to the performance of the structure in the physical and loading environments imposed on the structure during its lifetime. The key elements of each of the major areas are listed in Table 1.

TABLE I

SOME FACTORS AFFECTING STRUCTURAL PERFORMANCE

Design	Material	Manufacture	Environmental	
Stress levels	Composition	Material	Atmosphere	Conductivity
Dynamic response	Microstructure	Assembly methods	Humidity	pH
Configuration	Stiffness	Heat treatment	Air temperature	Calcerous deposits
Local geometry	Ductility	Joining	Sea temperature	Electrochemical potential
Materials selection	Strength	Mechanical finish	Dissolved oxygen	Wind and wave loadings
Method of joining	Toughness	Protective finish	Salinity	Tide
Finish selection	Weldability		Velocity	Biofouling
Corrosion protection	Homogeniety		Pressure	
Foundation properties	Fatigue strength			
Cost	Corrosion resistance			
	Creep behavior			

II. Types of Marine Structures

Large marine structures are of two basic types: ships and offshore platforms. A ship in a seaway is considered as an I beam or, more correctly, as a box girder, as shown in Figure 2. The analog for the deck is the top flange with the strakes carrying longitudinal and transverse I beams and stanchions loads down the web, past the neutral axes to the lower strakes, garboard, and keel. The supports are the buoyant forces of the sea, and the loads are the weight of the ship's structure and what the ship contains. In Figure 3a, the ship is supported by the waves, with the bow and stern each riding a wave crest and the midship region in a trough. The state of longitudinal bending stress is also shown in Figure 3a, and this condition, called sagging, places the deck in compression and the keel in tension. As the ship advances half a wave

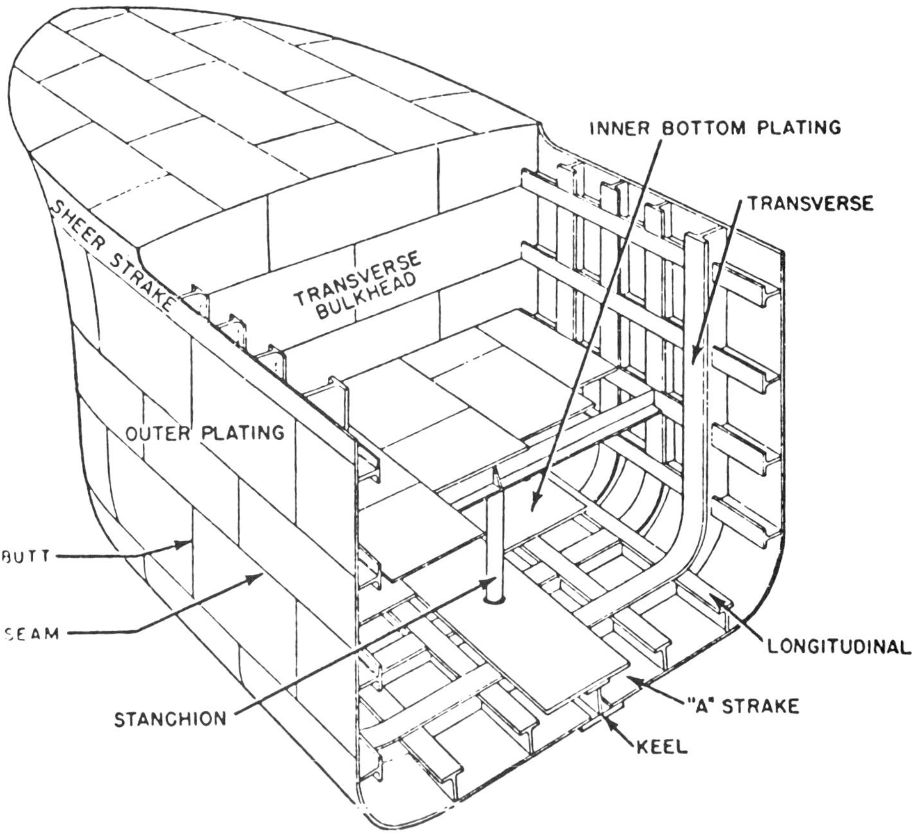

Fig. 2. Schematic of the cross section of a ship structure.

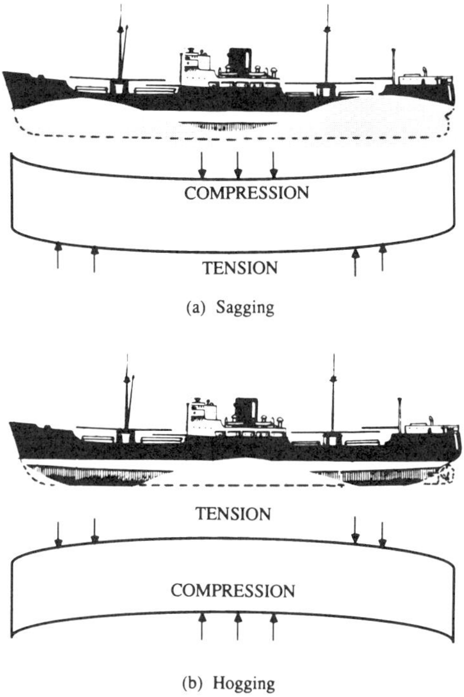

Fig. 3. Schematic showing the fatigue nature of wave loadings on ship structures.

length, another longitudinal bending stress state, called hogging, occurs that places the deck in tension, as shown in Figure 3b. This condition of alternating stresses from compression to tension is superimposed on the normal static stresses caused by the weight of the ship itself, unequal distribution of cargo, and nonuniform thermal expansion from temperature differences between air and water, and engine room heat. The above factors provide the complex stresses that are superimposed on all the stress risers in the ship such as hatches, cutouts, and passages and inherent weld flaws which can act as origins of fracture (*1*).

Details of ship hull design for U.S. ships are described in the American Bureau of Shipping (ABS) annual *Rules for Building and Classing Steel Vessels* (*2*). The specifications, including examples of calculations for structural details, are provided, and a section on welding and fabrication is also included.

Ship hull structures are examined after construction by surveys. Rules for surveys are also included in the ABS Rules. There is an annual survey for defects and structural integrity performed on the hull which includes such items as all hatchways, manholes, and scuttles in freeboard and superstructure decks, machinery castings, watertight bulkheads, bulkhead penetrations, crew rails, lack of improper alterations to the hull, and examination of hull thickness for deterioration.

The largest and most complex marine structures are offshore platforms. Offshore platforms are grouped into two categories: bottom supported platforms and floating platforms, as shown in Figure 4. Although over water drilling for crude oil took place along the California coast in the early 1900s, the first tower platform was not installed until 1947 off the Louisiana coast. The water depth was 20 feet. The structure was steel and weighted 1200 tons.

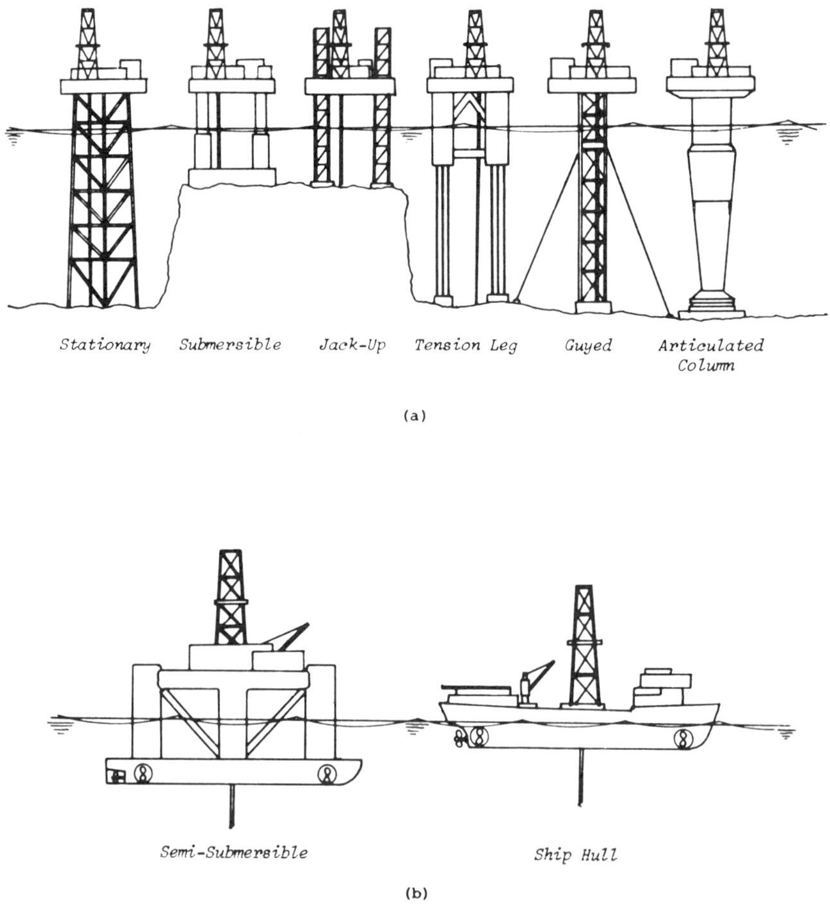

Fig. 4. Types of offshore platforms: (a) bottom supported platforms and (b) floating platforms.

As the large oil and gas fields in shallow and calm waters were identified and developed, platforms for deeper and more dangerous waters were constructed. Today there are over 10,000 offshore structures worldwide.

Offshore platforms are usually welded tubular steel space-frame structures that must have the ability to resist random fatigue loads from waves, large isolated loads from storms, and the constant aggressive corrosive attack of the environment. There are also some reinforced and prestressed concrete structures. The structural design aspects for offshore platforms are dictated by their location. The American Petroleum Institute (API) established detailed guidelines for the platforms located in the relatively shallow and calm waters of the Gulf of Mexico in the late 1960s, and API maintains the present U.S. standards (*3*). More recently, the ABS has been devising rules for offshore platform construction using a similar approach as for ship structures (*4*). Present developments in the North Sea for oil in the deeper Magnus field has initiated the need for new design considerations. The new requirements are directly related to the more severe environment of the North Sea compared with the Gulf of Mexico. This more severe weather in the North Sea is indicated by the height of the 100-year storm wave (i.e., the highest wave expected in 100 years). For the southern North Sea this wave height is 17 meters, while in the Gulf of Mexico it is 11 meters at the same water depth. For the North Sea Magnus oil field, the storm wave is 31 meters, whereas it would be only 26 meters at the same depth in the Gulf of Mexico. In addition, in the North Sea, the waters are both much deeper and exposed, and simultaneous wind gusts of 160 km/hr can be expected. Also the design water temperature is 22°C for the Gulf of Mexico and 10°C for the North Sea (*5*). The severity of the North Sea conditions has been responsible for specific structural requirements developed by Det Norske Veritas of Sweden (*6*) and the Department of Energy — Petroleum Engineering Division of Great Britain (*7*).

The modern offshore platforms and ships are designed and fabricated using the most advanced design techniques, alloys, and welding methods. In spite of this, complete structural reliability remains elusive.

III. Design Aspects

From the designer's point of view, the selection of materials for structural applications is based on the following five types of mechanical response (*8*):

(a) The strength and stiffness of undamaged, flaw-free material as reflected in the stress-strain response of the material to loads.

(b) Time-dependent material behavior such as creep, stress corrosion cracking, stress rupture, thermal fatigue, and material loss due to corrosion.

(c) The residual static strength of damaged, flaw containing material.

(d) Fatigue and corrosion fatigue of undamaged material.

(e) The life of damaged, flawed material.

The deformation of the structure during failure may follow any of five paths (*9*) shown schematically in Figure 5. These loading paths range from yielding dominated plastic collapse (path 4), graphically illustrated in the buckling failure of the M/T ALVENUS shown in Figure 6, to completely brittle failure (path 1) illustrated in the frequently published photograph of the USS Schenectady, which fractured while at dock following sea trials (Figure 7).

Until recently, the design of marine structures was based on the practice of using strength and stability criteria to configure the structure to resist yielding or buckling, with the emphasis on (a), (b), and (d) above. The primary material properties used in this approach are yield strength, modulus of elasticity, and fatigue endurance behavior (stress versus cycles to failure, S–N curves). In this regard, the material characteristics that control resistance to plastic flow are most important. These characteristics are grain size,

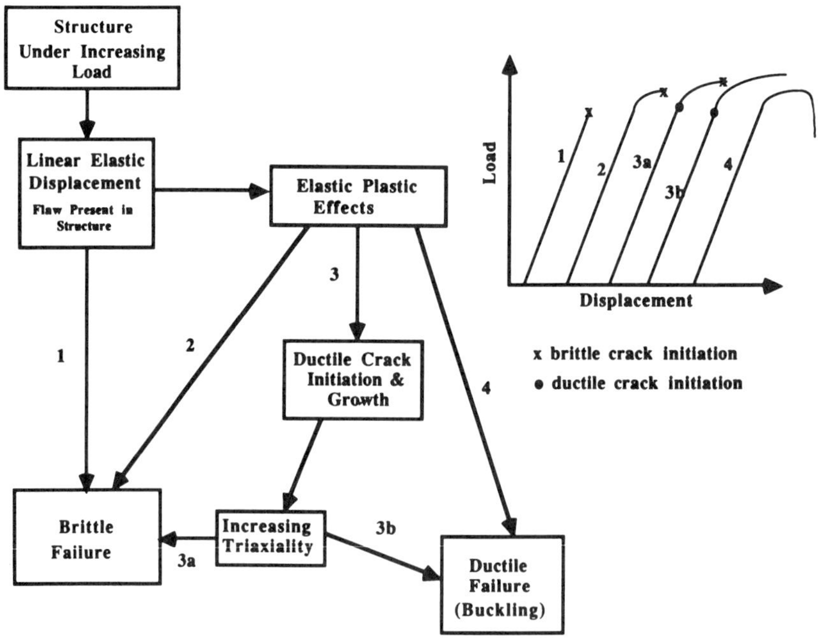

Fig. 5. The five loading paths to failure (adapted from ref. 9).

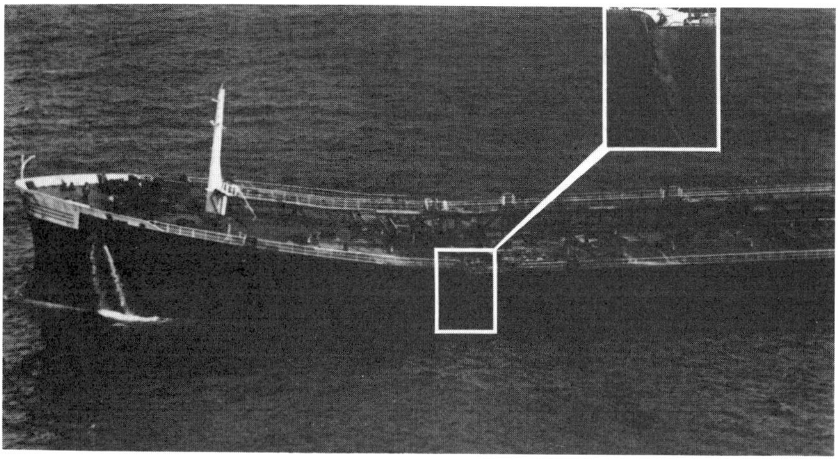

Fig. 6. Buckling failure of the M/T Alvenus (courtesy US Coast Guard).

precipitate size and distribution, dislocation structure, and point defect content—all characteristics connected with the microstructure of the material. If the structural member failed by yielding or buckling, as did the USS Alvenus, additional material was usually added to the load-bearing cross section to reduce the net section loads or decrease deflection of the structural member. If significant material was expected to be lost due to corrosion, section thickness was increased an additional amount to account for material loss with time.

Fig. 7. Brittle failure of the USS Schenectady (courtesy US Navy).

In the older design schemes, resistance to fracture was usually taken into account by assessing the impact properties of the materials. One common index used is the energy absorbed during impact of a Charpy-V notch specimen (*10*). The temperature dependence of the Charpy V-notch energy provides a means to establish the lower temperature limit for use of the material. Other common tests used to indicate fracture behavior are drop weight (*11*) and dynamic tear tests (*12*).

With the advent of fracture mechanics, however, more systematic approaches have been developed to account for fracture. In this regard, account is also taken of (c) and (e) above. For fracture dominated failure, macroscopic defects such as cracks, inclusions, weld flaws, porosity, forging laps, etc. are most important, because the essential feature of the fracture process is connected with the interaction of the local stress-strain field with the defect. Fracture mechanics provides a method to characterize failure based on stress analysis of a sharp crack. Although all flaws found in structures are not geometrically sharp, the worse cases occur for sharp cracks. Two analysis techniques are currently employed: linear-elastic and elastic-plastic. The linear analysis is used for materials with relatively low toughness for which the size of the crack tip plastic zone is small compared with the section thickness. The crack is characterized in the form of the stress intensity, K, which relates the stress to the crack size and geometry by a relation of the form:

$$K = \alpha\sigma\sqrt{a} \tag{1}$$

where α is a geometric constant, σ is the applied stress remote from the crack, and a is the crack length. Propagation of the crack leading to failure is assumed to occur at a critical value of the stress intensity, K_c, called the fracture toughness, which is a function of the section thickness. K_c is also related to the critical energy required for crack propagation:

$$K_c = (EG_c)^{1/2} \tag{2}$$

where G_c is the critical strain energy release rate for unstable crack propagation and E is Young's modulus of the material. The parameters K_c and G_c are functions of specimen thickness, but as the thickness increases and conditions of plane strain are established, K_c and G_c decrease to practically constant values that are characteristic of the material. For tensile loading of the crack (Mode I), these minimum values are called the plane-strain fracture toughness, K_{Ic}, and critical plane-strain energy release rate, G_{Ic}. The plane strain fracture toughness, K_{Ic}, therefore provides an index of the material's resistance to rapid crack extension. This index is independent of the crack length and section thickness. A standardized measurement of K_{Ic} is described in ASTM-E399-81 (*13*).

The elastic-plastic analyses are applicable to tougher materials which resist fracture. In this case, the extent of the plastic zone at the crack tip is larger with respect to the crack length and section thickness. This complicates the analysis because the fracture toughness index is not a single-valued function of the crack length and section thickness. Two methods to characterize elastic-plastic fracture are the *J* integral technique and the crack tip opening displacement technique (CTOD) used extensively by petroleum companies and shipyards to characterize toughness of welds.

The *J* integral technique assumes that deformation is reversible, either linearly or nonlinearly, and uses the energy integral:

$$J = \oint_{\Gamma} \left\{ W\,dy - T\left(\frac{\delta u}{\delta x}\right) \right\} ds \tag{3}$$

as the index of toughness. In this equation, W is the strain energy density evaluated over a contour, Γ, which surrounds the crack tip; T is the traction vector exterior to Γ; u is the displacement in the direction of T; and s is the arc length along the integration path. This integral, presented by Cherepanov (*14*) and Rice (*15*), can be related to the strain energy release rate, G, for nonlinear elastic deformation. For small-scale yielding in tensile load (Mode I), plastic deformation is approximated by nonlinear elastic deformation, thus

$$J_I = \frac{K_I^2}{E'} \tag{4}$$

where E' is Young's modulus, E, for plane stress, and $E/(1 - \nu^2)$ for plane strain and ν is Poisson's ratio of the material.

The CTOD technique is an extension of linear elastic fracture mechanics, which applies a correction for plastic deformation at the crack tip and uses the crack tip opening at the point of unstable crack propagation, CTOD, as the index of fracture toughness. The stress intensity at the crack tip is related to CTOD by:

$$\text{CTOD} = \left(\frac{K_I}{\sigma_y}\right)^2 e_y \tag{5}$$

where σ_y is the yield stress and e_y is the strain at yield. Methods to measure CTOD are provided in British Standard BS 5762: 1979 (*16*). Thus, a compatability exists in the toughness indices:

$$J_I = G_I = \frac{K_I^2}{E'}. \tag{6}$$

In relating fracture mechanics to the significance of defects, the defect tolerance parameters, CTOD/e_y and $(K_I/\sigma_y)^2$, are proportional to the critical flaw size a_c to cause failure by brittle fracture.

In marine construction, fatigue at welded joints is especially critical to structural integrity. Fatigue crack initiation, fatigue crack propagation, and residual strength of the material when cracks are present affect the failure of the structure. The performance of materials in fatigue is not a simple material behavior. The fatigue characteristics of materials are represented in an engineering sense by endurance limit curves where stress is plotted as a function of the log of the number of cycles to failure (S–N curves). The type of loading, the presence of flaws, and the microstructure of the material play a significant role is the material's performance.

The most common design approach to fatigue is based on the use of laboratory measured S–N curves and application of the Palmgren–Miner cumulative damage rule (*17*), where the loading spectra and the appropriate S–N curves are used to calculate the cumulative damage, D, from the relation:

$$D = \sum \left(\frac{n_i}{N_i} \right) \tag{7}$$

where n_i is the number of stress cycles corresponding to a discrete level of loading, and N_i is the number of cycles for the corresponding stress range of the S–N curve to cause failure. The criteria for failure is $D = 1$, so the structure is usually designed using a liberal margin of saftey for the total number of loading cycles expected over the design lifetime so that the effective D is >1.

A more modern design procedure, based on fracture mechanics, has been developed. The procedure is illustrated in Figure 8. The crack growth rate per loading cycle, da/dn, is measured as a function of applied stress intensity, ΔK, i.e., $da/dn = f(\Delta K)$, as schematically illustrated in Figure 8 step 2, and the total growth of the crack is computed by summing growth under each cycle of load:

$$a_n = a_0 + \sum f(\Delta K) \tag{8}$$

where a_n is the crack length after n cycles and a_0 is the initial crack length. A detailed review of these principles is given in reference (*18*).

The reliability of the Miner's rule and the fracture mechanics approach to the design of marine structures is controversial when converting laboratory measured S–N and da/dn versus ΔK curves to operational structural performance. This is because of the uncertainties associated with the presence of weld-related flaws, variability of the marine environment, uncertain loading from waves and storms, and other unknown or uncontrollable variables.

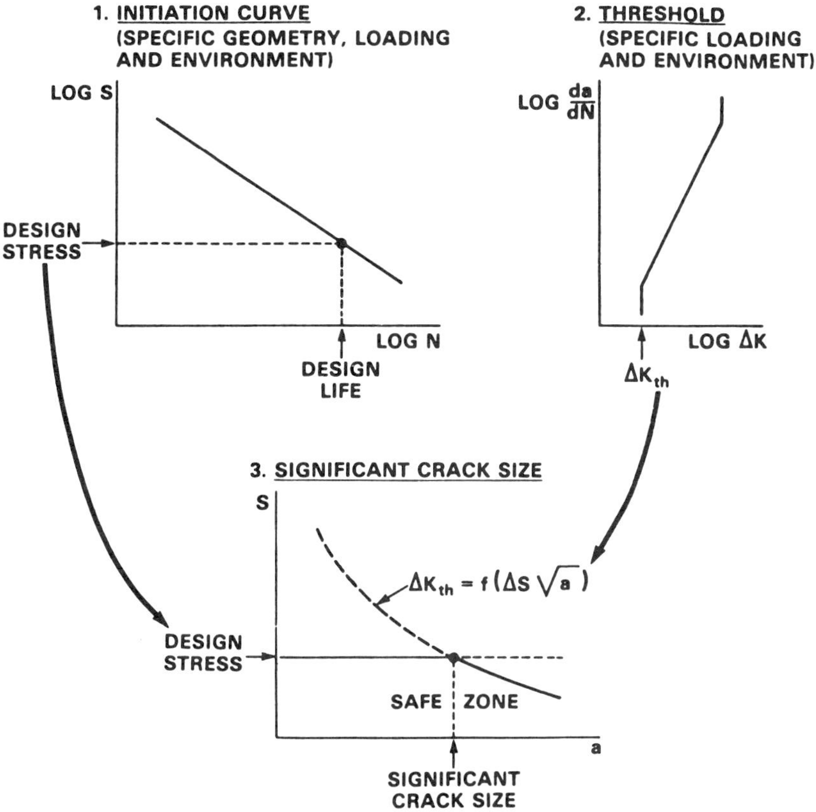

Fig. 8. Schematic showing the use of fracture mechanics in the design of structures for fatigue.

IV. Materials Selection

The selection of appropriate materials for fabrication and the implementation of appropriate corrosion protection plays an especially important role in the successful performance of a marine structure. Throughout history, extensive efforts have been made to find materials which can be used efficiently in marine structures. The earliest materials were woods which were used by prehistoric man in pilings for dwelling platforms raised over lakes. Wood satisfied the requirements for ship structures until the introduction of iron and steel in the nineteenth century. One of the earliest uses of metals in marine structures is traced to the Romans, who used lead to clad the bottoms of ships to reduce attack by wood-boring worms. It was years later, in the

1760s, that copper sheathing was used on the ship bottom of the HMS Alarm for the same purpose as lead was used by the Romans (*19*). It is interesting to note that some of the sheathing was attached to the hull of HMS Alarm by iron nails, which rapidly corroded from galvanic coupling to the copper, causing loss of some of the sheathing and degradation of other iron parts in contact with the copper. In subsequent uses of copper sheathing, the British Admiralty required use of copper nails for fastening the sheathing to the hull and lead sheathing to protect other iron parts in contact with the copper. Following this initial use, copper sheathing soon became commonplace, and it was quickly realized that the copper also stopped weeds and barnacles from attaching to the ship bottom, i.e., the copper, or more precisely the cuprous ion, acted as a fouling preventative.

In the nineteenth century, the industrial revolution resulted in an enormous increase in the use of metals for structural applications. By far the most common alloys used were of iron, and the use of steel is now commonplace. Iron barges were being built for use on inland waterways in England in the 1790s, and iron was being used for mooring cables around 1813. Iron vessels appeared around 1836 and the first publication of rules for the construction of iron ships was provided by Lloyd's Registry of Shipping in 1855 (*20*). By the mid nineteenth century iron was being used on warships for armor by France and Britain and the 172 foot ironclad gunboat, Monitor, possessing a unique revolving turret, provided a turning point in metal warship design and construction. By 1866 steels were being used in ship building.

Concrete has been used for over 2000 years as a structural material, often as a marine construction material. The nineteenth century saw the introduction of reinforced concrete by Joseph Lambot, a Frenchman, who fabricated a small ferrocement boat in 1848. Since this time, concrete has been used sporadically for floating marine structures and more extensively for pilings and bulkheads in harbor and coastal construction. During World War I, about 85,000 tons of sea-going shipping was constructed, and during World War II 488,000 tons of shipping was constructed. Since World War II, a number of ferrocement, prestressed concrete, and reinforced concrete barges, ship hulls, and pontoon structures have been built, but the major marine use of concrete has been in construction of fixed platforms for oil production and storage facilities, and in harbor and coastal construction (*21*).

Today, a large variety of materials are used to construct ships and offshore structures. The materials include various types of steels, non-ferrous alloys, concrete, and other materials. In the largest structures, the primary materials of construction are carbon-manganese steels of various grades. Also, reinforced and prestressed concrete is used in areas of greater water depth, such as in the North Sea, where severe environmental conditions exist and where proper ocean floor foundation characteristics exist which can support these massive structures. Specifications for the design and construction of prestressed and reinforced structures are found in *British Standard CP 110 Structural Use of Concrete* (*22*), and *Design and Construction of Concrete Structures*, published by Féderation Internationale de la Précontrainte (*23*).

Steel alloys are usually selected as materials of construction because of their relatively low cost, ease of fabrication by welding, and moderately good mechanical properties. The appropriate choice of structural steel should conform to the specifications for hull steels published by the classification societies. These steels are delivered from the steel mills with yield strength levels in the range of 270 MPa to 400 MPa in various grades within each strength level classification. The factors which affect the properties and quality of mill steels are shown in Table II. The various grades provide differing fracture toughness and weldability. Over the last ten years, however, major changes have occurred in the use of higher strength steels in construction of ships and offshore structures. The reasons for this increased usage at high stress levels is because of improvements in weldability achieved by thermal mechanical controlled processing (TMCP) of the steels. TMCP is a continuous method of producing steel plates that have reduced carbon equivalents. The carbon equivalent is an index which provides a means of assessing the weldability of steels based on the composition of the steel. Several carbon equivalent relations exist, the most common being the International Institute of Welding (IIW) index:

$$\mathrm{CE} = \frac{\mathrm{C} + \mathrm{Mn}}{6} + \frac{(\mathrm{Cr} + \mathrm{Mo} + \mathrm{V})}{5} + \frac{(\mathrm{Ni} + \mathrm{Cu})}{15} \tag{9}$$

where the concentration of each element (C, Mn, Cr, Mo, V, Ni, and Cu) is in weight percent. Carbon equivalents for normalized or controlled rolled steels are typically 0.40% to 0.42%. TMCP steels have carbon equivalents in the

TABLE II

FACTORS AFFECTING MICROSTRUCTURE OF MILL STEELS

Chemical composition
Melting practice
Type and extent of deoxidation
Special processing (vacuum degassing-desulphurization)
Grain refining
Rolling procedure (controlled—rolling—quench)
Heat treatment (normalizing—quench and tempering)

range 0.28% to 0.39%. The development of these fine-grained, clean, weldable structural steels have permitted their use at higher loadings over a wider temperature range. Today commercial treatments used in producing mill product forms usually assure that the steel will exhibit the properties attributed to that alloy when it is received by the customer. However, subsequent working or thermal treatments applied during secondary fabrication, especially during welding, can and often do alter the properties of the alloy and undesirable microstructures can result. The microstructure, in turn, is dependent upon the thermal and mechanical history of the alloy.

Rules for inspection and testing of hull materials are detailed in the American Bureau of Shipping Rules (*2*) along with the list and application of approved materials. Material details include steel grades, mill practices, compositions, fine grain practice, heat treatment, and tensile and impact specifications. For higher strength hull structural steel grades such as AH36, the frequency for Charpy-V notch impact testing of as rolled, normalized, controlled-rolled, or the new thermomechanical controlled process (TMCP) steels is specified. It is interesting to note that for commercial ship hulls, the Charpy-V notch test is still specified as the index of material toughness rather than the drop weight, dynamic tear, or fracture mechanics tests.

Improper selection of structural steels can lead to catastrophic consequences. This is especially true for ship structures, since ships are redundant stiffened plate structures in which brittle fracture, once initiated, may propagate to large lengths. The classic examples of this are the failures of the Liberty ships and T-2 tankers in World War II. Of 2500 Liberty ships built, 700 experienced failures, and 145 broke in two (Figure 7). These failures are attributed to using alloy compositions with high ductile to brittle transition temperatures and to welding-induced flaws. Structural integrity in ships can only be achieved by using crack-arresting steels.

Several important facts were learned from the ship failures and some of the more notable treatises are by Williams (*24*), who noted the interaction between stress concentration and notch sensitivity; Parker (*25*) and Biggs (*26*), who described the origin of the fractures at various ship locations, especially the older ships which had several crack origins at welded joints at the corner of square hatch openings; and Boyd (*27*), who summarized the lessons learned from the brittle fracture of the ships, with comments on the effects of stress, temperature, notches, thickness, welding, impact, and fatigue. Boyd's comment that thicker plates and sections are more liable to brittle fracture than thinner plates, due to the development of a triaxial stress state, appeared simultaneously with the early development of modern fracture mechanics for plane strain fracture conditions.

Other structurally efficient materials are aluminum and its alloys. Aluminum alloys are selected as materials of construction because of their high specific strengths, relative low cost, and to a lesser extent, their corrosion resistance. For these reasons, aluminum alloys have found considerable application in marine structures. The strength and corrosion resistance of aluminum alloys is also very composition and microstructure dependent. As a result, only a relatively few aluminum alloys are suitable for marine applications. One of the earliest uses of aluminum alloys in naval vessels was in the construction of the 60 foot torpedo boat, La Fonda, built for the French government by Yarrow and Company in England in 1894. The boat was constructed of an aluminum-6% copper alloy using triple riveting with aluminum rivets. Beginning in 1926, Al-Mg-Si-Cr alloys found increasing use in warship construction where there was, and still is, an urgent need for top-side weight savings (*28*).

As is now well known, the use of aluminum-copper alloys in sea water environments is not a good application because of the lack of corrosion resistance.

V. Welding

Because of relatively low fabrication costs and rapid construction, welding is extensively used in the fabrication of marine structures. This extensive use of welding has led to many welding related catastrophic failures which are a source of economic and human loss. An example of a modern welding related failure is the collapse of the mobile offshore platform Alexander Kielland in the North Sea on March 27, 1980. The Alexander Kielland was a drilling rig that had been converted to an accomodations platform. The rig capsized in heavy seas (20 m/sec winds and 6 to 8 m waves), killing 123 of the 212 on board. The cause of the disaster was attributed to a 6 mm fillet weld which joined a non-load bearing flange to one of the main braces (*29*).

The welding processes utilized for joining steels in the ship and offshore platform industries are shielded metal arc welding (SMAW): 50%; sub-merged metal arc welding (SAW): 15%; CO shielded gas metal arc welding (GMAW): 30%; with gas tungsten arc welding (GTAW) and electroslag welding (ESW) providing the remainder. GMAW is used more for the high strength quenched and tempered steels, while GTAW is used on a limited basis to remelt the weld toe to improve fatigue resistance of welds. ESW is used for high weld metal deposition rate in thick sections (*30*). Another process currently in development is narrow-gap welding. All aspects of welding fabrication processes, equipment, and filler materials are described in a multi-volume set of references from the American Welding Society (AWS) (*31*) and Volume 6-*Welding Brazing and Soldering* from ASM International (*32*).

Weldability is defined as the capacity of the metal to be joined satisfactorily. The criteria indicating the suitable joining of metals must be defined in metallurgical terms. The criteria are: (1) metallurgical compatibility for a specific process, i.e., how the weld region properties differ from the surrounding base metal; (2) ability to produce mechanical soundness, as indicated by lack of porosity, inclusions, or undesirable phases; and (3) serviceability under special requirements, such as fatigue and corrosion.

The weld consists of three principal regions, the weld metal, the heat affected zone (HAZ), and the parent metal (*33*). Figure 9 shows a schematic of the cross section of a typical weldment. Some of the metallurgical effects associated with welding are listed in Table III. The weld metal is a casting that reflects the alloy composition of the consumable filler material used and dissolved parent metal. It displays a deposit structure and segregation of alloying elements dependent on the welding conditions. The solidified weld metal contains large columnar grains, which have strength and toughness properties different from the base metal. The HAZ occurs because of the thermal history that the parent material experiences during welding. Material close to the weld will be heated from a peak temperature of the melting point at the fusion boundary to ambient temperature some distance away. The

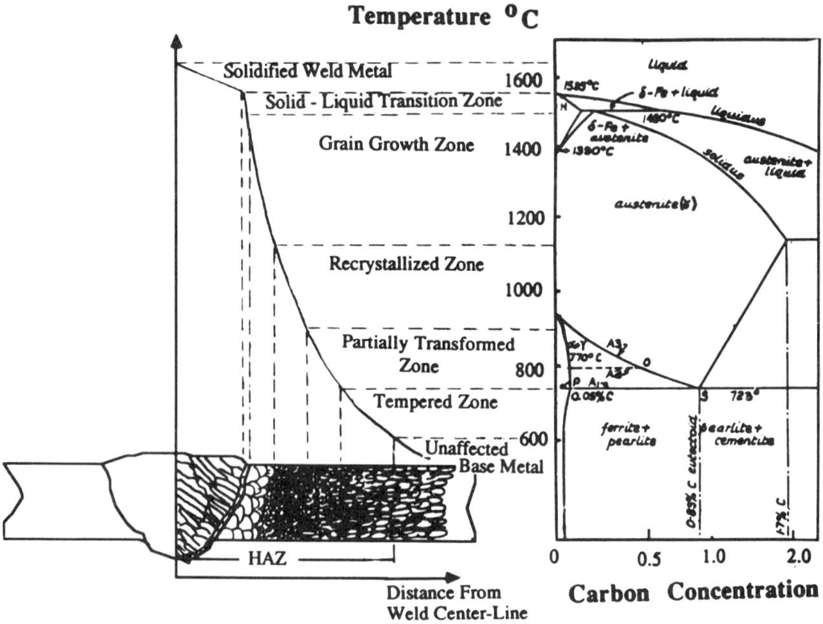

Fig. 9. Schematic of the cross section of a typical weldment and its relationship to the phase diagram.

TABLE III

TYPES OF CRACKING PHENOMENA ASSOCIATED WITH WELDING

Cracking phenomena	Position in weldment
Hydrogen attack	Weld deposit
Gas porosity	Weld deposit
Solidification cracking	Weld deposit
Liquation cracking, HAZ burning, or hot tearing	Fusion zone
Lamellar tearing	HAZ
Cold cracking or hydrogen cracking	HAZ
Reheat cracking	Weld deposit and HAZ

thermal history is position dependent, and the metallurgical changes occurring in the HAZ result from this thermal history. A fundamental understanding of the effects of the nonequilibrium thermal history which occur during welding is required to understand and achieve control of the welding process. The key to this understanding are the continuous cooling transformation curves (C–C–T) for the specific alloys which are being welded. An example of a typical C–C–T curve for a marine grade alloy is shown in Figure 10. Recrystallization and grain growth, phase changes, and precipitation produce the structure of the HAZ. The coarse grains in the HAZ of the base plate have higher hardenability than the base metal, and consequently demonstrate a greater tendency to form hard, brittle martensite. This may produce microcracking after welding. In certain carbon steels, formation of martensite can be avoided by preheating. This procedure causes slower cooling rates and provides sufficient time to form the softer transformation products of pearlite and proeutectoid ferrite in the microstructure, as depicted in Figure 10. In alloy steels with high hardenability, it is almost impossible to avoid formation of martensite, irregardless of plate thickness, thermal conductivity, weld geometry, and high heat input.

The local expansion and contraction associated with fusion welding usually results in the development of residual stresses on cooling to ambient temperatures. These residual stresses act in concert with those that are applied to the structure and therefore play a part in the mechanical response of the material to loads. Residual tensile stresses are deleterious and are avoided if possible. In instances where residual stresses present problems, proper post-weld heat treatment and/or mechanical working techniques to reduce the residual stress level should be applied.

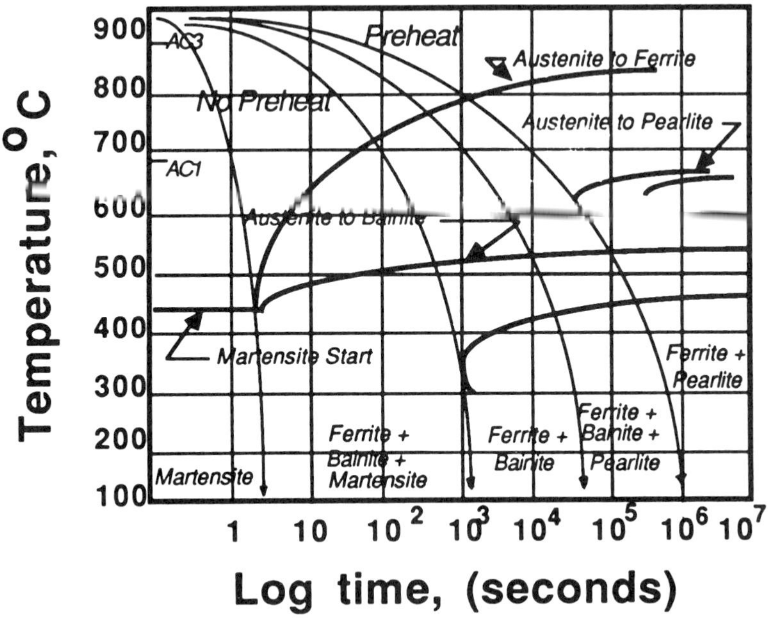

Fig. 10. Typical continuous cooling transformation (CCT) curve for a marine grade steel.

Good weldability is achieved by using steels with controlled alloy chemistry and controlled welding procedure. The composition dependence of weldability is expressed by the carbon equivalent (CE), equation (9). A value of CE < 0.43 is often considered necessary to achieve good weldability. The use of the carbon equivalent index reflects, in compositional terms, important metallurgical characteristics of the steel that affect the properties of the welded material. These properties include the martensite start temperature, Ms, and whether the pearlite and bainite start boundaries move to longer times due to the addition of substitutional elements in the steel alloy, as reflected in the C–C–T curve for the steel alloy. The carbon equivalent will therefore indicate the type of microstructural constituents to be expected in the HAZ. The undersirable microstructural constituent is martensite, which is the hardest and most fracture prone constituent formed in steel.

Good weldability under shipyard conditions is often difficult because control of the thermal history of the steel and its weldment is limited by the size of the structure, complex weld procedures, and difficult preheat conditions. If CE < 0.41, Lloyd's Rules do not require preheating of the weld area. When the CE is as high as 0.45, low hydrogen electrodes and preheating are required, especially for situations where high restraint occurs or where low service temperatures occur.

The effects of impurity elements in steel can be most deleterious to weldability. Phosphorous and sulphur, which seem to be inherent in steel making, are such troublesome impurities that strict control of their concentrations in the steel is required to maintain toughness. Other tramp elements such as antimony, arsenic, and tin have also been determined to cause embrittlement along with phosphorous in alloy steels with 3% to 5% nickel. These steels are typically stress relieved at temperatures that cause stress relief embrittlement (SRE). SRE is caused by enhanced diffusion due to nickel in the alloy of the impurity elements to the grain boundaries. This segregation of impurity elements causes low energy intergranular failure from impact loading.

Typical weld joint geometries that are utilized in marine applications are shown in Figure 11. Thin plate (less than 6.5 mm) can be welded by a single pass (a pass is called a "bead"), but in heavier plate, depending on the weld process, the weldment can require multiple passes. In the fabrication of tubular steel platforms, there are four types of welded joints namely, T, K, T-K, and Y joints, as illustrated in Figure 12. The welds around all these joints are of the fillet type, in which flaws such as undercuts, overbeading, and lack of penetration often occur.

The ease of producing sound welds depends to a significant degree on the welding position. Welding positions which are applicable to all types of

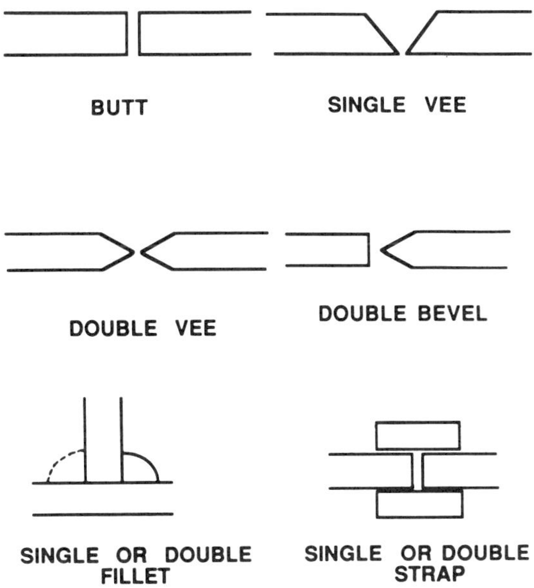

Fig. 11. Types of welds.

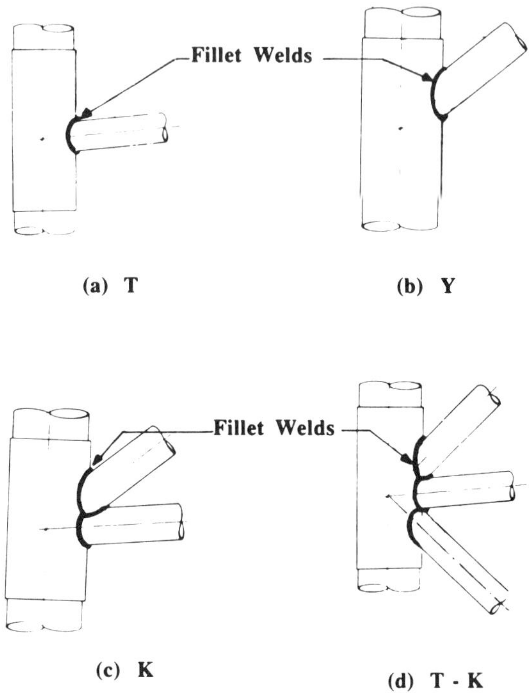

Fig. 12. Types of joints used in offshore platform construction. Joints are predominately of the fillet type.

welding are the flat, horizontal, vertical, and overhead. The flat welding position is the easiest and best welding position. The other three welding positions are considered out-of-position and usually require special procedures to ensure good welds. For ship building and offshore platform construction, out-of-position procedures are required. Shielded metal arc welding fluxes have been developed so that all position welding can be performed with fixed direct current, reversed electrode polarity (DCRP).

The amount of welding in offshore structures and ships is so large that weld defects can not be avoided. The problems encountered in welding carbon, carbon-manganese, and low alloy steels are listed in Table III. Of the various defects, four are from the heat-affected zone—specifically, liquidation cracking, lamellar tearing, cold or hydrogen cracking, and reheat cracking. Three types of cracking phenomena are associated strictly with the weld deposit, namely, hydrogen attack, gas porosity, solidification cracking, and inclusions from arc strikes or improper descaling. Other defects associated with welding are related to the weld profile. Undercutting reduces the cross section and

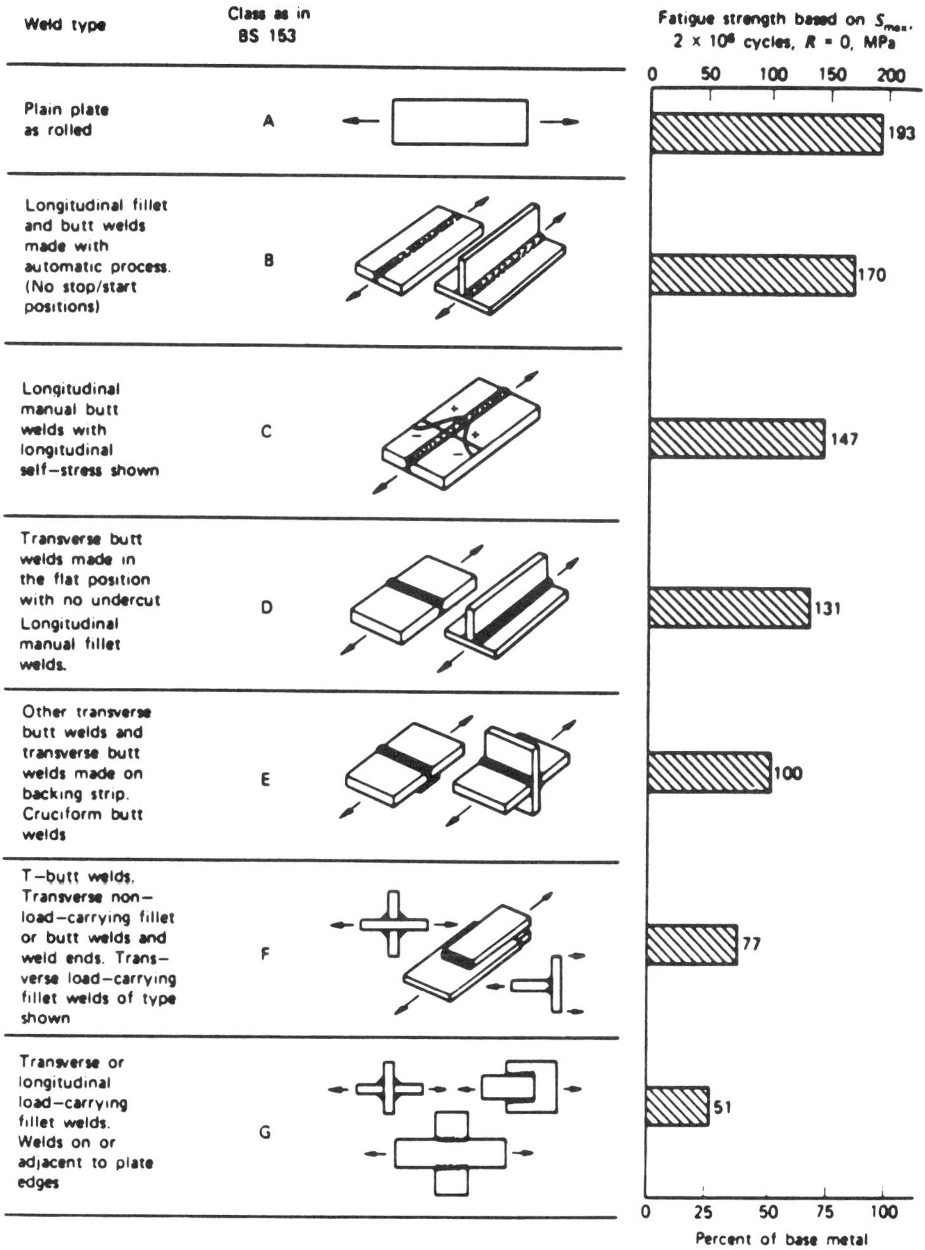

Fig. 13. Fatigue behavior of various welded joints (permission from © John Wiley Sons, Inc., ref 34).

also acts as a stress riser, and residual stresses can cause formation of cracks. These defects are especially critical to structural integrity.

The fatigue behavior of welded joints is critical for structural reliability. The region most subject to stress is in or near the toe of the weld, where the weld metal meets the parent metal. This region is usually the site of fatigue crack initiation. Reducing the reinforcement height, good weld contour, and/or GTAW remelting in the toe/base metal can improve fatigue resistance. Fatigue crack initiation, fatigue crack propagation, and residual strength of the material when cracks are present affect the failure of the structure. The type of loading, the presence of flaws, and the microstructure of the material play significant roles in the material's performance. The fatigue behaviour of various types of welded joints is illustrated by Richard's formula for BS 153 British Standard steel in Figure 13 from reference 34 in Fuchs and Stephens treatise on *Metal Fatigue in Engineering.*

For a sound weld, two key factors affecting weldment fatigue resistance are geometric stress concentration and residual stresses. A prime example of a geometric stress concentration at sites where fatigue cracks initiate are the toe of the weld at the intersection of the HAZ and the weld metal, as shown in Figure 14. Similar geometric stress concentrations are created from the reinforcement height in double V welds, as observed by Reemsnyder (*35*). His data, shown in Figure 15 for quenched and tempered carbon steel, showed that decreasing the height, *h*, of the transverse double V weld reinforcement

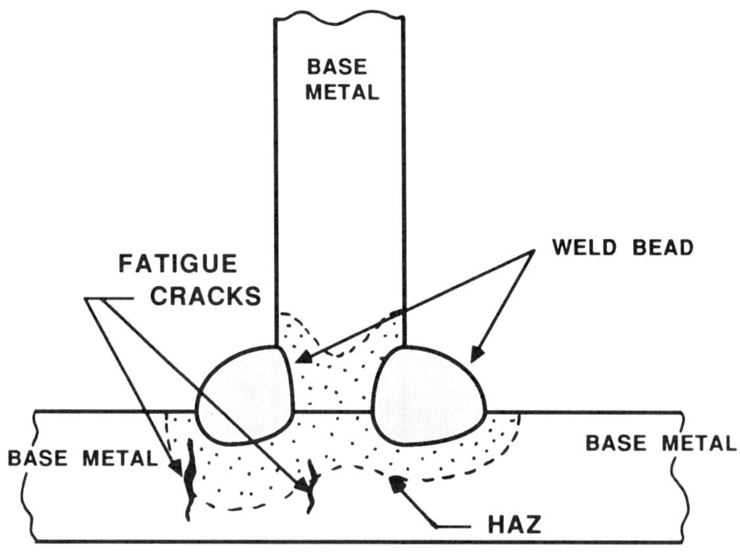

Fig. 14. Location of fatigue cracks in welds.

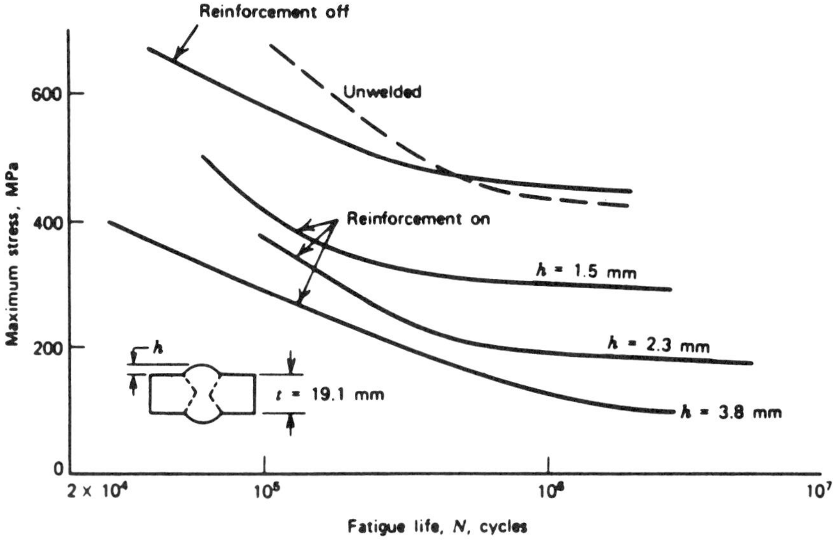

Fig. 15. Fatigue behavior of simple welded joints (permission from © John Wiley Sons, Inc., ref. 34).

provided an improvement in fatigue strength. This phenomena is due to the fact that as h increases, an effective notch forms between the reinforcement and the base plate, and this can act as a fatigue initiation site.

VI. Environmental Factors

Because the marine environment is so abundant and corrosive and because of unceasing wave action, considerable attention must be paid to integrating corrosion enginering principles into the design process. Many of the possible problems which affect structural safety and require maintenance operations are associated with corrosion. The causes and areas of corrosion on a stationary platform are shown in Figure 16. There are five major zones that provide sufficiently different environmental conditions to be treated separately: the marine atmosphere, which may also contain pollutants; the splash area; the tidal zone; the areas of full immersion; and the mud zone. The splash zone is generally considered to provide the severest environment because of the alternate wetting and drying which occurs in this area.

Steels are not very corrosion resistant in marine environments. Corrosion rates of structural steels in marine environments vary from about 0.2 mm/yr

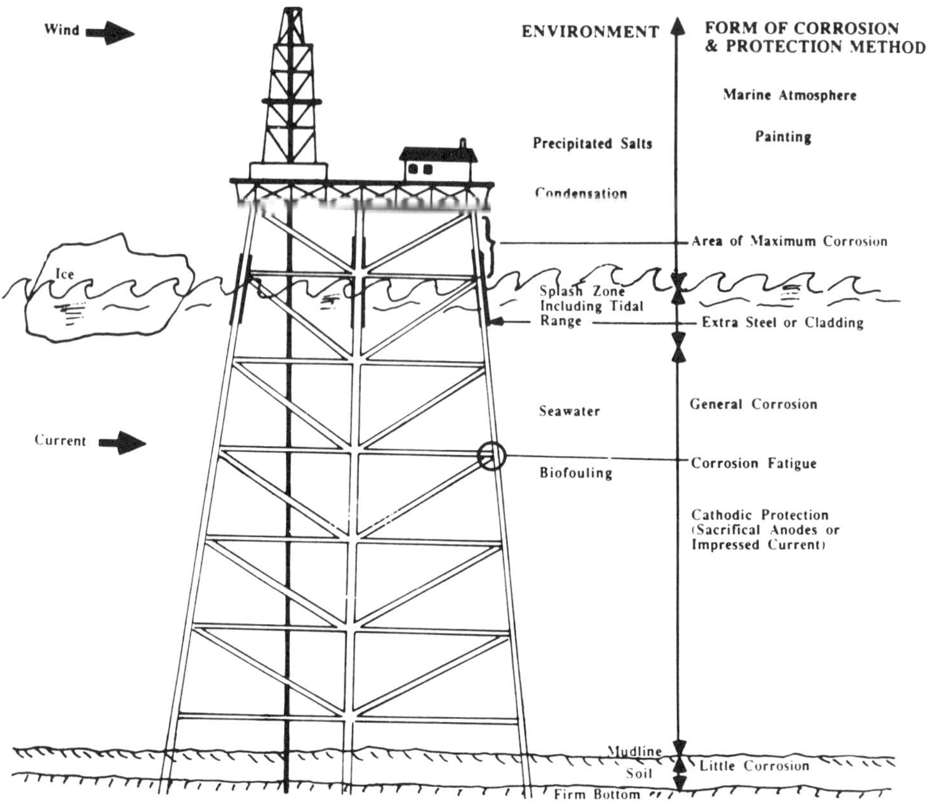

Fig. 16. Causes and areas of corrosion on an offshore platform.

in quiet, sheltered harbor conditions to about 5 or 6 mm/yr in more severe open sea locations. Nonmetallic coatings, along with sacrificial anodes and/or impressed current cathodic protection, are required to protect steel structures. While cathodic protection is used to reduce corrosion of immersed steel, it is not effective in the splash zone. In this area, special protective coatings, depositing calcerous deposits, or cladding with corrosion resistant 90/10 copper-nickel alloy are effective means of corrosion control. These corrosion protection methods are not completely passive, and maintenance is required to maintain structural safety and, hence, operation. Here again, modern numerical techniques promise significant benefits in designing corrosion protection into a structure based on the fundamental electrochemical behavior of the materials used in the structure.

The marine envrionment also has a profound effect on the fatigue behavior of steel. Figure 17 shows the results of S–N corrosion fatigue tests in air and

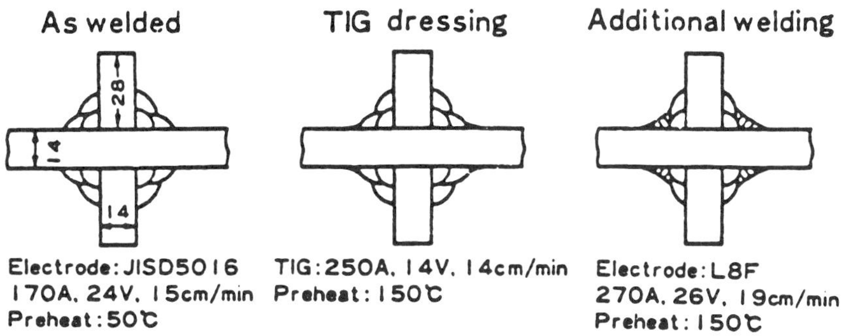

(a)

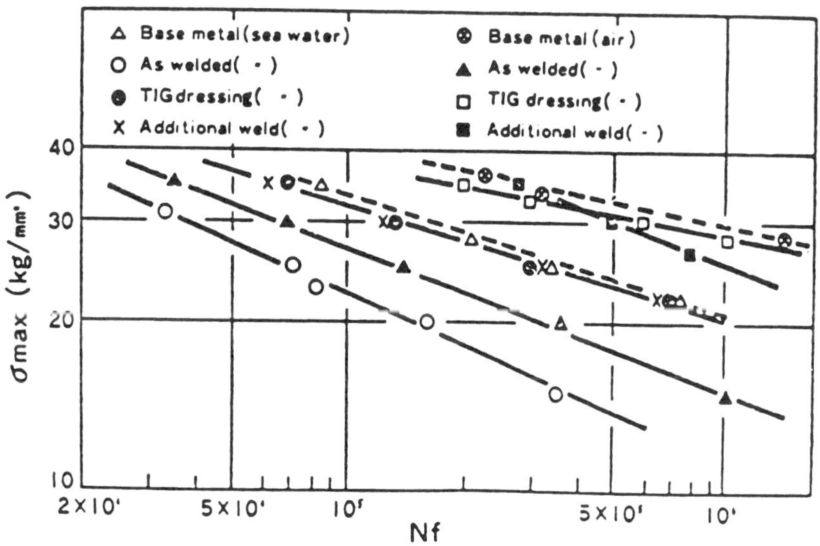

(b)

Fig. 17. Results of corrosion fatigue tests and weld profiling in air and seawater (permission from © International Institute of Welding, Great Britain, from ref. 36).

seawater and the effect of weld toe profiles (*36*). Besides a loss of fatigue resistance due to the environment, the steel exhibits further loss of fatigue resistance from welding. GTAW at the toe of the weld to profile the fillet improves the fatigue resistance, but the fatigue resistance of the steel never reaches the level of the material in air. Corrosion fatigue crack growth rate curves (see Figure 8) for Canadian pipeline steel are shown in Figure 18 (*37*).

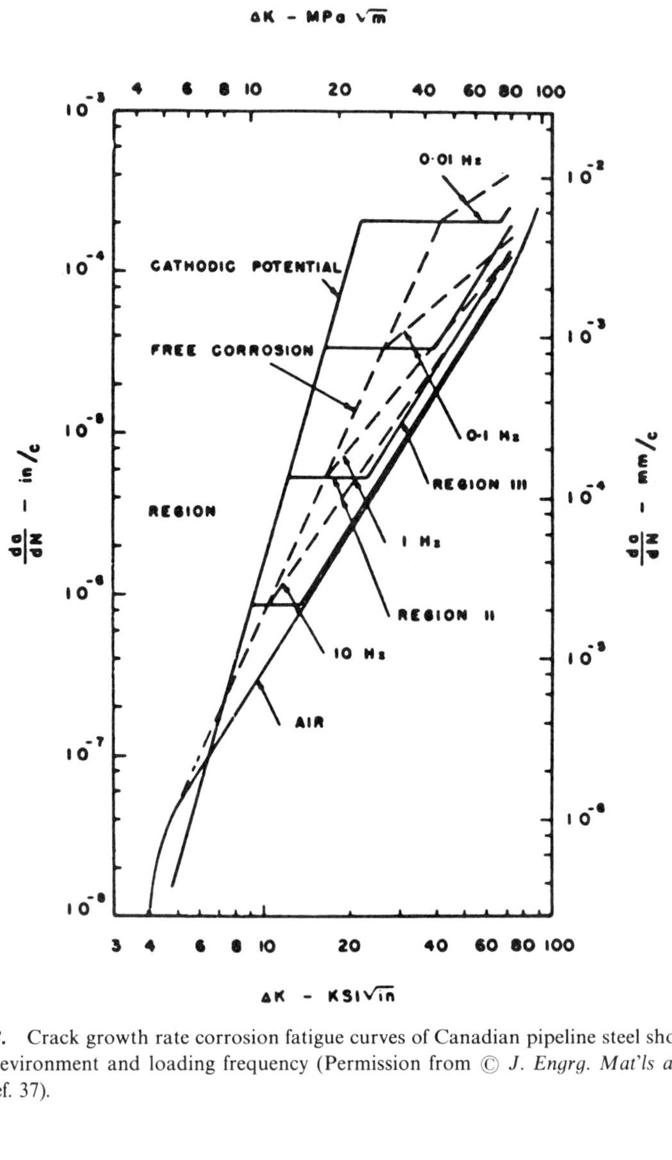

Fig. 18. Crack growth rate corrosion fatigue curves of Canadian pipeline steel showing the effects of evironment and loading frequency (Permission from © *J. Engrg. Mat'ls and Tech.*, ASME, ref. 37).

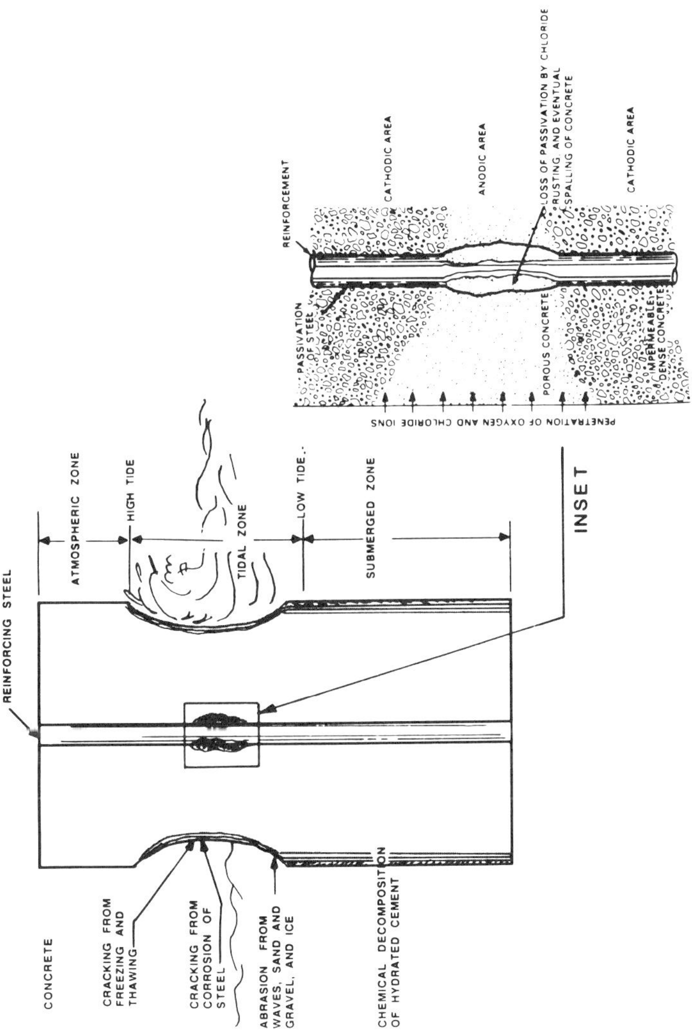

Fig. 19. Environmental degradation and failure in reinforced concrete.

This figure also shows the increased crack growth rate because of exposure of the steel to the environment and the effect of loading frequency.

The partial replacement of steel by more corrosion resistant metals could alleviate many of these problems, but a cost penalty may be involved. One such material is titanium and its alloys, which are resistant to almost all the corrosion problems shown in Figure 16. The structural efficiency of titanium, i.e., its strength-to weight ratio, is also high. The most recent production and cost figures for titanium, however, indicate that titanium would be impractical to replace the major structural elements of offshore platforms. But, the past concerns with the weldability of titanium alloys have been shown to be overstated, i.e., titanium and its alloys can be satisfactorily welded with proper care (*38*). Titanium alloys are then good choices for many applications such as marine risers, pumps, heat exchangers, and pressure vessels for the hulls of submersibles (*39*, *40*).

Properly formulated, high quality concrete is very resistant to marine environments. In this regard, the permeability of the concrete to seawater is an important factor. The permeability of the concrete can be controlled by the cement mix and choice of aggregate. Well hydrated Portland cements contain up to 30% CaOH, which maintains a highly alkaline environment (pH = 12.5) in the concrete. In reinforced and prestressed concrete, this highly alkaline environment produces a passive film on the steel that generally provides protection to the steel reinforcements. As illustrated in Figure 19, however, corrosion of the rebar produces voluminous corrosion products, which cause fracture and spalling of the concrete. This exposes bare steel to the marine environment (*21*).

VII. Closure

In recent years the trends in the use of materials in marine systems and structures have changed because marine structures have become larger, more complex in their fabrication, and more expensive. In addition, the structures are increasingly being used in more severe environmental conditions at lower temperatures and under higher loads, especially in structure designed for North Sea and arctic regions.

The various factors that affect materials performance interact in a synergistic manner to control the performance of the structure in the marine environment. In this chapter, the emphasis has been on delineating the interacting factors, especially the role of metallurgical and fabrication factors in structural integrity. The importance of microstructural factors in welding and welding-induced flaws are discussed, and the limitations of projecting laboratory measured fatigue data into real structural performance are noted. Fracture mechanics-derived tests, while providing new design approaches, must be used with caution in establishing structural performance, because the extrapolation of the laboratory test data to real structural performance is not rigorously established. None of the factors have been discussed in detail. However in the chapters that follow, much detail is provided, especially in regards to the dominant role that the environment plays in the performance of marine structures and materials.

Maintenance factors and nondestructive evaluation (NDE) also affect the performance of the structure. The need for nondestructive evaluations and maintenace programs is critical in finding and repairing fatigue-induced structural cracks to prevent catastrophic failures. Some papers dealing with maintenance, inspection, and NDE are given in references (*41*, *42*).

References

1. Anon., *Principles of Naval Engineering*. Bureau of Naval Personnel, NAVPERS 10788-B.

2. Anon. (1986). *Rules for Building and Classing Steel Vessels*, 1986. American Bureau of Shipping, Paramus, N.J.

3. Anon. (1982). *API Recommended Practice for Planning, Designing, and Construction of Offshore Platforms*, API RP-2A, 13th Edition. American Petroleum Institute, Dallas, Tx.

4. Anon. (1983). *Rules for Building and Classing Offshore Installations*, Special Committee on Offshore Installations, New York. American Bureau of Shipping, Paramus, N.J.

5. Collins, R., and Dover, W. D. (1984). Cracks in Offshore Oil Platforms. *American Scientist* **12**(6), 553–557.

6. Anon. (1977). *Rules for the Design, Construction, and Inspection of Offshore Structures*. Det Norske Veritas, Oslo, Norway.

7. Anon. (1984). *Offshore Installations—Guidance of Design and Construction*. Department of Energy—Petroleum Engineering Division of Great Britain, ISBN 0-11-7011185. Her Majesty's Stationary Office, London.

8. Hoeppner, D. W. (1972). "Corrosion Fatigue Considerations in Materials Selection and Engineering Design." In: *Corrosion Fatigue: Chemistry, Mechanics, and Microstructure*. National Association of Corrosion Engineers, Houston, Texas, pp. 3–11.

9. Harrison, R. P. and Milne, I. (1981). "Assessment of Defects: The C.E.G.B. Approach." *Philosophical Transactions of the Royal Society*, **A**, 299, pp. 145–150.

10. Anon. (1987). *ASTM Standard Test Method For Notched Bar Impact Testing of Materials*, ASTM E23-72. American Society For Testing and Materials, Philadelphia, Pa.

11. Anon. (1987). *ASTM Standard Test Method for Conducting Drop Weight Tests To Determine Nil Ductility Temperature of Ferritic Steels*, ASTM E-208-69. American Society For Testing and Materials, Philadelphia, Pa.

12. Anon. (1973). *Method For 5/8 Inch Dynamic Tear Testing of Metallic Materials*, Military Standard MIL-STD-1601(SHIPS). U.S. Department of Defense, Washington, D.C.

13. Anon. (1987). *ASTM Standard Test Method for Plane-Strain fracture Toughness of Metallic Materials*, E399-81. American Society For Testing and Materials, Philadelphia, Pa.

14. Cherepanov, G. P. (1967). *Applied Mathematical Mechanics* (PMM) **31**, 476.

15. Rice, J. R. (1968). *ASME Journal of Applied Mechanics*, **35**, 379–385.

16. Anon. (1979). *Methods For Crack Opening Displacement (COD) Testing*, BS 5762. British Standards Institution, London.

17. Miner, M. A. (1945). "Cumulative Damage in Fatigue." *Trans. ASME, J. Appl. Mech*, **67**, p. A159.

18. Burnside, O. H., Hudak, S. J., Oelkers, E., Chan, K. and Dexter, R. J. (1984). "Long–Term [illegible] U.S. Coast Guard.

19. Warren, N. (1980). *Metal Corrosion in Boats*, International Marine Publishing Co., Camden, Maine, pp. 189–191.

20. Lane, P. H. (1985). "Lloyd's Register—Surveying for Quality." Metals and Materials **1**, 484–488.

21. Burnside, O. H. and Pomerening, D. J. (1983). "Survey of Experience using Reinforced Concrete in Floating Marine Structure." *Ship Structure Committee Report SSC-321*, U.S. Coast Guard.

22. Anon. (1972). *The Structural Use of Concrete.* British Standards Institute, CP 110. Her Majesty's Stationary Office, London.

23. Anon. (1977). *Recommendations for the Design of Sea Structures in Reinforced or Prestressed Concrete*, 3rd Edition. Federation Internationale de la Precontraite, Paris.

24. Williams, M. L. (1953). "Analysis of Brittle Behavior in Ship Plates." In: *Symposium on the Effect of Temperature on the Brittle Behavior of Metals With Particular Reference to Low Temperatures.* ASTM Special Publication 158. American Society for Testing and Materials, Philadelphia, Pa.

25. Parker, E. R. (1957). *Engineering Structures*, John Wiley, New York.

26. Biggs, W. D. (1960). *The Brittle Fracture of Steel.* Pitman Publishing Corp., New York.

27. Boyd, G. M. (1970). *Brittle Fracture in Steel Structures.* Butterworth & Co, London.

28. Leveau, C. W. (1965). "Aluminum and Its Use In Naval Craft." *Naval Engineers J.*, Feb., 13–27.

29. Anon. (1981)., "The Alexander L. Kielland—Accident." *Norwegian Public Reports*, NOU 1981:11. Ministry of Justice and Police, Oslo.

30. Palko, W. A., Hasson, D. F. and Zanis, C. A. eds. (1985). *Electroslag Processing For Marine Applications*, Report EW-20-85, ADA 166926. U.S. Naval Academy, Annapolis, Maryland.

31. Kerns, W. H. ed. (1978)., *Welding Handbook*, 7th Edition. American Welding Society, Miami, Fla.

32. Anon, (1983). *Metals Handbook Vol.* 6: *Welding, Brazing, and Soldering.* ASM, Int'l, Metals Park, Ohio.

33. Easterling, K. (1983). *Introduction To The Physical Metallurgy of Welding.* Butterworth, London.

34. Fuchs, H. O. and Stephens, R. I. (1980). *Metal Fatigue in Engineering.* John Wiley and Sons, New York.

35. Reemsnyder, H. S. (1978). "Development and Application of Fatigue Data For Structural Steel Weldments." In: *Fatigue Testing of Weldments* (D. W. Hoeppner, ed.), ASTM STP 645. American Society For Testing and Materials, Philadelphia Pa, p. 3.

36. Todoroki, R. (1978). "Effects of Toe Profile Improvements on Corrosion Fatigue Properties of Welded Joints" IIW Doc. XIII-875-78. International Institute of Welding.

37. Vosikovsky, O. (1975). "Fatigue-Crack Growth in an X-65 Line-Pipe Steel at Low Cyclic Frequencies in Aqueous Environments." *Trans. ASME, J. Engin. Materials and Tech.*, **97**, 298–304.

38. Hasson, D. F. (1986). *A Review of Titanium Welding Processes*, Report EW-9-86, ADA166442. U.S. Naval Academy, Annapolis. Md.

39. Hasson, D. F. and Crowe, C. R. (1982). "Titanium for Offshore Drilling." *J. of Metals*, **34**, 23–28.
40. Crowe, C. R. and Hasson, D. F. (1981). "The Use of Titanium in Deep Sea Exploration." In: *Titanium for Energy and Industrial Applications* (D. Eylon, ed.) The Metallurgical Society of AIME, Warrendale, Pa, 93–110.
41. Brackett, R. L. (1986). *Sampling Criteria and Procedure for Underwater Inspection of Waterfront Facilities*, NCEL Report TN-1762. Naval Civil Engineering Laboratory, Port Hueneme, Ca.
42. Smith, A. P. (1986). *Underwater Nondestructive Testing of Concrete: An Evaluation of Techniques*, NCEL Report TN-1747. Naval Civil Engineering Laboratory, Port Hueneme, Ca.

2

The Marine Environment

STEPHEN C. DEXTER

College of Marine Studies
University of Delaware
Lewes, Delaware

and

ROBERT A. DALRYMPLE AND NOBUHISA KOBAYASHI

Ocean Engineering Group
Department of Civil Engineering
University of Delaware
Newark, Delaware

ISBN 0-12-341828-3

I. Introduction

The design of structures to be used in the ocean is far different than that for land-based structures, as the ocean environment is more adverse in many respects. Despite these design differences, there are also many points of similarity. A major component in the design analysis for both ocean and land-based structures is the period of time over which they are expected to function, called the design life. This factor affects not only the cost of the structure but also the intensity of the hazards it is likely to experience during its service life. A structure designed for a 20-year life will cost less to build, and statistically it will endure far less severe conditions than one designed for 100 years. Structures placed in the ocean environment are exposed to hazards not experienced on land. Storm waves, tidal currents, scour, and storm surges are uniquely ocean phenomena, and correct calculation of the loads they produce is of primary importance to the survival of the structure. Earthquakes at sea can be the direct cause of structural damage as can the induced seismic sea waves caused by the tremor. Another problem unique to the marine environment is caused by the increased weight and drag from growth of biofouling organisms on the submerged portions of marine structures.

Wind effects, on the other hand, are calculated in a similar way as on land, after taking into account the difference between the surface roughness of the land and the water. The effects of corrosion on structural integrity are also similar to those on land, except that the marine environment is more corrosive than are terrestrial fresh water environments. The presence of chlorides, biofouling from marine organisms, and damage to paint coatings from wave action, ice, and sediment scour can make marine corrosion control a difficult challenge.

The importance of the various environment factors lies, on the one hand, in their ability to produce loading on the structure, in terms of both the maximum forces and moments that might produce mechanical overloads, and on the periodic forces that might result in structural fatigue failure. On the other hand, the importance of environmental factors related to corrosion is that they lead to a reduction in the capability of the structure to withstand the forces discussed above.

The wind and ice loads that may occur are steady forces when compared to the more periodic wave forces or the more random earthquake excitation. Calculations of these forces and loading conditions are needed for two reasons. First, they are needed for the determination of the maximum design loads to be experienced by the structure, and second they are needed for the determination of the daily operating conditions. The number of days that wave heights exceed a certain value can be a critical factor in the cost of offshore operations, particularly if the operations become inefficient or must be shut down under large wave conditions.

The design of a structure ultimately comes down to an analysis of the loads on the structure and its foundation. The long-term integrity of a structure will depend on its ability to resist these loads and on the stability of the soil which underlies it. The ability of the bottom material to resist scour will be an important factor, as erodible material may require scour protection if the wave and current environment is severe. Additional structural loads not previously mentioned are the manmade live and dead loads created by use of the structure and also the impact loads from contact with support vessels.

This chapter discusses the nature of the environmental loads to be experienced in the ocean, as well as the environmental factors that decrease the ability of the structure to withstand the loads. Some indications are provided for general methods of calculating these loads; however, the design of offshore structures is so complex as to require far more detail than is provided here. The reader is therefore referred to more comprehensive engineering works for details, e.g., Sarpkaya and Isaacson (*1*), Wilson (*2*), the numerous *Proceedings of the Offshore Technology Conference* and the *International Conferences on Wind Effects on Buildings and Structures.* In addition, the American Petroleum Institute (*3*) provides recommended procedures for offshore platforms, as does Det Norske Veritas (*4*). For coastal structures, the U.S. Army Coastal Engineering Research Center's *Shore Protection Manual* is extremely useful. Information on marine corrosion and fouling can be found in numerous books including LaQue (*5*), Dexter (*6*), Uhlig and Revie (*7*), Fontana and Greene (*8*), Costlow and Tipper (*9*), and the *Metals Handbook* 8th edition and 9th edition, as well as the papers published yearly by the National Association of Corrosion Engineers, Houston, Texas.

II. Winds

Winds are generated due to variations in solar heating of the earth, leading to pressure gradients that drive the air currents. The orientation of the earth with respect to the sun results in cold poles and hot equatorial zones, providing the temperature differences to drive mean wind fields such as the Westerlies (lying to the north of 60° latitude) and the Trades in the Tropics. These winds are of interest for the physical oceanographer, as they are responsible for driving the major currents in the oceans.

Overwater winds differ from those on land principally due to the nature of the air-sea boundary, which is much smoother than its counterpart on land.

However, the structure of the wind velocity profile is treated similarly. The wind is retarded by the presence of the sea surface and the velocity increases with elevation. The usual expression for the profile is

$$V(z) = V_{z_0} \ln\left(\frac{z}{z_0}\right) \quad (1)$$

where $V(z)$ is the velocity at elevation z above the surface and V_{z_0} is a reference velocity measured at elevation z_0. The reference elevation is typically taken as 10 m. As an approximation, a power law relationship is often used:

$$V(z) = V_{z_0}\left(\frac{z}{z_0}\right)^k. \quad (2)$$

The power, k, varies with the surface over which the wind is blowing. For very rough surfaces, say, wooded areas or towns, $k = \frac{1}{7}$, while for open water conditions, $k = \frac{1}{10}$. The power law can be used over the entire elevation of the structure.

A. Wind Statistics

Wind velocities must be measured to obtain design data for a given site. Typically, long anemometer records are used to obtain information on the mean and extreme wind speeds and the gusts. More recently the spectral content of the wind has become useful in the dynamic design of compliant structures. This is especially important if the spectral components of the wind overlap the natural frequencies of the structure.

Useful measures of wind speed are the mean wind (both annual and hourly), the highest sustained wind, and the gust speeds. There are many ways to measure sustained wind speed. One of the most useful is to measure the average wind speed during the time it takes for a one mile long section of wind (known as a "mile of wind") to pass the anemometer. Thus the averaging time is inversely related to the wind speed. For example, a one minute averaging time is necessary for a 60 mile per hour wind. The highest average speed thus measured in a given time interval (say, one day or one week) is called the "fastest mile of wind" during that time interval. In addition to these measurements, the gust speeds give an indication of the severity of eddies in a turbulent wind. These give rise to the maximum wind speeds. A gust factor has been developed to relate the gust speed to the hourly mean winds. Over water, a gust factor of 1.3–1.5 is appropriate. Along with these measured wind speeds, it is vital to record the direction of the winds.

The values of wind speeds and directions at a site are extremely variable. This is true even for the mean annual wind speeds, which vary from year to year. To examine the winds for design purposes, it is important to have very long wind records from which the wind statistics can be extracted. Over short periods of time, the distribution of wind velocities can be taken as Gaussian. However, for extreme winds, other distributions such as the Gumbel or the Weibull distributions have been used to describe the wind statistics. These extreme value statistics are determined by ranking the maximum values of the wind over the length of the record from the largest to the smallest. The probability, p, of a given wind occurring is then the rank of the wind speed divided by the total number of measurements. Plotting these data on extreme value probability paper enables one to determine the probability of occurrence of different wind speeds. Given the probability, p, of a given wind speed occurring, the return period, T_r, can be determined.

$$T_r = \frac{1}{(1 - p)}. \tag{3}$$

The return period is often a deceiving number. A 100-year return period often gives a false sense of security. The encounter probability, P_L, is the probability that a wind of a given return period will occur in a given period of L years.

$$P_L = 1 - \left(1 - \frac{1}{T_r}\right)^L. \tag{4}$$

For example, a 100-year return period wind has an 18% probability of occurring in 20 years.

B. *Wind Forces*

The force of the wind on structures, F_w, is determined by the square of the wind speed, V.

$$F_w = \rho_a C V^2 A. \tag{5}$$

Here, ρ_a is the density of air, A the projected area of the structure, and C is the shape coefficient. For small structures, the above equation may be used with one constant wind speed, and A would be the entire projected area of the structure. For taller structures, or when the variation of wind speed with elevation is important, the force should be integrated over the height, with A being interpreted as the projected area per unit length of elevation, and V varying with elevation z.

The shape coefficient accounts for the amount of streamlining the structure possesses. For bluff objects, C is large (about 2), while for more streamlined objects, it can be less than 0.5. Examples of various shape coefficients can be found in Scruton and Newberry (*10*); Davenport (*11*), Sachs (*12*). An important parameter governing the magnitude of the shape coefficient, in addition to the structural geometry, is the Reynolds number, which is a dimensionless number composed of the wind speed, V, the air kinematic viscosity, ν, and a structural dimension. For a long circular cylinder, the Reynolds number is

$$\mathrm{R} = \frac{VD}{\nu} \tag{6}$$

where D is the cylinder diameter. For bluff-shaped objects, the wind direction also affects the shape coefficient. Wind loads for the entire structure need to be determined for all possible wind directions. Shielding coefficients have been determined to measure the amount of sheltering one object has on another. For objects within six diameters of an upwind object, shielding will be important for the force calculations.

Additional wind force considerations are due to the correlation of the wind along a structure. If the gust forces were correlated over the entire structure then the forces will be far greater than if the correlation distances were less. Good correlation in wind speeds occur over about 50 feet. After that the correlation drops off (*12*).

Dynamic effects of the wind forces can be very important, particularly when the natural frequency of the structure corresponds to an energetic portion of the wind spectrum. For example, shedding of vortices from alternate downwind sides of a structure causes a periodic shifting in the location of the velocity stagnation point at the front of the structure and leads to alternating transverse forces. This vortex shedding occurs at a frequency which can be predicted by the Strouhal number, S

$$S = \frac{nD}{V} \tag{7}$$

where n is the frequency of oscillation, D is a geometric measure of the structure, such as a characteristic diameter, and V is the local wind speed. The Strouhal number is a function of the Reynolds number and the shape of the structure. Galloping oscillations occur when the apparent wind direction of a moving structure creates a greater transverse force than for the static condition. Most structural shapes have this propensity, with the exception of a circular cylinder.

The designer is referred to the *ANSI Building Code Requirements for Minimum Design Loads in Buildings and Other Structures* (*13*) for more engineering details about wind forces. The book of Simiu and Scanlon (*14*) is another excellent reference.

III. Waves

Waves are created by the action of wind over water surfaces. As the surface of the ocean is so large, waves are always being generated at some location. The waves radiate away from their "birth place," with very little attenuation due to friction. A structure, then, no matter where it is located, is subjected to waves at almost all times, creating forces and induced motions.

A. Wave Effects

As water waves move past a fixed structure, individual particles of water are set in motion by the wave. The trajectories of the particles trace out nearly closed elliptical paths, as shown in Figure 1, thus producing periodic horizontal and vertical motions of water past the structure. The maximum horizontal velocities occur under the wave crest and are in the direction of wave travel. These horizontal and vertical water motions lead to two types of forces on the structure. The first is the viscous drag force, related to the horizontal and vertical velocities. The second force is the inertial force, related to the horizontal and vertical accelerations. The inertial force can be even larger than the drag force for large objects. If waves are breaking on the structure, a third type of force, related to the impact of the waves, can become important.

There are two methods for determining the wave forces on a structure: the design wave method and the design spectrum method. The design wave method was the first to be developed for offshore use. This method relies on the use of a single "maximum" wave to compute the design loads for the structure. However, since the character of the sea surface is influenced by

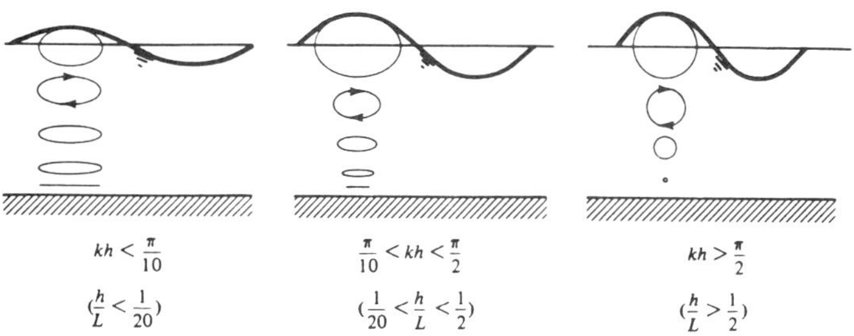

Fig. 1. Water particle trajectories in progressive water waves of different relative depths.

many wave trains, particularly during storm conditions, spectral methods have been developed to examine numerous frequencies and directions at the same time. Both methods have difficulties: The design wave method cannot handle the spectral and directional aspects of the sea surface, while the spectral method is limited to superimposing numerous small amplitude waves. The larger the wave, the more nonlinear, in a mathematical sense, the wave becomes, and spectral methods are limited to linear waves.

B. Wave Data

Measured data compiled over many years would provide the ideal data base for wave forecasting to determine either the design wave or the design spectrum. Unfortunately, only a few locations have had wave gages installed, and the length of the data set is very short. This is due to the difficulty of design for, and maintenance of, wave gages for very hostile environments. In deep water, most wave gaging is done by buoys, of which very few are capable of measuring a directional wave spectrum. Ship data also exist (15) for sea and swell conditions.

The American Petroleum Institute (*3*) has compiled wave heights for various locations around the U.S. In Table I, their reference and guideline wave heights and suggested deck elevations for offshore platforms are given for deep water conditions. The guideline wave height range allows for the variation of heights to be expected over a large area. Figure 2 shows the areas indicated in Table I and Figure 3 shows guideline and reference wave heights for the Gulf of Mexico. Ward, Evans, and Pompa (*16*) provide extreme wave data for the Atlantic coast of the U.S.

Water wave heights are approximately Gaussian distributed (*17*). To examine a historical record of N waves, the measured wave heights, H_i for $i = 1, N$, are ranked from the largest to the smallest.

The probability, p, that a wave exceeds a given height, $\hat{H}$, is denoted $P(H > \hat{H})$ and is found by

$$P(H > \hat{H}) = \frac{n}{N} \tag{8}$$

where n is the number of waves with heights greater than $\hat{H}$. From the Gaussian distribution,

$$P(H > \hat{H}) = e^{-(H/H_{\text{rms}})^2} = \frac{n}{N}. \tag{9}$$

TABLE I

WAVE PARAMETERS FOR TEN AREAS IN UNITED STATES WATERS

	Wave Height						
	Reference Level		Guideline		Wave Steepness	Reference Level Deck Clearance*	
	ft	m	ft	m		ft	m
Offshore Gulf of Mexico	70	21.3	(see Fig. 2.8-2)		$\frac{1}{12}$	48	14.6
Offshore Alaska							
1. Lower Cook Inlet	60	18.3	50–70	15–21	$\frac{1}{13}$	56	17.1
2. Icy-Bay Gulf of Alaska	100	30.5	90–120	27–37	$\frac{1}{15}$	80	24.4
3. Kodiak Shelf-Gulf of Alaska	90	27.4	80–110	24–34	$\frac{1}{15}$	72	21.9
4. Bering Sea/Bristol Bay	85	25.9	75–95	23–29	$\frac{1}{13}$	63	19.2
Offshore California							
1. Santa Barbara Channel	45	13.7	40–50	12–15	$\frac{1}{16}$	38	11.6
2. Outer Banks	60	18.3	55–70	17–21	$\frac{1}{15}$	44	13.4
Offshore Atlantic Coast							
1. Georges Bank	85	25.9	75–95	23–29	$\frac{1}{12}$	59	18.0
2. Baltimore Canyon	90	27.4	80–100	24–30	$\frac{1}{12}$	62	18.9
3. Georgia Embayment	75	22.9	65–85	20–26	$\frac{1}{12}$	53	16.2

* Above MLW in Atlantic; above MLLW in Pacific.

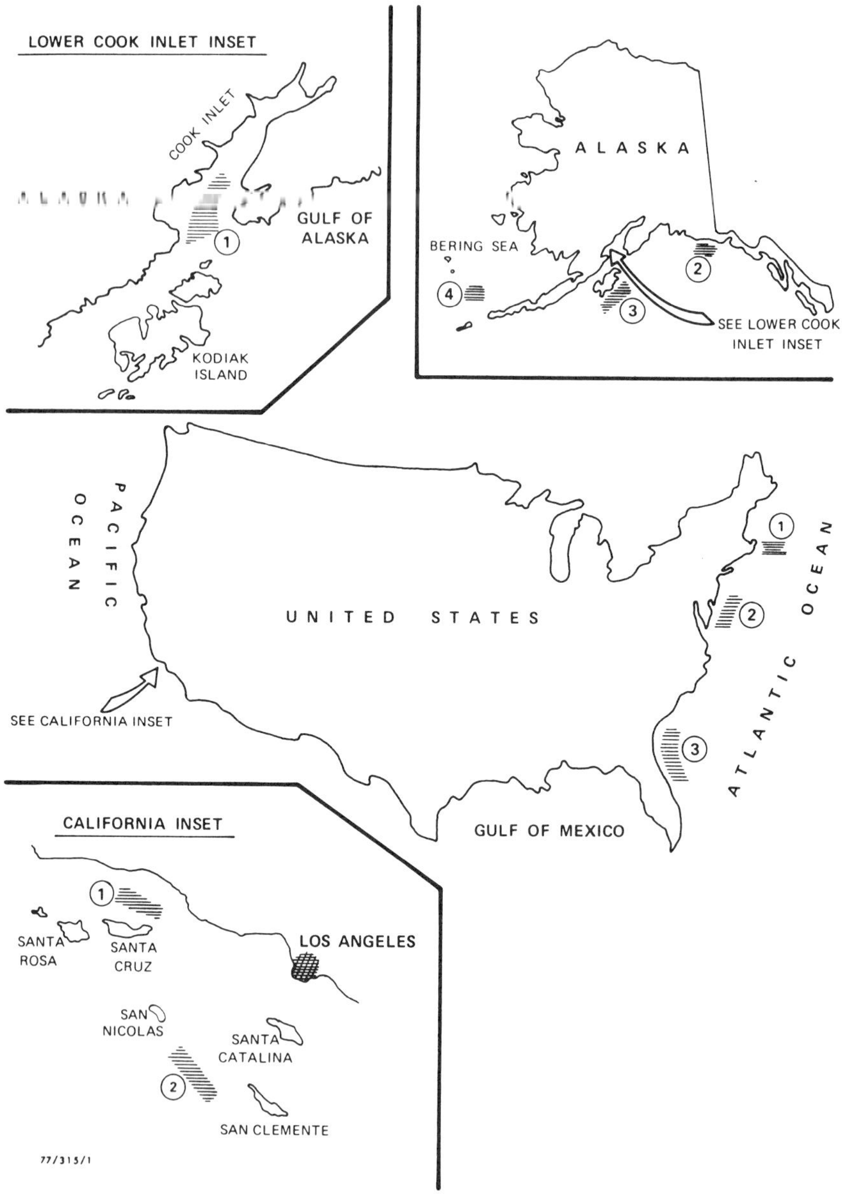

Fig. 2. Area location map.

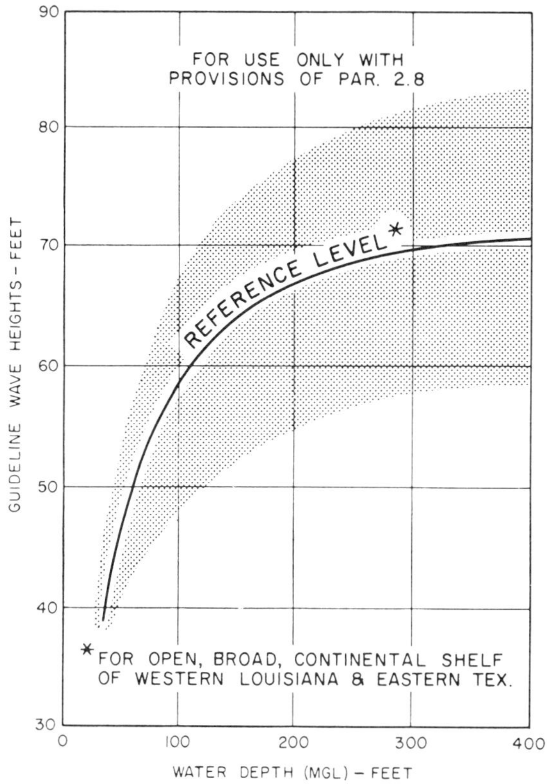

Fig. 3. Guideline wave heights for the Gulf of Mexico.

The wave height, H_{rms}, is the root-mean-square wave height determined from the measured waves by

$$H_{rms} = \left(\frac{1}{N}\sum H_i^2\right)^{1/2}. \tag{10}$$

Defining H_p as the wave height exceeded by pN of the waves, the relations in Table II hold from the Gaussian distribution.

The Gaussian relationship over-predicts the heights of the most extreme waves (*18*), as wave breaking will limit the heights of waves in all water depths.

Numerous techniques have been developed to "hindcast" waves from the more easily measured wind data. The models are quite sophisticated (see, e.g., Cardone et al.) (*19*).

TABLE II

RELATIONSHIP OF H_p TO H_{rms}

$H_{(1/100)}$	$= 2.36\ H_{\mathrm{rms}}$
$H_{(1/0)}$	$= 1.80\ H_{\mathrm{rms}}$
$H_{(1/3)}$	$= 1.42\ H_{\mathrm{rms}}$
$H_{(1/1)}$	$= 0.89\ H_{\mathrm{rms}}$

C. *Wave Theories*

For the design wave analysis, numerous wave theories exist for the determination of wave properties. The most widely used theory is the Airy (*20*) wave theory, due to its simplicity. It is not recommended for design as it does not correctly model the wave properties for large waves. The Airy theory is also known as linear wave theory, due to the mathematical fact that the theory is developed to only first order in wave steepness, ak, where a is the wave amplitude, k is the wave number, defined as $k = 2\pi/L$, and L is the wave length, the distance between two wave crests. Terms of higher order in wave steepness are neglected as being small. For the linear theory, the wave profile is given by the sinusoidal curve:

$$\eta = a\cos(\mathbf{k}\cdot\mathbf{x} - \sigma t) \tag{11}$$

The vector, **x**, is the horizontal vector (x, y, z), and **k** is the wave number vector, pointing in the direction of wave travel. The magnitude of k is found from the dispersion relationship

$$\sigma^2 = gk\tanh(kh) \tag{12}$$

where g is the acceleration of gravity; h is the local water depth; σ is the frequency of the wave, defined as $\sigma = 2\pi/T$, where T is the wave period or the time the wave takes to pass a point. A complete description of the Airy theory and its applications is given in Dean and Dalrymple (*21*).

The first major nonlinear theory was developed by Stokes (*22*) and extended up to fifth order by Skjelbreia and Hendrickson (*23*); see Fenton (*24*) for corrections. The advantage of higher order theories is that they are able to represent the wave more exactly by adding more harmonics to the wave form. The fifth order Stokes theory has been popular for design, but it fails to converge for shallow water conditions ($kh < \pi/10$). The stream function wave theory of Dean (*25*) has proven the most popular. Dalrymple (*26*) extended Dean's theory for the case of currents. The most exact numerical wave theory is that of Cokelet (*27*), which can be carried out to very high order. However, the Cokelet theory is difficult to use, and it is only in the very highest waves that it differs from stream function wave theory.

In very shallow water, analytical theories such as the solitary and the cnoidal (see, e.g., Wiegel (*28*)) wave theories have been used. The advantage of these theories is that fewer mathematical functions are used to describe the wave form than would be required by the Stokes expansions. The solitary wave theory does not represent a train of waves, only a single crest, and therefore is not applicable for offshore design. Svendsen (*29*) and Wiegel (*28*) have illustrated application of the cnoidal theory.

For convenience, Dean (*30*) has tabulated stream function results for 40 waves which span the cases from deep to shallow water. These 40 waves are shown in Figure 4. For each of these cases, given in dimensionless form, all relevant wave characteristics are determined, including simplified wave force and moment calculations.

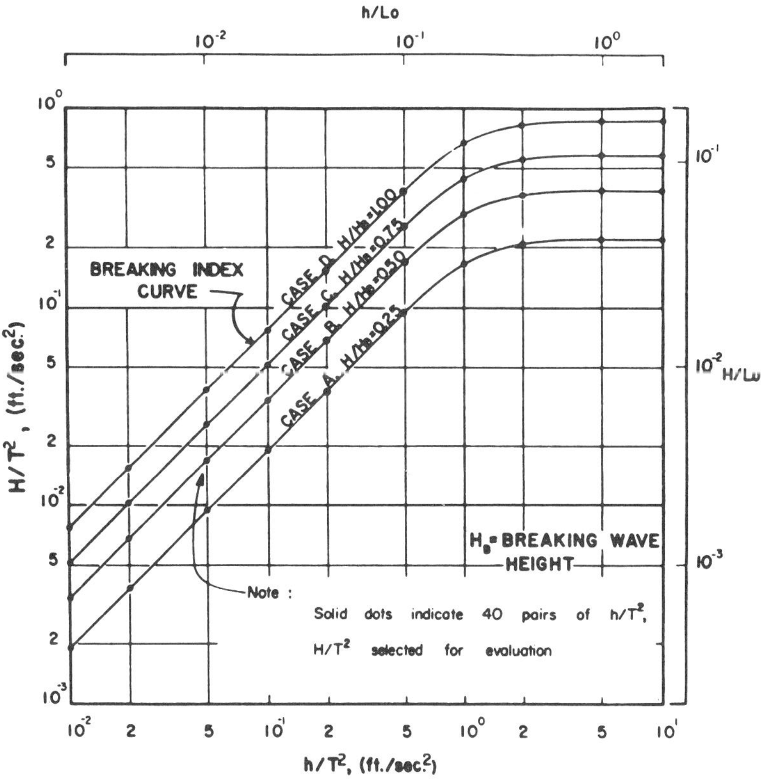

Fig. 4. Wave characteristics selected for evaluation.

Spectral models have been discussed by Borgman (*31*). Recently, directional spectral models have been used in design to examine the effects of the directionality of the waves, which in some instances can be more important than the nonlinearities. Berge (*32*) discusses the significance of offshore design with directional spectra.

D. *Wave-Induced Forces*

The basic equation for the calculation of wave forces is the Morison equation, developed by Morison *et al.* (*33*). This equation differs from the wind force equation by the inclusion of a term for the inertial forces due to accelerations in the wave. For horizontal forces, the Morison equation, in terms of the horizontal water particle velocity, u reads

$$F = \tfrac{1}{2}\rho_w C_D A|u|u + C_M \rho_w V \frac{Du}{Dt}. \tag{13}$$

Here, ρ_w is the density of sea water, C_D and C_M are the drag and inertia coefficients, respectively, which are dependent on the Reynolds number and the shape of the object. A and V are the projected area and volume of the structure. The absolute value sign in the drag force term is necessary to ensure that the drag force remains in the direction of the velocity.

The total acceleration of the fluid is given by four terms: the local change with time and the convective accelerations.

$$\frac{Du}{Dt} = \frac{\partial u}{\partial t} + u\frac{\partial u}{\partial x} + v\frac{\partial u}{\partial y} + w\frac{\partial u}{\partial z}. \tag{14}$$

When the velocities are small, the last three (nonlinear in velocity) terms may be dropped. For small objects, the Morison equation can be used to compute the total force on the structure, interpreting A and V as the total projected area and volume. For larger objects, A and V are interpreted as the area and volume per unit elevation, and the force, F, is integrated over the length of the structure. Dean (*30*) shows how to use the stream function wave theory to calculate the wave forces and moments on a structure. The total force, F, is a function of time and will vary over a wave period. The maximum force is found by determining the values of F as the wave advances past the structure.

In Figures 5a and 5b, the variation of the drag and inertial coefficients obtained from Gulf of Mexico field data are shown for cylinders with 2, 3, 3.71, and 4 foot diameters (*34*). Clearly, for this example, the inertial coefficient is not strongly dependent on the Reynolds number. For other water depths, pile diameters, and wave conditions, other data should be

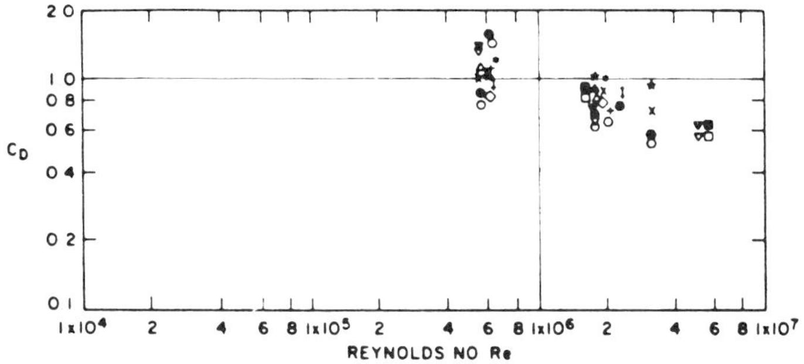

D/A Symbol			UF Symbol		Wave Project
(Ft)	Inline	Resultant	Inline	Resultant	
2■	○	●	◇	‡	I■
3■	▲	▲	+	•	
4■	▼	▼	○	●	
3 71‡	□	■	x	*	II†

■ Water Depth ~ 33 Ft
† Water Depth ~ 100 Ft
Note Wave Project I is a Combination of All Pile Dimensions

Fig. 5a. Data analysis results comparing UF results with Dean and Aagaar—drag coefficient versus Reynolds number (10% less data set).

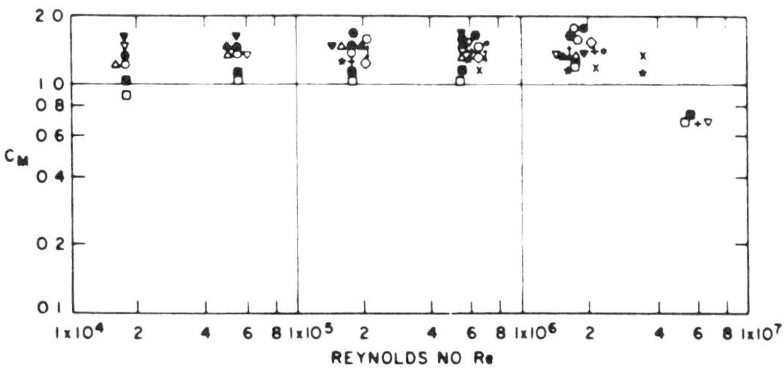

D/A Symbol			UF Symbol		Wave Project
(Ft)	Inline	Resultant	Inline	Resultant	
2■	○	●	◇	‡	I■
3■	▲	▲	+	•	
4■	▼	▼	○	●	
3 71‡	□	■	x	*	II†

■ Water Depth ~ 33 Ft
† Water Depth ~ 100 Ft
Note: Wave Project I is a Combination of All Pile Dimensions

Fig. 5b. Data analysis results comparing UF results with Dean and Aagaard—inertia coefficient versus Reynolds number (10% less data set).

examined for the drag and inertia coefficients. *The Proceedings of the Offshore Technology Conference* contains numerous papers which report these data. Examples are the papers resulting from the Ocean Test Structure, a 20′ × 40′ × 120′ instrumented platform placed in 66 feet of water in the Gulf of Mexico by Exxon Production Research Company (e.g., *35*, *36*, *37*).

Another parameter often used to compare drag and inertia coefficients is the Keulegan-Carpenter number, defined as

$$\text{K-C} = \frac{u_m T}{D} \tag{15}$$

where u_m is the amplitude of the local horizontal current, T is the wave period of the design (or measured) wave and D is a relevant diameter. For values of K-C less than 10, inertial forces dominate the drag force. If K-C exceeds 15, then the drag force is more important. Sarpkaya (*38*) provides curves of C_D and C_M versus the K-C number for a circular cylinder.

For complex structures with numerous legs and cross-bracing, the forces on each of the elements exposed to the wave must be computed at an instant in time and the results summed to determine the total force on the structure at that time. Various elements of the structure will be exposed to different phases of the wave at that instant, therefore the wave forces must be determined as a function of time (as the wave passes a structure) and the time for the maximum force must be found.

For very large objects, such as North Sea gravity platforms, the drag forces are small when compared to the inertia forces. In this situation, diffraction theories based on linear wave theory are appropriate, as the inertia force can be treated approximately linearly. The diffraction theory is based on the use of Green's functions, which are exact solutions of the Laplace equation that governs the irrotational motion of an incompressible fluid. The Green's functions, which can be envisioned as point wave sources, are distributed over the surface of the structure in order to ensure a no-flow boundary condition on the structure in the presence of an incident wave field. From pressure calculations, derived from the Green's function solution, forces on the structure are obtained. Examples of such methods are found in Garrison and Stacey (*39*), Garrison (*40*), Hogben and Standing (*41*), and Isaacson (*42*).

Recently, second order diffraction theory has been the focus of considerable research. Most of these methods are based on a Stokes expansion, using the first order diffraction results. To date, it has not been conclusively shown that these methods are accurate; in fact, no intercomparison of the different methods has been carried out. The reader is referred to the following works; Isaacson (*42*), Kurata and Ijima (*43*), Huang and Hudspeth (*44*), Hunt and Baddour (*45*), and Molin (*46*).

E. Currents

The presence of tidal or wind-driven currents with velocity, **U**, can greatly influence the loading on a structure. It is the drag force primarily that is affected, due to its quadratic dependency on velocity. For currents which are small with respect to the wave speed, the currents have only a small effect on the wave characteristics (*26*). In this case, it is reasonable to superimpose the currents, **U**, vectorially on the velocities, **u**, imposed by the waves. For the drag force, then,

$$F = \tfrac{1}{2} C_D A |\mathbf{u} + \mathbf{U}| (\mathbf{u} + \mathbf{U}) \tag{16}$$

where **u**, **U** are vector velocities. There is a slight modification of the inertia force due to the changes in the convective acceleration terms. For compliant structures, which move in response to the wave action, the velocity used in the Morison equation must be the relative velocity, as seen from the moving structure.

IV. Ice

Increasing coastal and ocean development in the arctic and other regions subjected to ice action has resulted in a rapid growth in the number of papers and reports dealing with ice and other arctic-related problems. The research results have been presented in various international conferences such as the Ice Symposia of the International Association for Hydraulic Research (IAHR), the Conferences of Port and Ocean Engineering under Arctic Conditions (POAC), the Offshore Technology Conferences (OTC), and the Symposia on Applied Glaciology. Furthermore, during the last two decades a series of large-scale projects culminating with the Arctic Ice Dynamics Joint Experiment (AIDJEX) (*47*) significantly improved our understanding of the growth, motion, and decay of sea ice in the interior of the Arctic Ocean. Long-term monitoring programs on the annual and inter-annual variability of the polar ice margins and their relationship with the atmospheric and ocean circulations have been in progress as well (*48*). These research results are normally published in the *Journal of Geophysical Research*. Since the literature is diverse and abundant, attempts have been made to report the state of the art of ice or arctic related problems (*49–54*). This very brief review of ice environments and associated engineering problems is mainly based on these references.

A. Ice Formation and Sea Ice Features

Ice is a complicated material because of the large number of processes that may intervene during its formation and growth from the liquid phase. If the water surface is very calm, it will become supercooled and a few crystals will appear to form plate ice of uniform texture. However, winds, current, and waves will generally mix and supercool a layer of water below the surface. Nucleation will initiate the formation of small frazil particles of discoidal shape. As they grow and agglomerate in slush they will float to the surface and refreeze to form the first layer of ice (*49*). Normal sea ice sheets grow as columnar-grained crystals. The strength and deformability of the ice is strongly dependent on the crystalline structure. The major difference between sea ice and fresh water ice is the amount of salts in the form of concentrated brine pockets in sea ice. Sea ice generally consists of ice crystals, brine, and precipitated salts. The liquid brine in sea ice tends to drain under the influence of gravity and migrate due to molecular diffusion. As a general rule, the older the sea ice, the fresher it will be. The salinity of sea ice that is less than one year-old typically ranges from 4 to 6 parts per thousand, while multi-year sea ice that has lasted through more than one melt season is normally less saline and mechanically stronger (*53*).

The thickness of first-year sheet ice depends on the rate of air-ice-sea heat exchange, the presence of insulating factors such as snow, and whether the ice remains fixed in one location or migrates. As an example, the first-year sheet ice will be typically 2 feet thick in Cook Inlet, 4 feet in Norton Sound, and 6 feet in the Southern Beaufort Sea along the south, west, and north coasts of Alaska, respectively (*53*). However, ice coverage varies from season to season and from site to site. In the Beaufort Sea freeze-up generally occurs in early October with complete ice coverage in mid-winter. On the other hand, mid-winter ice coverage in the Bering Sea is highly variable since sea ice formed in the northern regions tends to be transported towards the southern open-water regions. The first-year sheet ice may remain smooth and intact in protected bays and lagoons, as well as in the vicinity of shorelines as land-fast ice. It may also be transformed into other features such as rafted ice, pressure ridges, shear ridges, and rubble piles. Rafted ice consists of two or more ice sheet thicknesses as a result of overriding. Pressure ridges are formed from collisions between two ice features by relative motion perpendicular to their common boundary, whereas shear ridges are formed by shearing action between two ice features (*51*). Rubble piles are usually grounded on the sea bottom and are composed of several ice pileups. Floating rubble piles in deep water are also referred to as rubble fields or hummock fields. Multi-year ice features that have survived one or more melt seasons are mostly formed from first-year deformed ice features. It should be noted that icebergs are of fresh water origin, since they are separated from the ice cliff of a glacier reaching the sea.

Statistical ice data giving ice thickness, coverage, and movement, as well as occurrence and size of different deformed ice features, is typically obtained from satellite remote sensing, ship observations, and aerial reconnaisance. However, an available ice data base, if any, is usually too short in time, too sparse in space, and too meager in quality for practical engineering applications. Alternatively, a mathematical model is needed to generate or supplement the ice data base by hindcasting the ice conditions at a specific site using available historical meteorological data in a manner similar to wave hindcasting. Coupled nonlinear models for predicting the behavior of sea ice have been developed and calibrated in the AIDJEX program (*47*). However, these models deal with very large-scale sea ice behavior on a space scale of 100 km and a time scale of one day. The behavior of sea ice on much smaller space and time scales is generally required for engineering applications.

B. Mechanical Properties of Ice

Sea ice is a crystalline material whose behavior may be brittle, elastic, plastic, or viscoelastic, depending on a number of factors such as strain rate, temperature, brine volume, age, directionality of loading, and degree of confinement. Because of these factors, there is a wide variation in ice strength measurements quoted in the literature without adequate information on testing conditions and ice types (*49*). As a result, general relationships for the strength of ice have not yet been established.

For design purposes, the important mechanical properties of ice include compressive, flexural, shear, and tensile strengths as well as elastic modulus and Poisson's ratio. Compressive strength of unconfined ice is normally used for predicting the ice force resulting from ice crushing failure. In general, the compressive strength of unconfined ice increases as the strain rate is increased and reaches a peak at strain rates in the order of 10^{-3} sec^{-1}, after which an increase in strain rate may result in an apparent decrease in strength. This trend is frequently interpreted as a change in ice behavior from being ductile to brittle as the strain rate is increased (*53*). Under ductile conditions the ice continues to deform with corresponding lower stresses, whereas under brittle conditions a brittle crack is formed and propagates across an ice sample for a complete fracture (*49*).

The flexural strength of sea ice is approximately an order of magnitude less than its maximum compressive strength. As a result, inclined structures have been proposed to induce bending failures in sea ice before crushing failures occur. The flexural strength of sea ice determined from beam tests decreases as the brine volume is increased (Vaudrey in reference *52*). Only limited testing has been made for other mechanical properties of sea ice. In short, the mechanical properties of sea ice are not well established since so many factors influence its strength. In addition to these strength properties, the design of ice resistant structures needs to account for forces from the large scale processes such as ridge building, rafting, and rubble piling, which are also not well understood.

C. *Ice Forces on Structures*

One of the earliest ice problems was the ice force exerted on dams caused by the thermal expansion of a continuous ice sheet constrained in a reservoir. In addition, bridge piers and lighthouses were placed in moving ice. These structures typically employed vertical cylindrical elements passing through the waterline. As the moving ice sheet was forced past a vertical-sided structure, the ice sheet failed in crushing and the structure appeared to be indenting the ice sheet. Since ice features in the marine environment tend to move under the action of winds and currents, the ice force on vertical-sided marine structures is generally predicted using an indentation formula such as that recommended by API (*51*). The indentation force is normally assumed to be proportional to the product of the unconfined compressive ice strength and the projected structural area struck by the ice. The proportionality constant depends on the crystallography and the failure pattern across the width of the indentor. The average drift velocity of the surrounding ice field determines the strain rate which strongly affects the compressive strength of sea ice. The indentation formula is simple and easy to apply except for some uncertainties associated with its coefficients. However, estimation of the design ice force on a vertical-sided structure requires statistical information on different ice features, specifically, their thickness and strength, which the structure will encounter during its service life. The governing design ice feature may comprise a double- or triple-thickness of rafted ice or a floating hummock field with a consolidated refrozen core.

Inclining the cutting edge of a pier usually produces an important decrease in the horizontal component of the ice force and the appearance of a vertical component (*49*). The sheet ice force exerted on a conical-shaped structure may be separated into the fracture and ride-up forces. The fracture force is that force required to fracture the ice sheet in bending as it is pushed against the structure. The ride-up force is that force required to push the already broken pieces of ice up the inclined slope of the structure and around it. The formula developed by Ralston on the basis of a plastic limit analysis is widely used to predict the sheet ice force (*52*). The ice force caused by a ridge has been analyzed regarding the ridge as a beam on elastic foundation (*53*). Ice phenomena to be investigated for the design of marine structures in ice infested waters also include ad-freeze, pileup, ride-up, and ice jamming. Ad-freeze is the phenomenon of ice becoming frozen fast to a structure. Vertical loads of ad-frozen ice caused by tidal fluctuations will be important if the tidal range is large. Ice pileup and ride-up are associated with the onshore movement of floating ice sheets on inclined slopes. Shore ice ride-up can extend over 100 m inland and destory structures (Sodhi and Kovacs in reference *54*). Ice jamming is the accumulation of unconsolidated ice pieces against structures. Multiple-legged structures are especially likely to become jammed, thus enhancing the risk of blocking large ice features with an accompanying major increase in horizontal loads. To reduce the total force and the possibility of local damage, a structural configuration should be selected so as to limit the number of structural components in the ice zone.

D. Navigation in Ice

The design of a transportation system can be based on two approaches. The first is utilization of an icebreaker supported system for which the breadth of ships assisted by an icebreaker should not exceed that of the icebreaker. The second approach is use of ice-strengthened ships which navigate on their own. The design of icebreaking ships is strongly dependent on the characteristics of the sea ice to be broken, such as ice thickness, strength, drift, and features. The total resistance of a ship progressing in the continuous icebreaking mode may be expressed as the sum of the forces needed to: (1) break the ice in bending and submerge the broken ice pieces, (2) impart a lateral velocity to the ice pieces and overcome ice friction on the hull, and (3) propell the ship in open water (*49*). In general, physical modeling using model ice in a laboratory is performed to verify and calibrate the preliminary design based on analytical modeling and design experiences.

V. Chemical and Biological Characteristics

A. Seawater Chemistry — The Major Features

1. Major Ions

The concentrations of the major constituents of full strength seawater are shown in Table III. Major constituents are considered to be those whose

TABLE III

CONCENTRATIONS OF THE MOST ABUNDANT IONS AND MOLECULES IN SEA WATER OF 35‰ SALINITY. (DENSITY = 1.023 g/cm^3 AT 25°C)

Ion or Molecule	Concentration (m moles/kg-sw)	(g/kg-sw)
Na^+	468.5	10.77
K^+	10.21	0.399
Mg^{+2}	53.08	1.290
Ca^{+2}	10.28	0.412
Sr^{+2}	0.09	0.008
Cl^-	545.9	19.354
Br^-	0.84	0.067
F^-	0.07	0.0013
HCO_3^-	2.30	0.140
SO_4^{-2}	28.23	2.712
$B(OH)_3$	0.416	0.0257

concentrations are greater than 1 mg/L and are not greatly affected by biological processes. The behavior of these major ions and molecules is said to be "conservative" because their concentrations bear a relatively constant ratio to each other over a wide range of dilutions. Although most of the known elements can be found dissolved in seawater, the eleven ions and molecules listed in Table III account for over 99.85% of the total dissolved solids. Moreover, the conservative nature of these major ions means that all of their concentrations can be calculated if the concentration of any one of them is measured.

The most commonly measured chemical property of seawater is its salinity. Historically, salinity, S, in parts per thousand (‰), has been defined as: "the total weight in grams of inorganic salts in one kilogram of seawater, when all bromides and iodides are replaced by an equivalent quantity of chlorides, and all carbonates are replaced by an equivalent quantity of oxides" (*55*, *56*). Salinity is usually determined by measuring either the chlorinity or the electrical conductivity of the seawater. Chlorinity (Cl) is defined as: "the mass in grams of silver required to precipitate the halogens in 0.3285234 kg of seawater" (*56*). This is nearly equal to the mass of chloride in the seawater sample. Chlorinity is related to salinity by the relationship:

$$S(‰) = 1.80655\,\mathrm{Cl}(‰) \quad (17)$$

Provided that pure water is the only substance added to or removed from seawater, the concentration of any ion, χ, from Table III at a salinity other than 35‰ can be calculated from the following relationship:

$$\text{(Concentration of ion } \chi \text{ at salinity S‰)} = \text{(Concentration of } \chi \text{ at 35‰ salinity} \times \text{(S‰)/(35‰)}. \tag{18}$$

For example, using equation (18) and Table III, the concentration of sodium ion in seawater of 20‰ salinity is:

$$\begin{aligned}(Na^+ \text{ at } 20‰) &= (0.468 \text{ moles/kg} - \text{sw}) \times (20‰)/(35‰) \\ &= 0.2667 \text{ moles } Na^+/\text{kg of seawater}\end{aligned} \tag{19}$$

This relationship may break down at very low salinities (less than 10‰ for Ca^{+2} and HCO_3^- and less than 5‰ for the other major ions). The relationship may also break down in grossly polluted seawater. In this latter case, the concentration of the ion of interest must be known in both seawater and in the solution being added to it.

The conservative nature of the major ions is due to the fact that the chemical and geological processes that affect their concentrations are slow relative to the 1500-year mixing time (approximate) of the world ocean (*57*). In the short term, the only processes that affect the concentrations of the major ions are evaporation, precipitation, and river discharge. Since water is the only substance added to or removed from seawater by the first two of these processes, they affect the absolute concentrations of each ion but have no effect on the concentration ratios.

River discharge does add constituents other than pure water. Hence, the conservative behavior of the major ions may not hold in the low salinity regions near river outfalls. To a first approximation, river water is a 0.4 millimolar solution of calcium bicarbonate (*58*) as shown in Table IV. At a salinity of 10‰, the calcium concentration from equation (18) and Table III, but ignoring the river input, will be 10% too low. The error for bicarbonate ion will be even larger. Errors in concentrations of the other major ions calculated by ignoring the river input, however, will be much less due to their relatively high concentrations in seawater compared to those in river water.

2. Dissolved Gases and Other Minor Ions

Many of the minor ions that are important to corrosion processes are dissolved gases such as carbon dioxide and oxygen. Their concentrations are not conservative, as they are influenced by air-sea exhange as well as by

TABLE IV

CONCENTRATIONS OF THE MOST ABUNDANT IONS AND MOLECULES IN AVERAGE RIVER WATER

Ion or Molecule	Concentration	
	m moles/L	mg/L
Na^+	0.274	6.3
K^+	0.059	2.3
Mg^{+2}	0.171	4.1
Ca^{+2}	0.375	15.0
Cl^-	0.220	7.8
HCO_3^-	0.958	58.4
SO_4^{-2}	0.117	11.2
NO_3^-	0.016	1.0
Fe^{+2}	0.012	0.67
$Si(OH)_4$	0.218	20.9

biochemical processes. The concentration of dissolved oxygen in surface waters will usually be within a few percent of the equilibrium saturation value with atmospheric oxygen. The solubility of oxygen in seawater varies inversely with both temperature and salinity, but the effect of temperature is larger. If the absolute temperature (T, °K) and salinity (S‰) are known, the solubility of oxygen can be calculated from the relation:

$$\ln O_2 \text{ conc.}\left(\frac{\text{mL}}{\text{L}}\right) = A_1 + A_2\left(\frac{100}{T}\right) + A_3\ln\left(\frac{T}{100}\right) + A_4\left(\frac{T}{100}\right) + S‰\left[B_1 + B_2\left(\frac{T}{100}\right) + B_3\left(\frac{T}{100}\right)^2\right]. \quad (20)$$

The constants A_1 through B_3 are given in Table V, and the equilibrium oxygen saturation levels in mL/L as a function of temperature and salinity calculated from Equation (20) and Table V are show in Table VI. More detailed tables are given by Kester (*59*). The preferred units for dissolved oxygen among chemical oceanographers are μ mol/kg. Thus, values of the constants used in equation (20) in those units are shown in Table V in parentheses.

Generally, surface ocean waters are in equilibrium with the atmosphere. Two sets of conditions, however, can lead to the waters becoming substantially supersaturated with oxygen. The first of these conditions is oxygen production due to photosynthesis by microscopic marine plants called

TABLE V

CONSTANTS FOR USE WITH EQUATION (20)*

A_1	A_2	A_3	A_4
−173.4292 (−173.9894)	249.6339 (255.5907)	143.3483 (146.4813)	−21.8492 (−22.2040)

B_1	B_2	B_3
−0.033096 (−0.037362)	0.014259 (0.016504)	−0.0017000 (−0.0020564)

* To yield O_2 concentration in mL/L relative to air at 760 mm Hg total pressure and 100% relative humidity. (Values in parentheses yield O_2 concentration in μmol/kg).

phytoplankton. During the spring phytoplankton bloom, or in upwelling regions rich in nutrients, intense photosynthesis can produce concentrations as high as 200% saturation for periods of up to a few weeks. Such oxygen supersaturation is most often found in near-shore regions as a transient phenomenon. Once the bloom is over and photosynthesis returns to normal levels, the surface waters quickly reequilibrate with the atmosphere.

The second condition that may cause oxygen supersaturation is the entrainment of air bubbles due to wave action. This factor will not usually cause supersaturations greater than about 10% because vigorous wave action also promotes reequilibration with the atmosphere.

TABLE VI

SOLUBILITY OF OXYGEN (mL/L) IN SEA WATER AS CALCULATED FROM EQUATION (20)

	Salinity (‰)					
T(°C)	0	8	16	24	31	36
0	10.22	9.70	9.19	8.70	8.27	7.99
5	8.93	8.49	8.05	7.64	7.28	7.04
10	7.89	7.52	7.14	6.79	6.48	6.28
15	7.05	6.72	6.40	6.10	5.83	5.65
20	6.35	6.07	5.79	5.52	5.29	5.14
25	5.77	5.52	5.27	5.04	4.84	4.70
30	5.28	5.06	4.84	4.63	4.45	4.33

The concentration of carbon dioxide is less affected by air-sea interchange than is that of dissolved oxygen, due to the fact that the carbon dioxide system in seawater is buffered by the presence of bicarbonate and carbonate ions. Carbon dioxide is a weak acid and undergoes two ionizations in aqueous solutions (*60*).

$$CO_2 + H_2O = H^+ + HCO_3^- \qquad \text{First ionization} \tag{21}$$

$$HCO_3^- = H^+ + CO_3^{-2} \qquad \text{Second ionization} \tag{22}$$

Equations (21) and (22) can be represented by the appropriate equilibrium constants:

$$K_1 = \frac{(H^+)(HCO_3^-)}{(CO_2)} = 1 \times 10^{-6} \qquad \text{First equilibrium constant of carbonic acid} \tag{23}$$

$$K_2 = \frac{(H^+)(CO_3^{-2})}{(HCO_3^-)} = 8 \times 10^{-10} \qquad \text{Second equilibrium constant of carbonic acid} \tag{24}$$

At pH = 8, where the hydrogen ion concentration is 1×10^{-8}, calculations using the above equations show that 93% of the total inorganic carbon is present as HCO_3^-, 6% as CO_3^{-2}, and 1% as CO_2. Bicarbonate accounts for at least 85% of the total inorganic carbon under all naturally occurring conditions. However, the relative concentraions of CO_2 and CO_3^{-2} vary greatly depending on pH. Surface seawater characteristically has a pH value greater than 8, due to the combined effects of air-sea exchange and photosynthesis. Consquently, the carbonate ion concentration is relatively high in surface waters, and surface waters are nearly always supersaturated with respect to the calcium carbonate phases, calcite and aragonite.

The concentrations of carbon dioxide and oxygen are closely coupled and related to the pH of seawater through the processes of photosynthesis and biochemical oxidation as represented in the following general reaction:

$$CH_2O + O_2 \underset{\text{Biochemical oxidation (respiration)}}{\overset{\text{Photosynthesis}}{\longleftrightarrow}} CO_2 + H_2O. \tag{25}$$

CH_2O represents a typical organic carbohydrate molecule. During decomposition of this organic material in seawater, equation (25) runs from left to right, dissolved oxygen is consumed, and CO_2 is produced. Production of

CO_2, in turn, makes the water more acidic (i.e., lower pH) and decreases the saturation state with respect to carbonates.

The exact concentrations of CO_2, HCO_3^- and CO_3^{-2} depend on the total concentration of inorganic carbon (TCO_2):

$$TCO_2 = CO_2 + HCO_3^- + CO_3^{-2} \tag{26}$$

and on the values of the first and second equilibrium constants of carbonic acid, which are functions of temperature, salinity, and pressure (*60*). The data in Table VII illustrate the effects of temperature and pressure on the equilibrium constants in seawater. It should be noted that chemical equilibria are pressure dependent. This is especially true for reactions, such as the dissolution of calcium carbonate, involving solids which ionize upon solution.

B. Seawater Chemistry—Its Variability

Now that we have a general picture of the major chemical characteristics of seawater, let us examine another feature that has an effect on corrosion. That is, the variability of seawater. In the following sections, data will be presented on both the horizontal and vertical variations in seawater. Data will first be presented on the general large scale distribution of temperature, salinity, dissolved oxygen, and pH in surface waters. These four variables are the only ones important to corrosion for which there exists a large enough data base to permit general surface water mapping. From a corrosion point of view, it is not fruitful to do this over a large area in greater detail than is presented here because local conditions change so rapidly with time. These general horizontal distribution plots will serve only to illustrate trends with latitude and longitude. Following the horizontal distribution plots, variations with season and depth will be illustrated for several near-shore and deep-water locations for which data are plentiful.

TABLE VII

TEMPERATURE AND PRESSURE EFFECTS ON K_1 AND K_2.

Ionization Constant	Ratio of Const. at 25°C to Value at 2°C K_{25}/K_2	Ratio of Const. at 1000 atm to value at 1 atm K_{1000}/K_1	Approx. Value of Equilibrium Const. at 25°C 35 ppt, 1 atm
K_1	1.76	2.86	1.0×10^{-6}
K_2	2.43	2.05	8.0×10^{-10}

1. Horizontal Variations

The Pacific Ocean will be used as an example of large scale mapping of surface water variations in temperature, salinity, dissolved oxygen, and pH. The data presented in Figures 6 through 17 were taken from the *Russian Atlas of the Pacific Ocean* (*61*). Several additional sources of similar information are given by Dexter and Culberson (*62*).

Temperature

Figures 6 and 7 show the average surface temperatures of the Pacific Ocean for the months of February and July, respectively. The maximum tempera-

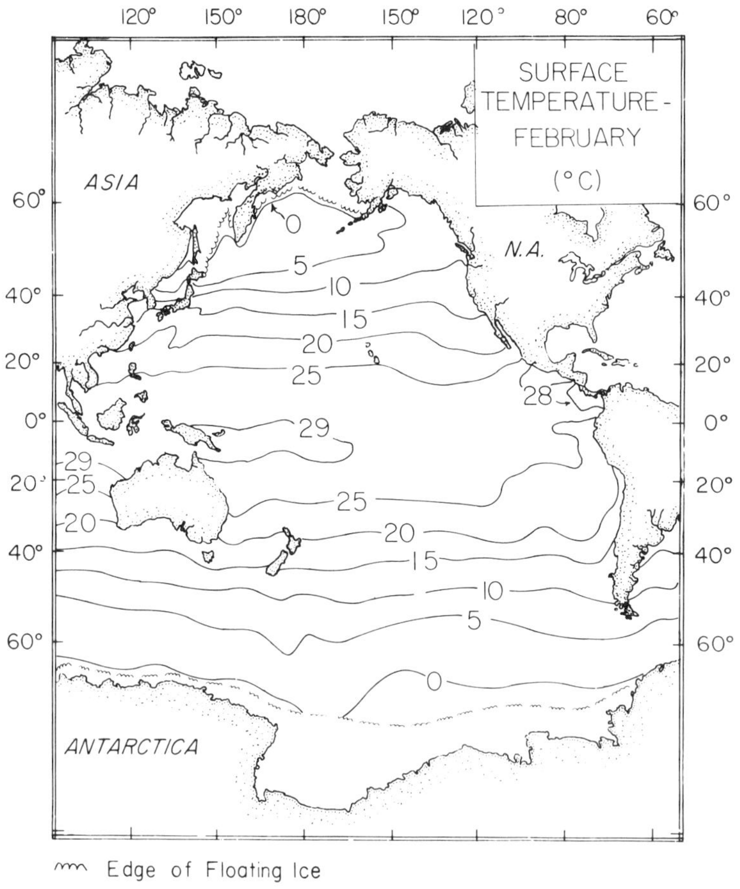

Fig. 6. Pacific Ocean surface temperature (C) distribution for February.

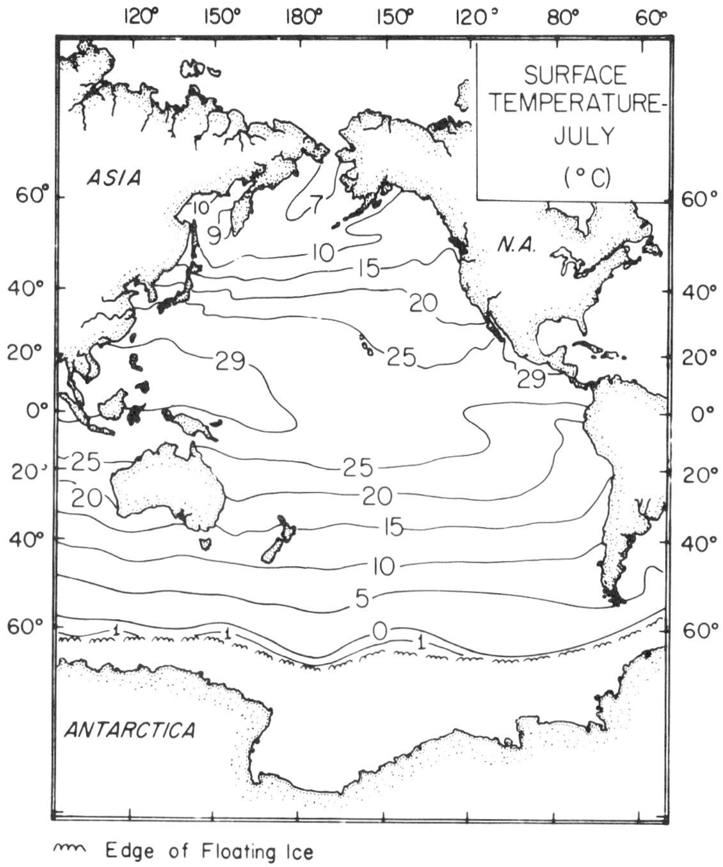

Fig. 7. Pacific Ocean surface temperature (C) distribution for July.

ture difference shown on these two charts is about 30°C. If all other factors were held constant, this would mean that the average corrosion rates should be roughly two times higher at the equator than in high latitudes. This general effect, however, is complicated and often overshadowed by increases in the saturation level of dissolved oxygen with decreasing temperature and by short-term local temperature fluctuations which do not appear on these figures. Differences in biofouling rates at various locations can also be an important factor.

Salinity

The large-scale salinity distribution for the Pacific Ocean is shown for the months of February and August in Figures 8 and 9, respectively. The

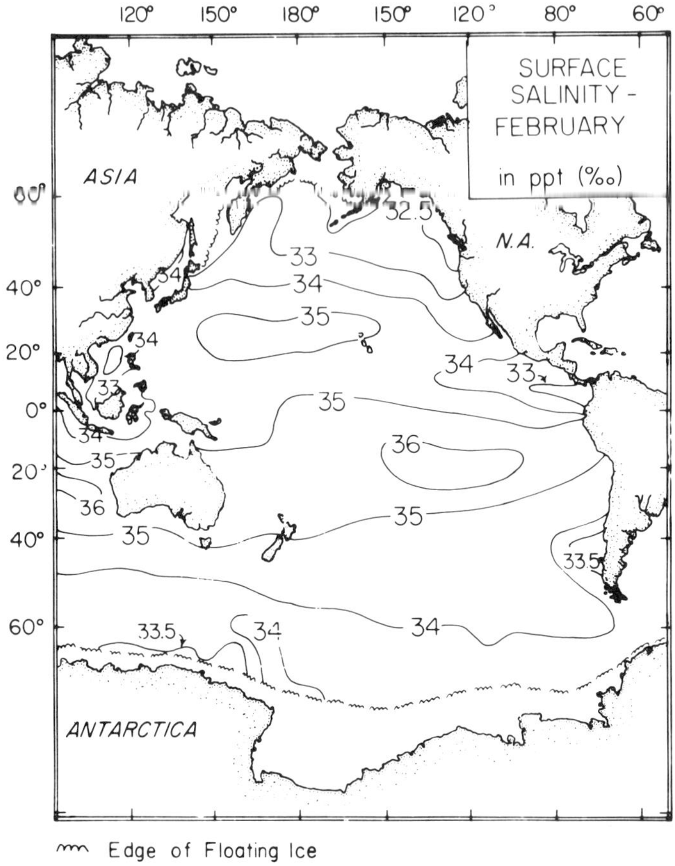

Fig. 8. Pacific Ocean surface salinity (‰) distribution for February.

maximum difference shown here (31.0‰ to 36.0‰) is secondary in importance to differences in dissolved oxygen and temperature for determining corrosion rates.

Dissolved Oxygen

The distribution of dissolved oxygen in the surface waters of the Pacific Ocean is shown in Figure 10 for the months of January through March. Comparison with Figures 6 and 7 reveals that the highest concentrations of oxygen coincide with the lowest temperatures. The surface distribution of oxygen for July through September is shown in Figure 11. Note particularly the lower oxygen values at high latitudes in the Northern Hemisphere as

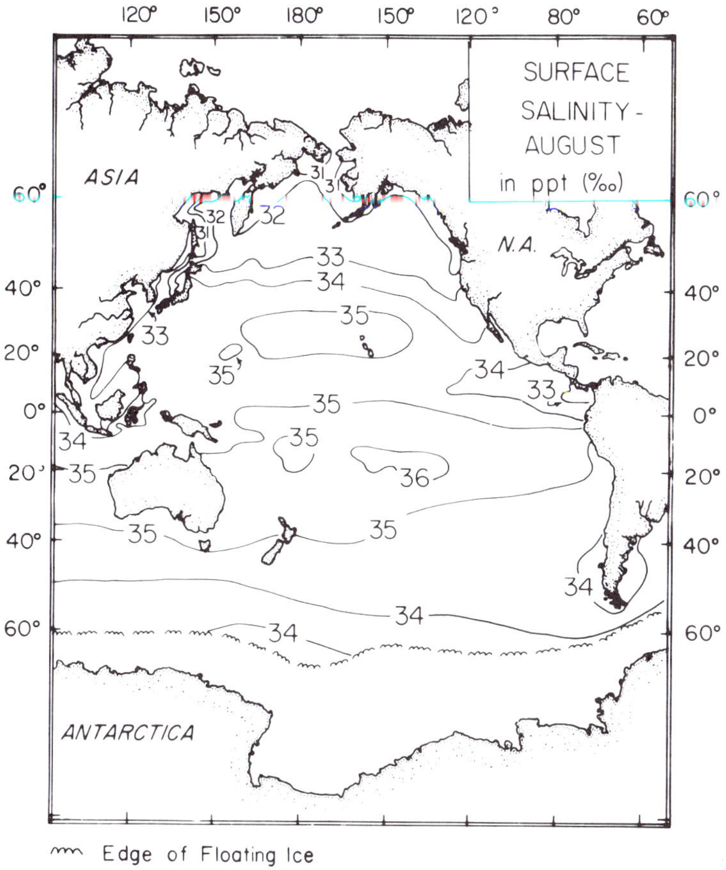

Fig. 9. Pacific Ocean surface salinity (‰) distribution for August.

compared to those in Figure 10. This seasonal reduction in oxygen is in response to warmer water temperatures and to the consequent decrease in oxygen solubility (refer to Table VI). As discussed previously, these surface waters are either saturated or supersaturated with oxygen at atmospheric conditions.

In contrast, deep waters are often undersaturated due to consumption of oxygen during the biochemical oxidation of organic matter. Figures 12 and 13 show horizontal maps of dissolved oxygen concentration in the Pacific Ocean at depths of 500 and 1000 m, respectively. Note the generally lower level of oxygen at these depths, especially in the northeastern Pacific. As will be discussed in more detail in a later section, the dissolved oxygen concentration in the Pacific Ocean goes through a minimum at intermediate depths.

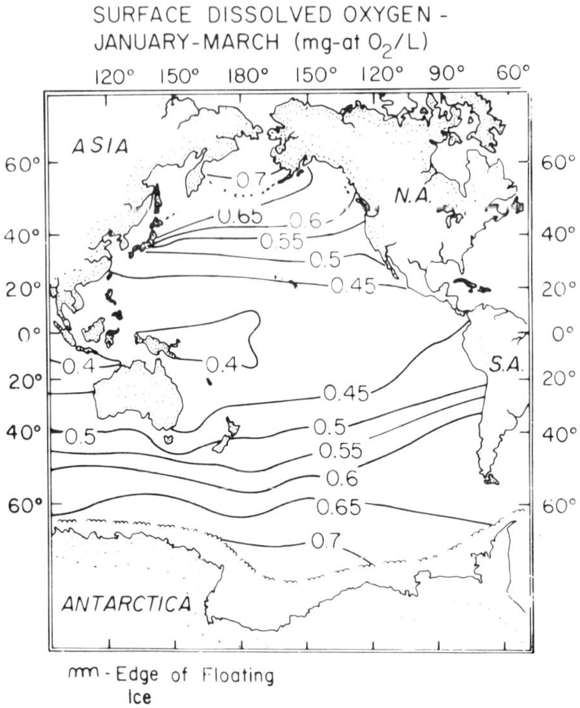

Fig. 10. Pacific Ocean surface dissolved oxygen (mg atm/L) distribution for January through March.

Figure 14 presents the depth of that minimum (solid contours) and the concentration of oxygen at that depth (dotted contours). The depth of the oxygen minimum ranges from 400 m in the eastern Pacific near the equator to over 2400 m in the central South Pacific. The concentration of oxygen at the depth of the minimum ranges from 0.01 to 0.40 mg-atm/L (mg-atm/L × 11.2 = mL/L).

pH

The pH of the Pacific surface waters ranges from 8.1 to 8.3, and its general distribution for the months of January to March is shown in Figure 15. Distributions of pH at depths of 500 and 1000 m are shown in Figures 16 and 17. In comparing these to the dissolved oxygen distributions at the same depths (Figures 12 and 13), note that the trends for the two variables are similar. For example, the dissolved oxygen maximum centered on 180°

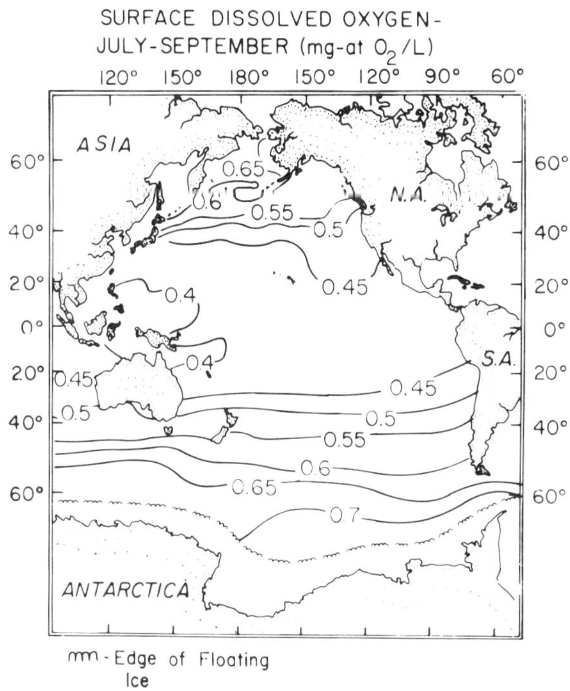

Fig. 11. Pacific Ocean surface dissolved oxygen (mg atm/L) distribution for July through September.

longitude between 20° and 40° north latitude in Figure 12 is reproduced closely for pH in Figure 16. Likewise, the gradual decrease in oxygen at 1000 m from southwest to northeast in Figure 13 is reproduced for pH in Figure 17. These two comparisons illustrate the closely coupled relationship between dissolved oxygen and pH mentioned in an earlier section. This relationship is found in all areas of the world ocean. At any given location, the shape of a pH profile with depth will be similar to that of the dissolved oxygen profile, and the minima (or maxima) of the two variables generally will occur at the same depth.

2. Variability with Depth

The variability of seawater with depth will now be illustrated for several deep-water and near-shore locations for which data are plentiful. These locations are shown in Figure 18 and Table VIII.

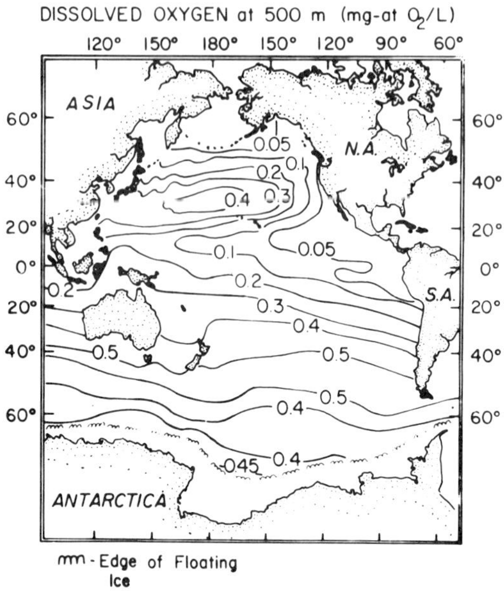

Fig. 12. Pacific Ocean dissolved oxygen (mg atm/L) distribution at 500 m depth.

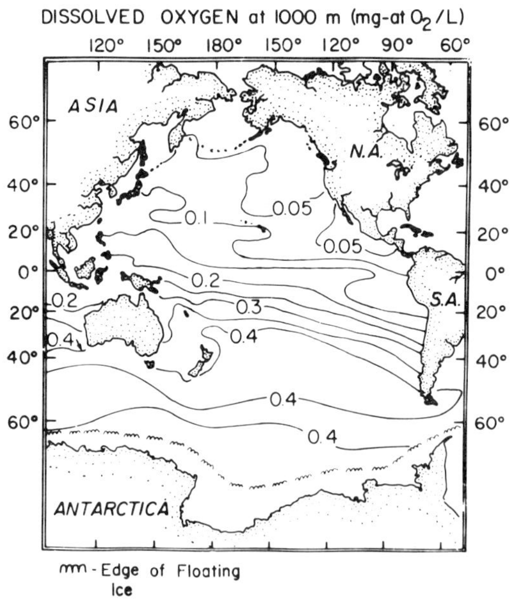

Fig. 13. Pacific Ocean dissolved oxygen (mg atm/L) distribution at 1000 m depth.

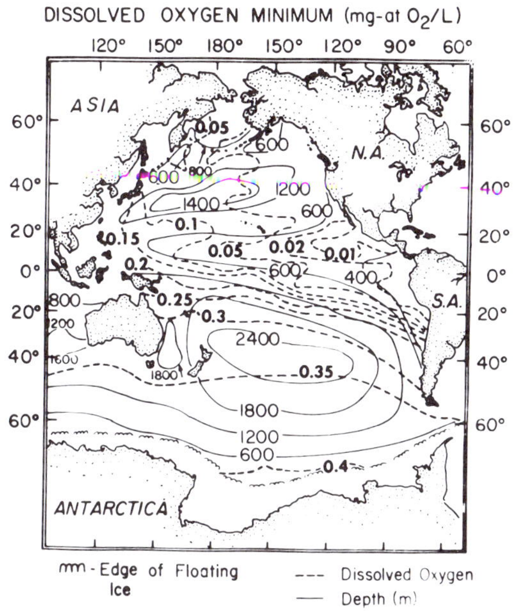

Fig. 14. Depth in meters of the dissolved oxygen minimum in the Pacific Ocean (solid contours) and the value of the minimum in mg atm/L (dashed contours).

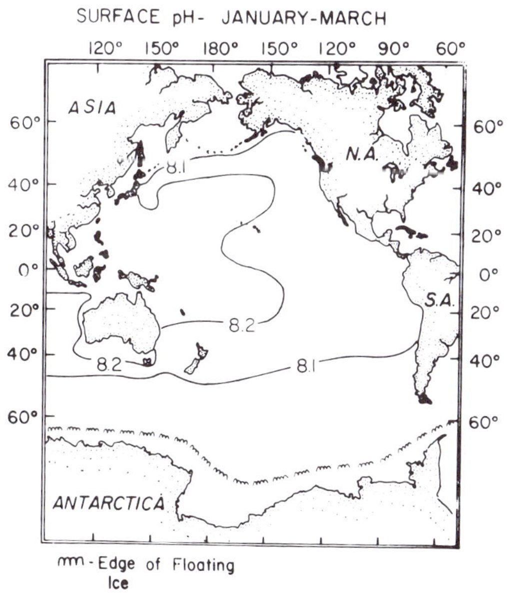

Fig. 15. Pacific Ocean surface pH distribution for January to March.

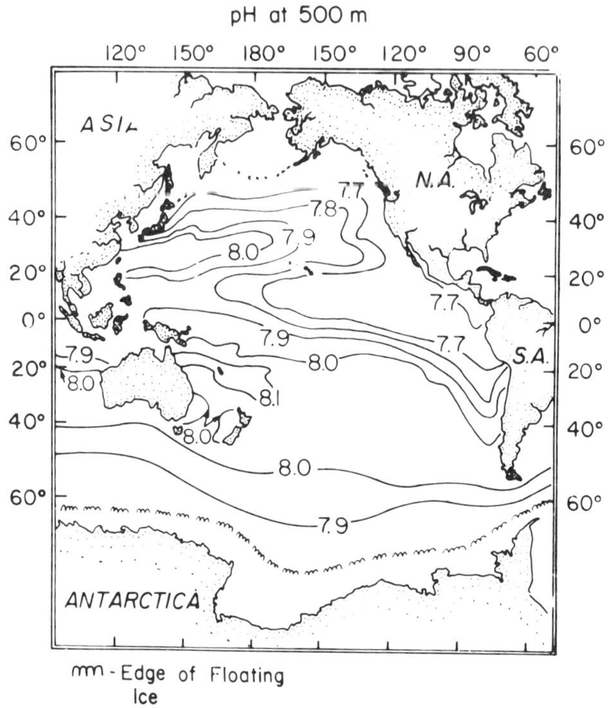

Fig. 16. Pacific Ocean pH distribution at a depth of 500 m.

In the Open Ocean

The two open ocean stations, location 2 in the northeastern Pacific and location 6 in the northwestern Atlantic, as shown in Figure 18, will be considered first. Typical profiles of temperature and salinity versus depth (*62*) are compared for these two sites in Figures 19 and 20. The North Pacific waters are colder and less saline than those in the North Atlantic. At depths greater than about 1500 m, however, the temperatures and salinities in both oceans are nearly the same from a corrosion point of view. This relative consistency of temperature and salinity in the deep waters of the world oceans is due to the fact that deep water is formed in only two locations: the North Atlantic north of Iceland (3°C, 34.9‰) and the Wedell Sea in the Antarctic sector of the South Atlantic (−2°C, 34.6‰). The deep waters of the Atlantic, Indian, and Pacific Oceans are composed of a mixture of these two water types. The slight differences in temperature and salinity noted above are important to chemical and physical oceanography but have little significance to corrosion.

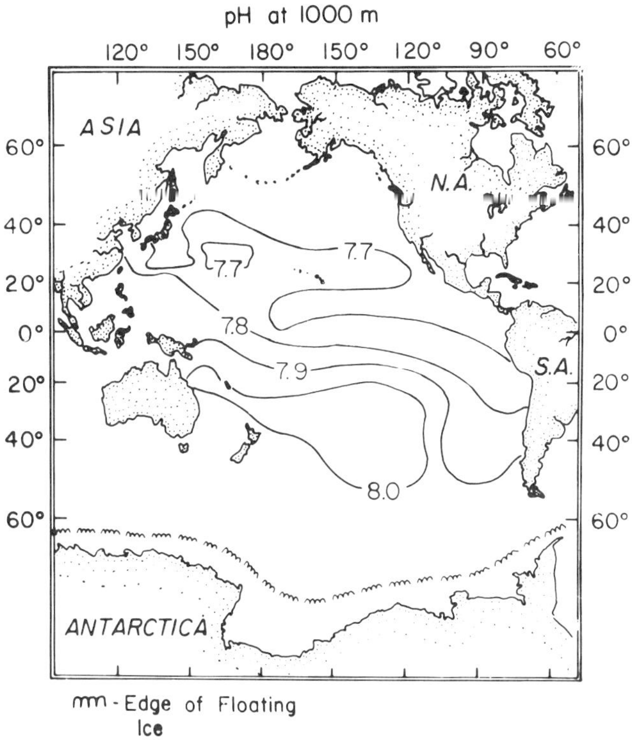

Fig. 17. Pacific Ocean pH distribution at a depth of 1000 m.

In terms of dissolved oxygen (Figure 21), the differences between the North Atlantic and Pacific Oceans are much greater. Surface values are higher at the Pacific site due to colder temperatures. At depths below 400 m, however, the Atlantic values are consistently higher. The lower oxygen concentrations in the deep North Pacific are due to the fact that deep North Pacific water has a considerably longer residence time than that in the North Atlantic. Consequently, the deep North Pacific water has lost more dissolved oxygen through biochemical oxidation of organic matter.

The pH profiles for these two open ocean locations are shown in Figure 22. A comparison of the corresponding pH and oxygen profiles from Figures 21 and 22 reveals the closely coupled nature of their relationship through the carbon dioxide system as discussed previously. Note that the oxygen and pH minima are reached at the same depth for a given location as was predicted. Note also that deep North Pacific water is from 0.15 to 0.40 pH units more acidic than that in the North Atlantic, due primarily to the increased oxidation of organic matter in the North Pacific.

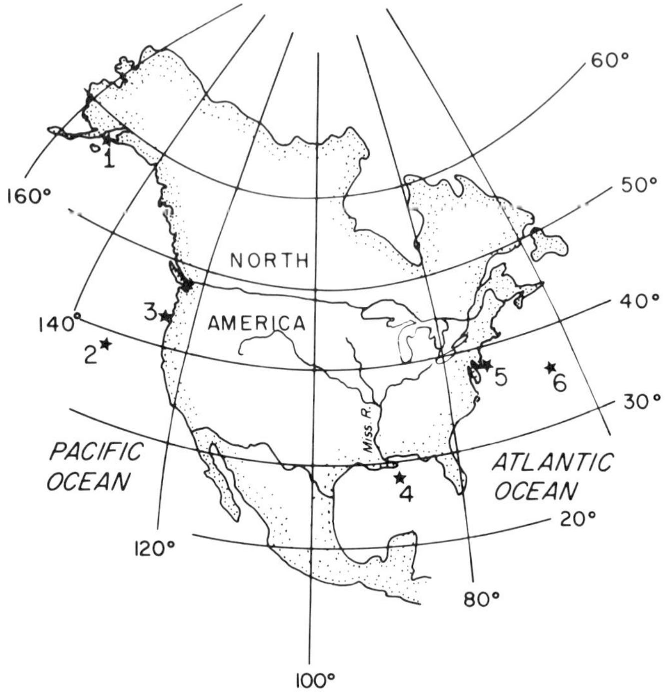

Fig. 18. Station positions for Figs. 19 to 30 (Table VII).

TABLE VIII

STATION POSITIONS FOR FIGURE 18.

Station no. (see Figure 18)	Chart coordinates	Reference
1	58° 57′ N	Kinney et al. (63)
Cook Inlet	152° 59′ W	
2	38° 21′ N	Dexter and Culberson (62)
N.E. Pacific	133° 38′ W	
3	44° 39′ N	Barstow et al. (64)
West Coast USA	124° 39′ W	
4	28° 12′ N	Leipper (65, 66)
Gulf of Mexico	88° 26′ W	
5	39° 20′ N	Dexter and Culberson (62)
East Coast USA	73° 40′ W	
6	36° 44′ N	Dexter and Culberson (62)
N.W. Atlantic	64° 28′ W	

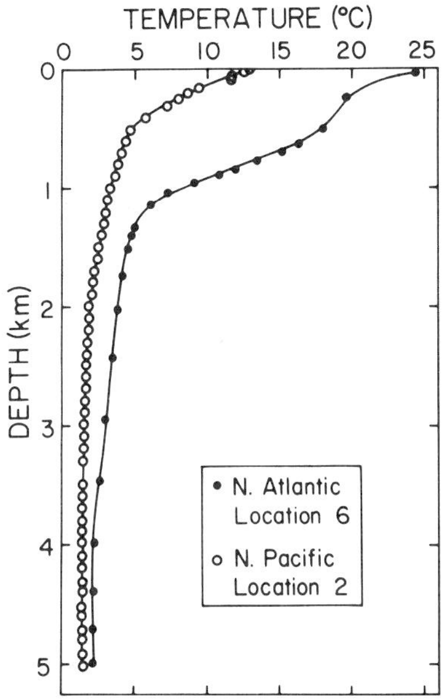

Fig. 19. Comparison of temperature–depth profiles for open ocean sites 2 and 6 of Fig. 18.

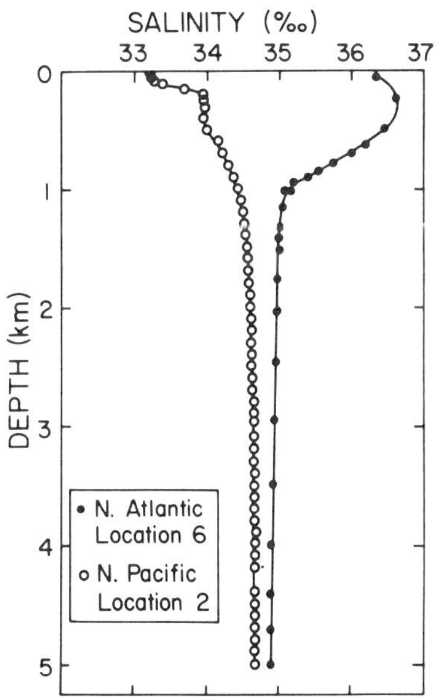

Fig. 20. Comparison of salinity–depth profiles for open ocean sites 2 and 6 of Fig. 18.

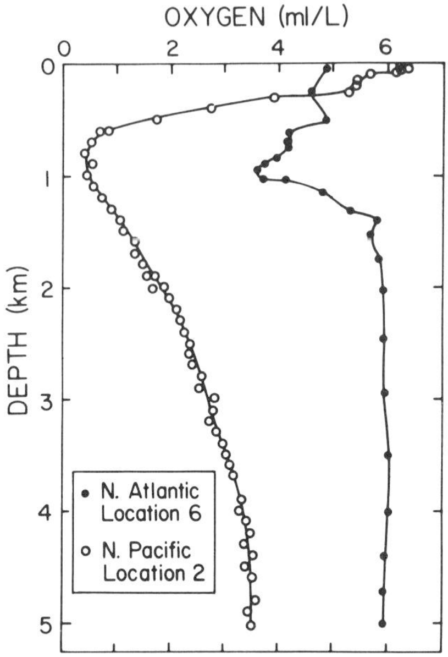

Fig. 21. Comparison of dissolved oxygen–depth profiles for open ocean sites 2 and 6 of Fig. 18.

In Coastal Waters

Temperature, salinity, and dissolved oxygen profiles (*63*) for the mouth of Cook Inlet (station 1, Figure 18) are shown for the month of May, 1968 in Figure 23. No reliable data are available for the winter months. Referring to Table IV, the deep waters between 75 and 150 m are saturated with oxygen at their temperature and salinity, while the surface waters, especially the top 25 m, are supersaturated.

The coastal waters off Oregon (site 3, Figure 18) are characterized for February and July (*64*) in Figures 24 through 26. The close correlation between the shapes of the oxygen and pH profiles in both winter and summer is particularly striking here. Compare Figures 25 and 26. In the Gulf of Mexico, Figures 27 to 29 (site 4, Figure 18), rather large seasonal differences occur within the first 50 to 100 m below the surface. These gradually disappear in the deeper water (*65*, *66*). Values of salinity, temperature and dissolved oxygen in the vicinity of location 4 may also be influenced intermittently by the plume from the Mississippi River.

In the shallow continental shelf waters off the east coast of the United States (location 5, Figure 18), seasonal differences persist from the surface to the bottom (*62*) as shown in Figure 30. Seasonal variations of temperature

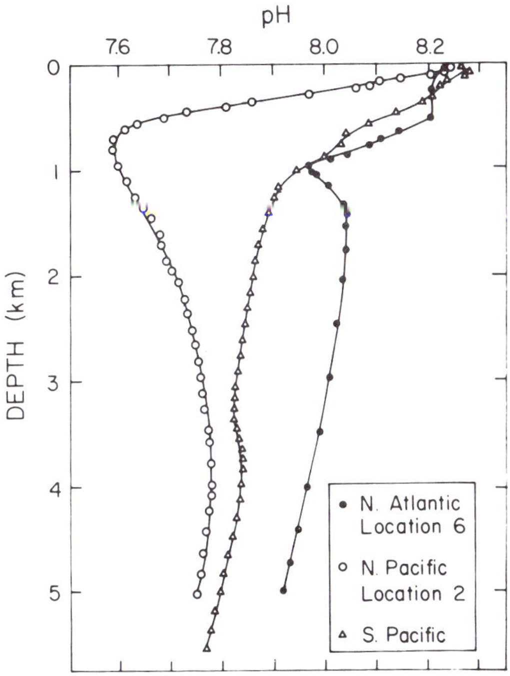

Fig. 22. Comparison of pH–depth profiles for open ocean sites 2 and 6 of Fig. 18. Note that data for the South Pacific is highest at the surface, but intermediate at depths greater than 500 m.

and salinity have been given for other coastal locations in the United States by McIlhenny and Zeitoun (*67*). Upon close examination, the oxygen and pH profiles in Figure 30 do not appear to be closely related in the manner seen earlier. In March, the water column is well mixed down to the bottom, and the changes with depth of all four variables are small. In August, however, the dissolved oxygen profile is nearly independent of depth, while the pH and temperature profiles show substantial changes. Based on salinity and temperature, the oxygen saturation levels during August are about 5.2 mL/L in the surface waters and 6.5 mL/L in the deep water. The oxygen profile for August shows that the surface waters are nearly saturated, while in the deep waters, biological activity has used up enough oxygen and produced enough CO_2 to decrease the pH, but not enough to produce a strong oxygen minimum. This points out the danger inherent in assuming that the pH minimum can always be located by measuring an oxygen profile. The two profiles will correspond closely in shape only when the biological demand for oxygen is intense enough to produce a strong oxygen minimum as seen in Figures 21 and 22, or when there is not a strong temperature gradient.

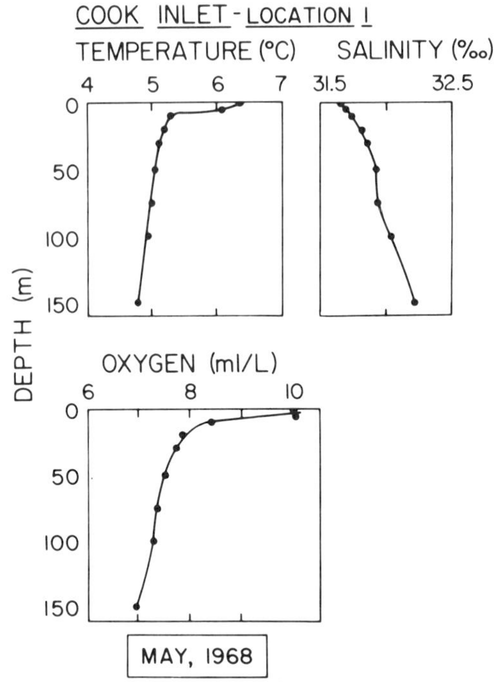

Fig. 23. Temperature, salinity, and dissolved oxygen variations with depth at Cook Inlet (site number 1, Fig. 18) for the month of May.

The surface waters of most of the world ocean, including the sites in Figure 18, are 200% to 500% supersaturated with respect to the calcium carbonate species, calcite and aragonite *(60)*. This means that precipitation of carbonate type scales is likely to be an important part of any corrosion reaction in surface water at most locations. Carbonate precipitation is most likely to occur in the elevated pH regime adjacent to cathodic surfaces where OH^- ions are produced during reduction of dissolved oxygen. Deep waters, below about 300 m in much of the Atlantic and central Pacific, and below 200 to 1000 m in the North Pacific, are often undersaturated. Even here, however, a cathodic surface may provide a micro-pH environment basic enough for precipitation to occur, especially in the absence of strong water currents.

Effect of Pollutants

Let us now consider the effects of pollutants on the aspects of the marine environment discussed in the previous sections. The ratios of the major ions are not affected by pollutants as long as the salinity remains above 5‰ to

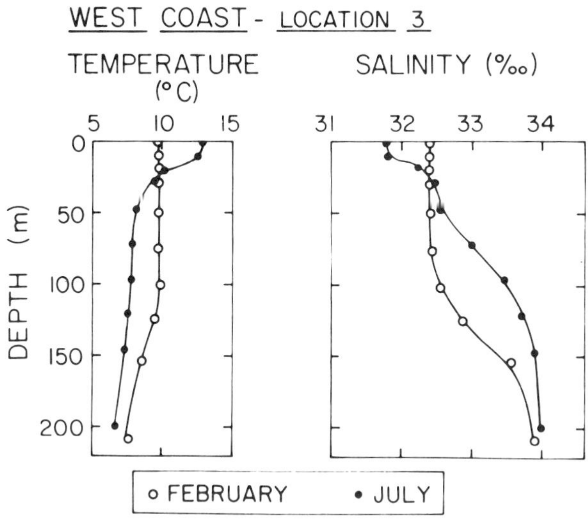

Fig. 24. Variation of temperature and salinity with depth and season off the Oregon coast (site number 3, Fig. 18).

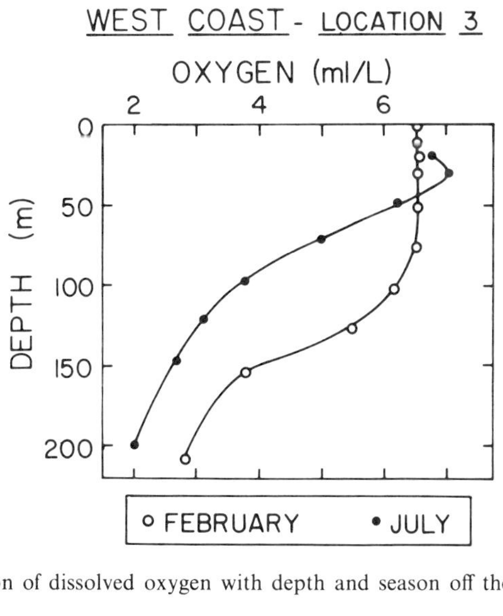

Fig. 25. Variation of dissolved oxygen with depth and season off the Oregon coast (site number 3, Fig. 18).

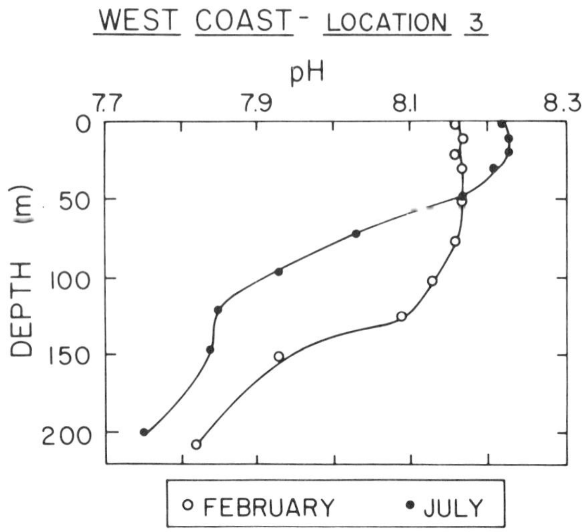

Fig. 26. Variation of pH with depth and season off the Oregon coast (site number 3, Fig. 18).

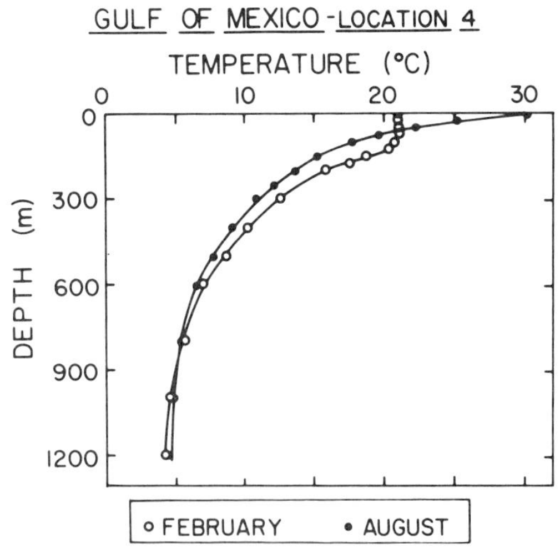

Fig. 27. Variationof temperature with depth and season in the Gulf of Mexico (site number 4, Fig. 18).

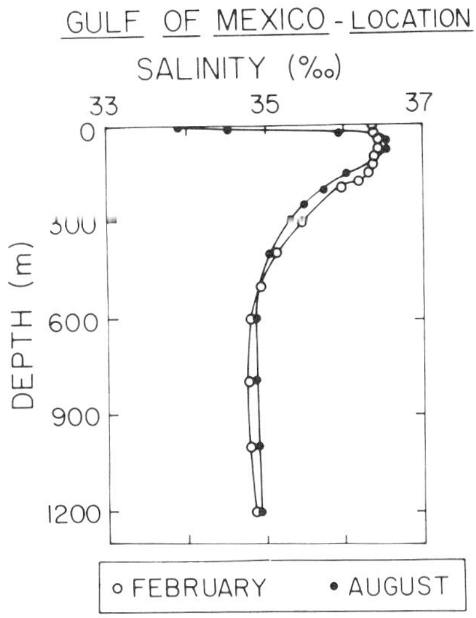

Fig. 28. Variation of salinity with depth and season in the Gulf of Mexico (site number 4, Fig. 18).

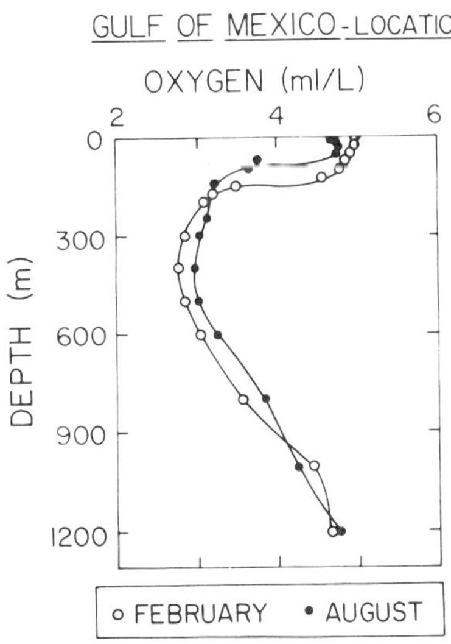

Fig. 29. Variation of dissolved oxygen with depth and season in the Gulf of Mexico (site number 4, Fig. 18).

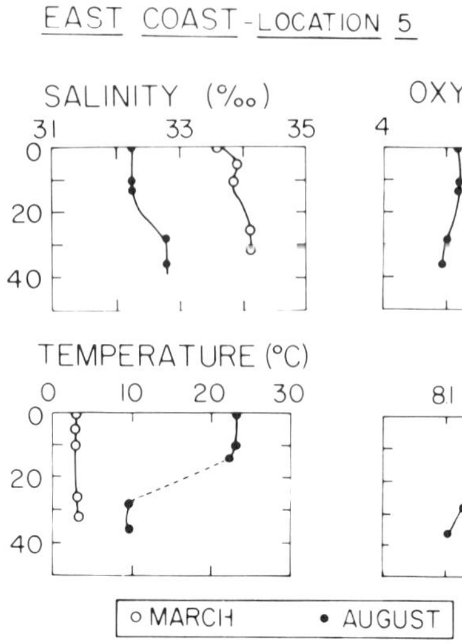

Fig. 30. Variation of temperature, salinity, dissolved oxygen, and pH with depth and season off the New Jersey and Delaware coasts (site number 5, Fig. 18).

10‰. The relationship between the major ions should hold except, perhaps, in a confined waterway with poor tidal flushing in which a pollutant containing a large concentration of one of the major ions was introduced in quantities approaching that of the waterway itself.

In contrast, the concentrations of the minor constituents of seawater may be radically changed by pollution. Concentrations of heavy metals, nutrients such as nitrates and phosphates, dissolved organics, dissolved gases such as oxygen and carbon dioxide, sulfides, and the pH of the water are particularly sensitive to pollution (*62*). Many pollutants, especially those containing organic material, increase the utilization of dissolved oxygen in seawater. As the organics become oxidized, oxygen concentrations fall, CO_2 concentrations rise, and the water becomes more acidic. These conditions often result in a decrease in the corrosion rate of steel structures. The reverse can be true, however, if certain types of microorganisms are present in the water as explained in the next section. For active-passive metals such as aluminum and stainless alloys, a decrease in dissolved oxygen can produce either an increase or decrease in the corrosion rate.

It should be carefully noted that it is usually the minor elements in seawater that affect the corrosion of structural metals and alloys, rather than the major elements. Thus, while changes in the minor elements are often unimportant to a subject like physical oceanography, they can be of major importance to corrosion.

C. *Influence of Biological Organisms*

Seawater is a biologically active medium in which dwell a large number of microscopic and macroscopic organisms. Many of these organisms are commonly observed in association with solid surfaces in seawater (*68*, *9*). As this subject will be dealt with in detail in a later chapter of this volume, only a brief description will be given here.

Immersion of any solid surface in seawater initiates a continuous and dynamic process, beginning with adsorption of nonliving, dissolved organic material and continuing through the formation of bacterial and algal films, and the settlement and growth of various macro-invertebrates (*68*). This process, by which the surfaces of all structural materials immersed in seawater become colonized, adds to the variability of the ocean environment in which corrosion occurs.

1. Bacterial Films

The process of colonization begins immediately upon immersion with the adsorption of a macromolecular organic film. This "conditioning film" is nearly complete within the first two hours of immersion (*69*), at which time the initially colonizing bacteria begin to attach in substantial numbers. The bacterial, or "primary," film develops over a period of 24 to 48 hours in most natural seawaters, although further changes in the film can often be observed over more than a two week period. The reader is referred to the excellent monograph by Marshall (*68*) and the proceedings volumes edited by Berkeley et al. (*70*) and by Costlow and Tipper (*9*) for further information.

The bacterial film changes the chemistry at the solid-liquid interface in a number of ways that have an important bearing on corrosion (*71*, *72*). As the biofilm grows, the bacteria in the film produce a number of metabolic byproducts. Among these are organic acids, hydrogen sulfide, and protein-rich polymeric materials commonly called "slime." The first effect of the composite film of bacteria and associated polymer, an example of which is depicted in Figure 31, is to create a barrier between the metal-liquid interface and the bulk seawater. The barrier itself is over 90% water, so it does not truly isolate the interface, but rather supports strong concentration gradients for various chemical species. Thus, the water chemistry at the interface may be different than that in the bulk, although the two are closely coupled through diffusive processes.

Two chemical species, oxygen and hydrogen, that are often implicated (or even rate controlling) in corrosion, are also important in the metabolism of the organisms. A given biofilm can be either a source or a sink for either oxygen or hydrogen. Moreover, biofilms are rarely continuous. Usually, they

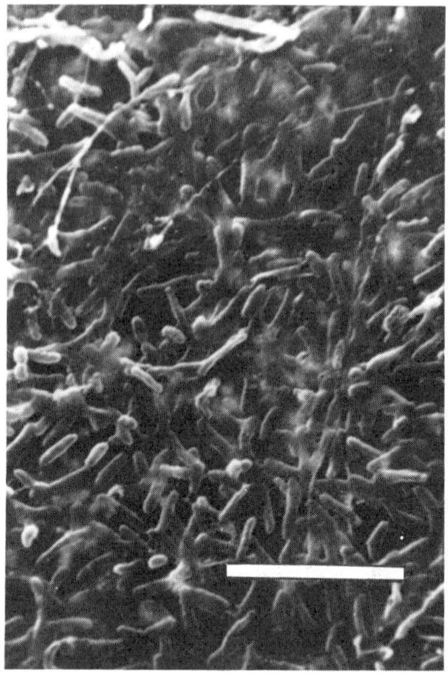

Fig. 31. Scanning electron micrograph of a nearly continuous film of rod shaped marine bacteria and slime on a metal surface that had been coated with a copper based antifouling paint and exposed in seawater at Woods Hole, MA for 7 days at a depth of 5 m. The magnification marker indicates 10 micrometers.

provide only spotty coverage of the metal surface. Thus, they are capable of inducing oxygen (or other chemical) concentration cells. Bacterial action on decaying organic matter in the biofilm can also result in the production of ammonia and sulfides. Ammonia causes stress corrosion cracking of copper alloys, and sulfides have been implicated in accelerated localized and/or uniform corrosion of both copper alloys and steels.

Under anaerobic conditions, such as those found in many marshy coastal areas, where all the dissolved oxygen in the water is used in the decay of organic matter, the corrosion rate of steel is expected to be very low. Under these conditions, however, the action of the sulfate-reducing bacteria of the genus *Desulfovibrio* utilize hydrogen produced at the metal surface in reducing sulfates from the decaying organic material to sulfides, including H_2S. The sulfides combine with iron from the steel to produce an FeS film, which is itself corrosive. The bacteria, thus, transform a benign environment into an aggressive one in which steel corrodes quite rapidly (*73*). Even under open ocean conditions at air saturation, the presence of a microbiological film can result in anaerobic conditions at the metal surface. Oxygen utilizing bacteria in the initial film may eventually increase sufficiently in numbers that they use all the oxygen diffusing through the film before it can reach the metal surface. This creates an anaerobic layer right next to the metal surface, and provides a place where the sulfate-reducing organisms can flourish. In all these examples, the biofilm is able to substantially change the chemistry of the electrolyte at the water/metal interface. Thus, the corrosion rate may depend more on the details of the electrolyte chemistry at the interface than it does on the ambient bulk seawater chemistry. Additional details about many aspects of microbial corrosion can be found in two recent proceedings on the subject (*74*, *75*).

2. Macrofouling Films

Within about the first two or three days of immersion, the solid surface, already having acquired both conditioning and bacterial films, begins to be colonized by the larvae of the macrofouling organisms. A heavy encrustation of these organisms can have a number of undesirable effects on marine structures. Both weight and hydrodynamic drag on the structure will be increased by the fouling layer. Interference with the functioning of moving parts may also occur. In terms of corrosion, the effects of the macrofouling layer are similar to those of the microfouling layer. If the macrofoulers form a continuous layer, they decrease the availability of dissolved oxygen at the metal-water interface. If the layer is discontinuous, they may induce oxygen or chemical concentration cells, leading to various types of localized corrosion as discussed above. Fouling films may also break down protective paint coatings by a combination of chemical and mechanical action. An excellent collection of papers on these subjects appears in the proceedings volume edited by Costlow and Tipper (*9*).

D. Summary

Throughout this discussion on the chemical and biological aspects of the marine environment, we have seen that several things can contribute to a variability in seawater properties. This variability may in turn influence corrosion of metallic structures in that environment. Variations in the chemistry of open ocean seawater tend to take place slowly (over times of three to six months), and over horizontal and vertical distances large compared to the dimensions of most marine structures. Such gradual changes may produce an equally gradual change in the corrosion rate of structural materials as the seasons progress, but such changes are unlikely to produce sharp changes in either corrosion rate or mode. Moreover, such gradual changes are relatively easy to measure and to monitor.

On the other hand, the changes brought about by micro- and macrofouling films, tend to take place over periods of hours to days, and over distances of centimeters to meters. These types of changes take place, not in the bulk water, but rather, directly at the metal-water interface where the corrosion takes place. This means that the chemical environment in which the electrochemical reactions take place, in the presence of a biofilm, may bear little resemblance to that of the bulk water. It is these types of changes, which can be produced quickly and lead to sharp chemical gradients over short distances, that often result in the onset of localized corrosion. Crevice corrosion beginning under the base of a barnacle is a good example of this type of influence. Whether the biofilm is composed of microscopic bacteria or large sedentary fouling organisms is often less important than whether the film provides complete or spotty coverage of the metal surface. Almost invariably, a spotty film, or one that forms in discrete colonies of organisms with bare metal in between, will be more likely to induce structurally significant corrosion than a film that produces a continuous layer.

References

1. Sarpkaya, T., and Isaacson, M. St. Q. (1981). *Mechanics of Wave Forces on Offshore Structures.* Van Nostrand Reinhold Co., New York, 651 pp.
2. Wilson, J. F., editor (1983). *Dynamics of Offshore Structures.* John Wiley and Sons, New York, 546 pp.
3. American Petroleum Institute (1982). *Recommended Practice for Planning, Designing and Constructing Fixed Offshore Platforms.* American Petroleum Institute, Washington, 96 pp.
4. Det Norske Veritas (1975). *Rules for Construction and Classification of Mobile Offshore Units.* Det Norske Veritas, Oslo, Norway.
5. LaQue, F. L. (1975). *Marine Corrosion*, Wiley-Interscience, New York.
6. Dexter, S. C. (1979). *Handbook of Oceanographic Engineering Materials.* Wiley-Interscience, New York.
7. Uhlig, H. H., and Revie, R. W. (1985). *Corrosion and Corrosion Control*, Third Edition. Wiley-Interscience, New York.
8. Fontana, M. G., (1986). *Corrosion Engineering*, Third Edition. McGraw-Hill, New York.
9. Costlow, J. D., and Tipper, R. C., editors (1984). *Marine Biodeterioration, Proceedings of the Symposium.* Naval Institute Press, Annapolis, MD.
10. Scruton, C., and Newberry, C. W. (1963). "On the Estimation of Wind Loads for Building and Structural Design." *Proc. Inst. Civ. Eng.* **25**, 97–126.
11. Davenport, A. G. (1960). *Wind Loads on Structures.* Div. Building Res., Paper No. 88, Ottawa.
12. Sachs, P. (1978). *Wind Forces in Engineering.* Pergamon Press, Oxford, 400 pp.
13. *Building Code Requirements for Minimum Design Loads in Buildings and Other Structures*, A58, pp. 1–72. American National Standards Institute, New York.
14. Simiu, E., and Scanlan, R. (1986). *Wind Effects on Structures.* 2nd ed. Wiley-Interscience, New York.
15. U.S. Naval Weather Service Command (1975). *Summary of Synoptic Meterological Observations*, National Technical Information Service, Springfield, VA.

16. Ward, E. G., Evans, D. J., and Pompa, J. A. (1977). "Extreme Wave Heights Along the Atlantic Coast of the United States." *Proc. Offshore Tech. Conf.* (OTC-2846), 315–324.
17. Longuet-Higgins, M. S. (1952). "On the Statistical Distribution of the Heights of Sea Waves." *J. Mar. Res.* **11**, 245–266.
18. Forristall, G. Z. (1978). "On the Statistical Distribution of Wave Heights in a Storm." *J. Geophys. Res.* **83**, 2353–2358.
19. Cardone, V. J., Pierson, W. J., and Ward, E. G. (1975). "Hindcasting the Directional Spectra of Hurricane Generated Waves." *Proc. Offshore Tech. Conf.* (OTC-2332).
20. Airy, G. G. (1845). "Tides and Waves." *Encyclopaedia Metropolitana* **5**, Art. 192, 241–396.
21. Dean, R. G., and Dalrymple, R. A. (1984). *Water Wave Mechanics for Engineers and Scientists.* Prentice-Hall, Englewood Cliffs, NJ, 354 pp.
22. Stokes, G. G. (1847). "On the Theory of Oscillatory Waves." *Trans. Camb. Phil. Soc.* **8**, 441–455.
23. Skjelbreia, L., and Hendrickson, J. A. (1960). "Fifth Order Gravity Wave Theory." *Proc. 7th Coastal Engrng. Conf.*, ASCE, The Hague, 184–196.
24. Fenton, J. (1985). "A Fifth-Order Stokes Theory for Steady Waves." *J. Waterway, Port, Coastal and Ocean Engrg.*, ASCE **111**(2), 216–234.
25. Dean, R. G. (1965). "Stream Function Representation of Nonlinear Ocean Waves." *J. Geophys. Res.* **70**, 4561–4572.
26. Dalrymple, R. A. (1974). "A Finite Amplitude Wave on a Linear Shear Current." *J. Geophys. Res.* **79**, 4498–4504.
27. Cokelet, E. D. (1977). "Steep Gravity Waves in Water of Arbitrary Uniform Depth." *Phil. Trans. Roy. Soc. Lond.* **A**(286), 183–230.
28. Wiegel, R. L. (1964). *Oceanographic Engineering.* Prentice-Hall, Englewood Cliffs, N.J.
29. Svendsen, I. A. (1974). "Cnoidal Waves Over a Gently Sloping Bottom." Series Paper 6, ISVA, Tech. Univ. Den., Lynbgy.
30. Dean, R. G. (1974). *Evaluation and Development of Water Wave Theories for Engineering Application.* SR-1, U.S. Army Coastal Engrg. Res. Center.
31. Borgman, L. E. (1972). "Statistical Models for Ocean Waves and Wave Forces." *Advances in Hydroscience* **8**, 139–181.
32. Berge, B. (1981). "Design of Offshore Structures: Directional Spectra." *Proc. Directional Wave Spectra Applications*, ASCE, Berkeley, 495 pp.
33. Morison, J. R., O'Brien, M. P., Johnson, J. W., and Schaaf, S. A. (1950). "The Force Exerted by Surface Waves on Piles" *Petrol. Trans., AIME* **189**, 149–154.
34. Dean, R. G., Dalrymple, R. A., and Hudspeth, R. T. (1981). Co-efficients from Wave Project I and II Data, Including Free Surface Effects." *Soc. Petrol. Engrgs. J.* **21**(6), 779–786.
35. Haring, R. E., and Spencer, L. P. (1979). "The Ocean Test Structure Data Base." *Civil Engrg. in the Oceans IV*, ASCE, 669–683.
36. Heideman, J. C., Olsen, O. A., and Johansson, P. I. (1979). "Local Wave Force Coefficients." *Civil Engrg. in the Oceans IV*, ASCE 684–699.
37. Dean, R. G., Lo, J-M., and Johansson, P. I. (1979). "Rare Wave Kinematics vs. Design Practice." *Civil Engrg. in the Oceans IV*, ASCE, 1030–1049.
38. Sarpkaya, T. (1976). "In-line and Transverse Forces on Cylinders in Oscillatory Flow at High Reynolds Numbers." *Proc. Offshore Technology Conf.* (OTC-2533), 95–108.
39. Garrison, C. J., and Stacey, R. (1977). "Wave Loads on North Sea Gravity Platforms: A Comparison of Theory and Experiment." *Proc. Offshore Tech. Conf.*, (OTC-2794).
40. Garrison, C. J. (1978). "Hydrodynamic Loading of Large Offshore Structures: Three Dimensional Source Distribution Methods." In: *Numerical Methods in Offshore Engineering.* (O. C. Zienkiewicz, R. W. Lewis and K. G. Stagg, eds). Wiley, Chichester, pp. 97–140.
41. Hogben, N., and Standing, R. G. (1975). "Experience in Computing Wave Loads on Large Bodies." *Proc. Offshore Tech. Conf.* (OTC-2189), 413–431.

42. Isaacson, M. St. Q. (1977). "Nonlinear Wave Forces on Large Offshore Structures." *J. Waterways Port Coastal and Ocean Div., ASCE* **101**, 166–170.
43. Kurata, K., and Ijima, T. (1979). "Finite Element Analysis of Wave Forces on Structures." *Civil Engrg. in the Oceans IV, ASCE*, 172–186.
44. Huang, M-C., and Hudspeth, R. T. (1982). "Nonlinear Diffraction by Eigen-Function Expansions." *J. Waterway Port Coastal and Ocean Div.*, ASCE, 108, 306–325.
45. Hunt, J. N., and Baddour, R. E. (1981). "The Diffraction of Nonlinear Progressive Waves by a Vertical Cylinder." *Q. J. Mech. Appld. Math.* **34**, 69–87.
46. Molin, B. (1979). "Second Order Diffraction Loads Upon Three-Dimensional Bodies." *Appld. Ocean Res.* **1**, 197–202.
47. Pritchard, R. S., editor (1980). "Sea Ice Processes and Models." *Proceedings of the Arctic Ice Dynamics Joint Experiment.* University of Washington Press, Seattle.
48. Untersteiner, N. (1983). *Air-Sea-Ice Interaction: Research Programs for the* 1980*s*. Applied Physics Laboratory, University of Washington, Seattle.
49. Michel, B. (1978). *Ice Mechanisms.* Les Presses de L'Université Laval, Québec.
50. Wadhams, P. (1980). "Ice Characteristics in the Seasonal Sea Ice Zone." *Cold Regions Science and Technology*, 2, 37–87.
51. American Petroleum Institute (1982). *Bulletin for Planning, Designing and Constructing Fixed Offshore Structures in Ice Environments. API Bul.* 2*N*, API Production Department, Dallas.
52. Caldwell, S. R., and Crissman, R. D., editors (1983). *Design for Ice Forces.* Technical Council on Cold Regions Engineering, American Society of Civil Engineers, New York.
53. Watt, B. J. (1983). "Hydrocarbon Extraction in Arctic Frontiers." In: *Proceedings of the Third International Conference on the Behavior of Offshore Structures*, Vol. 1 (C. Chryssostomidis and J. J. Connor, eds). Hemisphere Publishing Corporation, New York.
54. Schwartz, J. et al. (1984). *State of the Art Report of the IAHR Working Group on Ice Forces.* 7th International Symposium on Ice, International Association for Hydraulic Research, Hamburg.
55. Sverdrup, H., Johnson, M., and Fleming, R. (1942). *The Oceans.* Prentice-Hall, Englewood Cliffs, NJ.
56. Wilson, T. R. S. (1975). "Salinity and the Major Elements of Sea Water." In: *Chemical Oceanography*, 2nd edition, Volume 1 (J. P. Riley and G. Skirrow, eds). Academic Press, New York, p. 365.
57. Broecker, W. S. (1974). *Chemical Oceanography.* Harcourt Brace Jovanovich, New York.
58. Livingstone, D. A. (1963). "Chemical Composition of Rivers and Lakes." In: *Data of Geochemistry* (M. Fleischer, ed.). U.S. Geol. Survey Prof. Paper 440, Chapter G.
59. Kester, D. R. (1978). "Dissolved Gases Other Than CO_2." In: *Chemical Oceanography*, 2nd edition, Volume 1 (J.P. Riley and G. Skirrow, eds). Academic Press, New York, p. 498.
60. Skirrow, G. (1975). "The Dissolved Gases — Carbon Dioxide." In: *Chemical Oceanography*, 2nd edition, Vol. 2 (J. P. Riley and G. Skirrow, eds). Academic Press, New York, p. 1.
61. Gorshkov, C. G. (1974). *Atlas of the Oceans — Pacific Ocean.* Ministry of Defence of the USSR, Military Sea Transport, Leningrad. (In Russian; See also their *Atlas of the Mediterranean Sea.*)
62. Dexter, S. C., and Culberson, C. H. (1980). "Global Variability of Natural Sea Water." *Materials Performance* **19** (9), 16–28.
63. Kinney, P. J., Groves, J., and Button, D. K. (1970). *Cook Inlet Environmental Data R/V Acona Cruise 065. May 21–28, 1968.* Report No. R-70-2. Institute of Marine Science, University of Alaska, Fairbanks.
64. Barstow, D., Gilbert, W., Park, K., Still, R., and Wyatt, B. (1968). *Hydrographic Data from Oregon Waters, 1966*, Data Report No. 33, Reference 68–34. Department of Oceanography, Oregon State University, Corvallis, OR.

65. Leipper, D. F. (1968). *Hydrographic Station Data, Gulf of Mexico. February to March Nansen Casts, 1965–1968.* Reference 68-15T. Department of Oceanography, Texas A & M University, College Station, TX.
66. Leipper, D. F. (1968). *Hydrographic Station Data, Gulf of Mexico. August to September 5, 1968. Nansen Casts and STD.* Reference 68–17T. Department of Oceanography, Texas A & M University, College Station, TX.
67. McIlhenny, W. F., and Zeitoun, M. A. (1969). *Chemical Engineering.* **76**, Part 1, p. 81; Part 2, p. 251, McGraw Hill, New York.
68. Marshall, K. C. (1976). *Interfaces in Microbial Ecology.* Harvard University Press, Cambridge, MA.
69. Loeb, G. I., and Neihof, R. A. (1975). "Marine Conditioning Films." In: *Applied Chemistry at Protein Interfaces,* Advances in Chemistry Series no. 145 (R. E. Baier, ed). American Chemical Society, Washington, DC, p. 319.
70. Berkeley, R. C. W. et al., editors (1980). *Microbial Adhesion to Surfaces.* Ellis Horwood Limited, Chichester, West Sussex, England.
71. Gerchakov, S., and Udey, L. R. (1984). "Microfouling and Corrosion." In: *Marine Biodeterioration, Proceedings of the Symposium* (J. Costlow and R. Tipper, eds). Naval Institute Press, Annapolis, MD, p. 82.
72. Dexter, S. C. (1985). "Fouling and Corrosion." In: *Condenser Biofouling Control Symposium,* EPRI CS-4339 (W. Chow and Y. G. Massalli, eds). Electric Power Research Inst, Palo Alto, CA.
73. Iverson, W. P., and Olson, G. J. (1983). "Anaerobic Corrosion by Sulfate-reducing Bacteria." In: *Microbial Corrosion Proceedings,* The National Physical Laboratory, Teddington, England, March 1983. The Metals Society, London, p. 46.
74. The Metals Society (1983). *Microbial Corrosion Proceedings.* The National Physical Laboratory, Teddington, England, March 1983, The Metals Society, London.
75. Dexter, S. C., editor (1986). *Biologically Induced, Corrosion, Proceedings of the International Conference.* National Bureau of Standards, Gaithersburg, MD, June, 1985; National Association of Corrosion Engineers, Houston, TX.

3

Marine Biofouling

BRENDA J. LITTLE

Naval Ocean Research and Development Activity
NSTL, Mississippi

and

JOHN R. DEPALMA

U.S. Naval Oceanographic Office
NSTL, Mississippi

Exposure of virtually any engineering material to naturally occurring seawater initiates a series of sequential and parallel biological and chemical events that culminate in the formation of a complex layer of inorganic, organic, and cellular components known as fouling (*1*, *2*). Biofouling is a general term used to describe both the microbiological and macrobiological growths that develop on exposed surfaces. Microfouling is the accumulation of such microbial forms as bacteria, fungi, and microalgae and their secretions. Macrofouling is the accumulation of such macroscopic organisms as oysters and barnacles on submerged surfaces. The relationship between micro- and macrofouling has been reviewed by Mitchell and Benson (*3*).

ISBN 0-12-341828-3

I. Marine Microfouling

The interaction between microorganisms and solid surfaces can be anticipated on the basis of colloidal theory; however, microorganisms are more complex than simple colloids. Microorganisms are capable of independent locomotion, of growth and colonization, of forming tenacious films of exuded extracellular polymeric materials, and of chelating inorganic ions to their exudates.

A. Stages of Microbial Film Formation

Microfouling films may require a few hours or weeks to form, depending on flow rate, substratum composition, and nutrients in the water column; however, all microfouling can be described by four sequential events, as outlined in Table I (*4*).

1. Adsorption of Conditioning Film

Adhesion of microorganisms to surfaces is always preceded by adsorption of multilayers of natural polymers (*5*). This adsorption of polymers from the liquid phase to a material surface makes the surface wettable and may provide a primary site of nutrient concentration for microorganisms.

2. Attachment and Colonization of "Pioneer" Species

As water moves across a surface, microorganisms approach the solid-liquid interface. Attachment may take several hours, since the process is complicat-

TABLE I

STAGES OF MICROBIAL FILM FORMATION

Stage	Comments
1. Conditioning	Chemical adsorption; organisms not directly involved; instantaneous reaction.
2. Colonization by "pioneer" bacterial species	Fast-growing bacterial periphytes become established in a matter of hours.
3. Colonization by other microorganisms	Other bacteria, fungi, and protozoa may develop extensively in a matter of days.
4. Accumulation	Adhesion of particles, dead cells, detritus. Chelation of heavy metals from water.

ed by electrostatic forces at the interface. Bacteria are held at some distance from the surface by electrical, double-layer repulsion energies (*6*). In this reversible phase, the bacteria are only weakly held at the solid surface. They exhibit Brownian motion and can be easily removed. Irreversible attachment is possible as a result of bridging extracellular polymers produced by bacteria. The microbial polymers appear to be polysaccharide in nature and provide strong adhesive for the bacteria. The first colonizing species are often small, rod-shaped bacteria (Figure 1). The small rods have a selective advantage over other bacterial groups in this irreversible sorption (*7*). Small rods are able to reproduce at low nutrient levels and to produce extracellular polymeric fibrillar material (*8*). Hendricks (*9*) reported that bacteria sorbed to surfaces appeared to be metabolically more active than suspended ones. Furthermore, bacteria anchored by polysaccharide fibers are able to withstand enormous shear forces (*10*). Once attached to the surface, microorganisms assimilate nutrients and synthesize new cellular and extracellular material, which accumulates in the surface deposit. The rate of biofilm production is a function of nutrient concentration, fluid velocity, and temperature (*11*).

3. Colonization by Other Microorganisms

The attachment of a primary film of bacteria to a surface is normally followed by the attachment of other groups of microbes in rapid succession

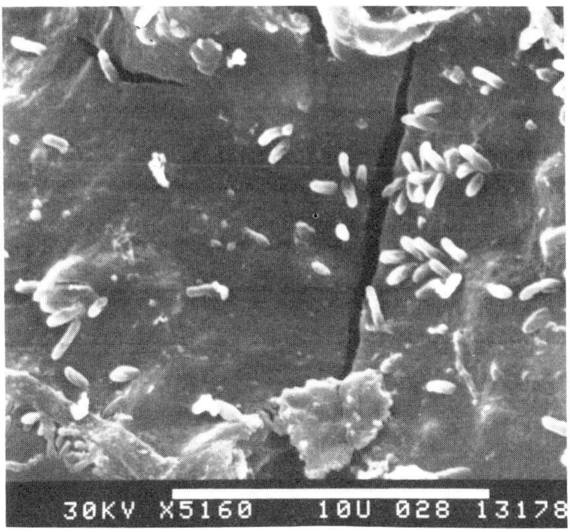

Fig. 1. Rod-shaped bacteria on aluminum surface exposed to flowing seawater for 8 days.

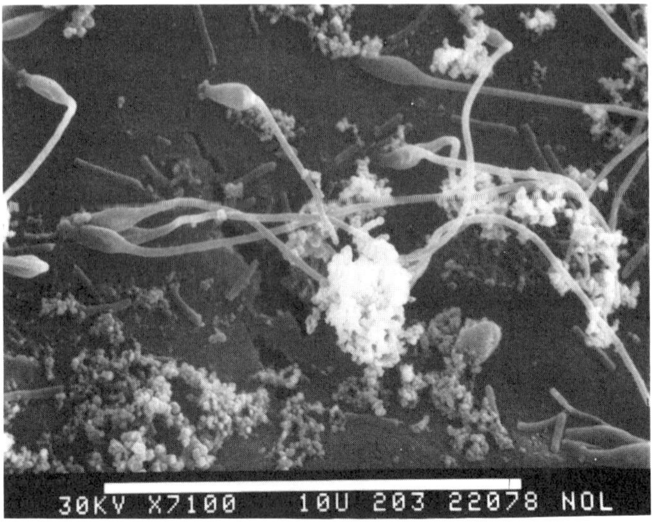

Fig. 2. Filamentous bacteria on aluminum surface exposed to flowing seawater for 14 days.

(*1*). Secondary colonizers are usually stalked or filamentous bacteria, diatoms, other microalgae, and protozoa (Figures 2 and 3).

4. ACCUMULATION

The microbial film composed of cells and polymers is very different from the original surface. The biofilm surface is highly adsorptive, which is partially due to its polyelectrolyte nature, and can collect quantities of particulates and cellular debris. In addition, the polysaccharide secretions of microorganisms possess a net negative charge and are capable of chelating inorganic ions from the passing fluid. Although microorganisms and their remains make up the most conspicuous components of fouling deposits, varying amounts of organic secretions, trapped detritus, inorganic precipitates, and corrosion products compose the bulk of the fouling layer (Figure 4).

B. *Factors Influencing Microfouling*

A wide range of environmental factors affects the fouling rate and dominance of the organisms involved in the fouling process. Water chemistry, water temperature, pressure, shear stress, substratum composition, and design features can affect the settlement and attachment of organisms. Thus,

Fig. 3. Pennate (elongate) and centric (cylindrical) diatoms enmeshed in biofilm on stainless steel surface exposed to seawater for 21 days.

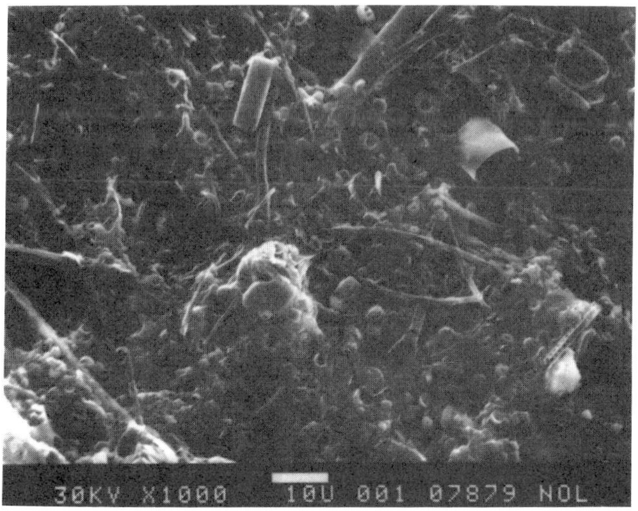

Fig. 4. Coccoliths, diatom frustules, oxide and debris on titanium surface exposed to estuarine water for 18 days.

biofilm properties are system specific because their growth conditions are different. A biofilm developing in the presence of a hydrated corrosion layer will have a different chemical and biological composition and may require control of different frequency or intensity as compared to a biofilm developing in the absence of such a layer. Similarly, a biofilm resulting from a cold-water source will have unique properties not found in films of warm-water origin.

1. WATER CHEMISTRY

Microfouling kinetics are closely related to the water chemistry. Biofilm growth has been described by a wide variety of rate expressions whose rate constants are functions of pH, temperature, limiting nutrient concentration, nutrient type, terminal electron acceptor, and organism concentrations.

To grow, organisms must draw from the environment all of the substances required for the synthesis of their cell materials and for the generation of energy. These substances are collectively called nutrients. Microorganisms are extraordinarily diverse in their specific physiological properties and in their specific growth requirements. The chemical composition of cells indicates the major requirements for growth. Water accounts for some 80% to 90% of the total cellular weight and is therefore the major essential nutrient. The solid matter of cells contains hydrogen and oxygen (derived from water), carbon, nitrogen, phosphorous, and sulfur, in order of decreasing abundance. Many other elements are included in the residual fraction. For example, bacteria of marine origin always require sodium (NA^+) for growth. All of the required metallic trace elements can be supplied as nutrients in the form of the cations of inorganic salts. One nonmetallic element, phosphorous, can also be used as a nutrient when provided in an inorganic form as, for example, phosphate salts. The needs for carbon, nitrogen, sulfur, and oxygen cannot be so simply described because microorganisms differ with respect to the specific chemical form under which these elements can be provided as nutrients.

Heterotrophic organisms require carbon from an organic source. Since most organic substrates (carbon sources) are at the same oxidation level as organic cell constituents, they can be used directly without further oxidation or reduction. In addition to meeting the biosynthetic needs of the cell for carbon, organic substrates must also supply the energetic requirements of the cell. Microorganisms differ with respect both to the kind and the number of organic compounds that they can use as a principal source of carbon and energy—many microorganisms can use a single organic compound for both needs, while others require a variable number of compounds. Autotrophic microorganisms obtain energy from sources other than organic compounds. For example, lithotrophic microorganisms derive energy from the oxidation of such inorganic compounds as H_2S, NH_3, Fe^{2+}, and Mn^{+1}. Both heterotrophs and autotrophs are found in the sea.

Reduced forms of nitrogen and sulfur occur in organic compounds of the cell as amino and sulfhydryl groups. Some bacteria can meet the needs for nitrogen and sulfur from nitrates and sulfates; however, some are unable to reduce these anions and must be supplied with the elements in a reduced form. A reduced nitrogen source is available in ammonium salts. The need for reduced sulfur can be met by a sulfide or an organic compound containing a sulfhydryl group. Nitrogen and sulfur requirements can also be met by organic nutrients that contain both elements in the reduced form, such as amino acids and complex protein degradation products.

In addition to these nutrients, any organic compound that a microorganism requires as a precursor or constituent, but which it cannot synthesize from simpler carbon sources, must be provided as a nutrient. These materials are known as growth factors and are typically amino acids, purines and pyrimidines, and vitamins.

Many bacteria are obligate aerobes that are completely unable to grow in the absence of oxygen. Oxygen is required for two purposes: It is a terminal electron acceptor in respiration, and it is necessary for the biosynthesis of sterols and unsaturated fatty acids. Microorganisms that can grow only in anaerobic environments are called obligate anaerobes. The physiology of obligate anaerobes is poorly understood. Many organisms can grow in both aerobic and anaerobic environments; these are called facultative. Aerobes and anaerobes coexist in natural environments.

Nutrient-rich waters yield thick biofilms at rapid rates, whereas nutrient-deficient waters yield very thin biofilms and the film formation rate is very slow. In addition, the attachment of microorganisms to surfaces is closely rated to the production of biopolymers used to anchor microorganisms to surfaces. In waters that are rich in carbon and deficient in nitrogen, the microbes tend to produce large quantities of polymers. These polymers also stimulate colonization of other species and can concentrate nutrients. The chemistry of the water is an important factor in the initial growth of microbial films; however, the process is probably independent of the nutrient status of the water in later stages. Inorganic constituents of the fluid can also affect film properties. For example, calcium, magnesium, and iron have been implicated in film bonding strength (*12*).

It is generally accepted that the incidence of bacteria in the water column decreases significantly with depth. One of the most important factors governing the distribution of bacteria in the sea is the availability of nutrients. In the upper waters of the sea and in the sediments, there are relatively more organic nutrients than in the intermediate waters and, proportionally, more heterotrophic bacteria at those extremes.

2. Water Temperature

Temperature is an important environmental factor controlling both the number of microorganisms available for attachment to a substratum and the rate of growth after attachment. Both have been shown to increase at higher temperatures (*13*). Tests with the highly polluted Rhine River water in Ludwigshafen, West Germany and estuarine water from Oresund in Landskroma, Sweden (*14*) indicated that the fouling rate in both instances was controlled to the greatest extent by the temperature of the water.

3. Pressure

The effect of pressure and pressure changes on microbial species is unknown. The marine biosphere is characterized by hydrostatic pressures of 1 atm at the surface, with pressure increasing roughly 1 atm for every 10 meters of depth. Evidence of rich microbial populations is found near the marine sediment water interface. Jannasch et al. (*15*) found reduced microbial activity in the deep sea and attributed this finding to the coeffects of temperature and pressure. However, Berger (*16*) proposed that microbial activity in ocean depths proceeded very slowly due to the inhibitory effects of pressure on protein synthesis.

4. Shear Stress

Nutrient concentration affects the rate of initial biofilm formation, but not the extent of biofouling. The extent of microfouling is primarily controlled by wall shear stress in turbulent flow (*17*). The relationship is a complex one, which involves the adsorption of nutrients onto the substratum and the nature of the microbe attached to the surface so that the exact extent of shear stress on biofilm growth cannot be predicted. Biofilms formed under high shear stress are more compact and tenacious than those formed under low shear stress. Furthermore, microbial fouling cannot occur to any significant extent when shear stress exceeds 200 g/cm sec^2 (*17*).

5. Substratum Composition

Substratum, the surface to which the microorganism attaches, can also impact the fouling rate and the dominant species. Both qualitative and quantitative differences in the attached microbial communities have been reported to occur in response to differing types of metal surfaces. Copper surfaces are colonized more slowly than nontoxic metal surfaces, and the colonizing species appear to exude copious quantities of polymer with which to insulate themselves from the copper surface (*18*).

6. Mechanical Factors

Mechanical factors that influence the attachment and growth of microfouling organisms include geometry, orientation, layout, surface finish, and polarization. The major effects of geometry and orientation on microfouling involve tolerance rather than prevention. Geometry indirectly impacts biofouling by influencing velocity and turbulence. Surface roughness has been shown to be important in initial fouling . Polished surfaces are initially more resistant than roughened surfaces; however the effect is rapidly overcome as proliferation takes place. Crevices from bolted or gasketed joints, discontinuous welds or welds without full penetration, and other situations where fluids can be trapped provide sites for colonization.

C. Impact of Microbial Film Formation

1. Increased Fluid Frictional Resistance

Changes on the surfaces of well-defined solids exposed to seawater from diverse sources have been observed by microelectrophoresis, ellipsometry, contact angle measurements, and pyrolysis/mass spectrometry These observations indicate that the adsorbed organic film which forms rapidly on all exposed surfaces lowers surface energy and imparts a moderately negative electrical charge (*19*). This molecular fouling is instantaneous. Based on "thickness" measurements, molecular fouling can have no significant effect on fluid flow or heat transfer. Nevertheless, the surface properties resulting from adsorption of an organic film may affect the sequence of microbial events that follow and that are known to impact both of these engineering parameters (*20*). Thin, viscoelastic biofilms developing on conduit walls cause unexpectedly large increases in fluid frictional resistance and hydrodynamic drag. Measurements of the effect of several different types of microbial slimes on the drag of rotating discs of different materials with differing surface roughness have been made (*21*). Results of these measurements showed that friction drag increases of 10% to 20% could be caused by tenacious microbial slimes that were persistent even at high speeds.

2. Decreased Heat Transfer Efficiency

Biofilm buildup on heat exchanger surfaces seriously limits heat transfer efficiency. The relationship between biofilm thickness and the inhibition of heat transfer is discussed in Characklis and Zelver (*22*).

3. Biodeterioration

Biofilms also contribute to the biodeterioration of materials in aquatic environments. Bacteria and fungi attack materials such as fiber nets, lines, corks, gasoline, kerosene, lubricating oil, crude oil, rubber, plastics, and metals.

The most frequently cited mechanisms for microbiologically induced corrosion of metal surfaces are: (1) production of corrosive metabolic products, such as acids or hydrogen sulfide; (2) formation of discontinuous deposits on the surface, resulting in differential aeration and concentration cells; (3) disruption of natural and other protective films; (4) breakdown of corrosion inhibitors and coatings; and (5) deplolarization of cathodic and anodic reactions.

By-products of metabolism such as H_2S from *Desulfovibrio desulfuricans*, CH_3COOOH from *Clostridium aceticum*, and H_2SO_4 from *Thiobacillus thiooxidans* are obvious contributors to corrosion processes. Staffeldt and Calderon (*23*) demonstrated that organic acids of the Krebs cycle promoted the corrosion of copper. Burnes et al. (*24*) demonstrated that solutions of various organic acids, including citric, fumaric, ketoglutaric, glutaric, maleic, malic, and succinic acids, formed metallic salts of copper, tin, and zinc in sterile solutions in the presence of oxygen. The metals were oxidized electrochemically and the organic acids removed the protective oxide film. Ashton et al. (*25*) found acetic, ketoglutaric, succinic, and lactic acids in a culture of *Escherichia coli* where carbon-steel corrosion was enhanced. They suggest that acidic metabolites were responsible for accelerating the corrosion by dissolution of iron oxide. Webb (*26*) and Kramer (*27*) demonstrated accelerated corrosion of copper in the presence of amino and dicarboxylic acids.

Within the biofilm layers, oxygen-deficient and even oxygen-free conditions arise, since oxygen is taken up by respiring organisms near the surface of the slime as rapidly as it can diffuse in. This situation will give rise to potential differences with subsequent corrosion currents between the bacterial colony sites and the surrounding metal surface (*18*). Growth of filamentous iron bacteria, *Sphaerotilus*, *Crenothrix*, and *Leptothrix*, and the stalk-forming unicellular bacterium, *Gallionella*, also gives rise to differential aeration cells.

These organisms oxidize ferrous to ferric ions to obtain energy for growth reactions and cause thick deposits of ferric hydroxide to accumulate between the organisms. The efficient oxygen scavenging by iron-oxidizing bacteria sets up a microhabitat suitable for the growth of the highly corrosive, anaerobic, sulphate-reducing bacteria.

The enhancing effect of sulfate reducers of the genus *Desulfovibrio* on corrosion processes has been attributed to the removal of hydrogen from cathodic sites of corrosion cells and the production of FeS (*28*). Similar observations for cathodic depolarization were made in the case of nitrate-reducing bacteria (*29*).

Pitting potential was used by Salvarezza et al. (*30*) to assess the aggressiveness of the biological species *Cladosporium resinae* in microbial corrosion of aluminum alloys. Chloride anions in their system controlled the pitting process. Nitrates and phosphates, used as nutrients in the culture media, acted as inhibitors displacing the pitting potential value to high anodic regions at low chloride/inhibitor ratios. In these conditions, microbial uptake of pitting inhibitors was the principal cause of the pitting potential decrease during microbial growth. At high chloride/inhibitor ratios or in the absence of inhibitors, metabolite production was the most important factor in the aggressiveness increase. The acidic metabolites produced by *Cladosporium resinae* are able to facilitate the breakdown of the passive oxide film by chloride anions, decreasing the pitting potential value.

D. Methods for Quantifying Microbial Film Formation

Methods routinely used to analyze the biofilms developing on surfaces can be roughly divided into biochemical techniques and microscopy.

1. Biochemical Techniques

Adenosine triphosphate (ATP) is a constituent of all living organisms and can be used to quantify the presence of living microbes on surfaces (*31*, *32*, *33*). Muramic acid, a component found only in cell walls of Cyanophytes and bacteria, can also be related to cell number (*34*, *35*). The quantification of these cellular extracts requires relatively complex chemical manipulations. White et al. (*36*) have developed a technique for measuring extractable lipid phosphate to estimate microbial biomass. In addition, examination of the lipid components can be used to identify major features of the microbial community structure.

2. Microscopy

The scanning electron microscope (SEM) provides a means for qualitatively and semiquantitatively measuring the types and numbers of microorganisms that attach to surfaces. The microorganisms must be properly fixed and dehydrated for viewing (*2*). Microorganisms attached to surfaces can also be stained with fluorescent dyes and viewed with an epifluorescent microscope. Such a technique has been described by Hobbie et al. (*37*).

E. Methods of Control

Microfouling has traditionally been controlled by chemical and/or mechanical cleaning. The main factors to be considered in selection of microfouling control agents are effectiveness, toxicity, effect on equipment, and ease of implementation.

1. Chemical Cleaning

Removing air from cooling water for heat exchangers has been proposed as an environmentally safe way to control microfouling in closed systems. It can be achieved by physical (hot well and flash evaporation) and chemical (sulphite and hydrazine) methods. Deaeration is independent of design and does have the advantage of reducing corrosion in many alloys. However, anaerobic conditions encourage the growth of anaerobic bacteria and the pitting of copper, nickel, and other alloys sensitive to pitting in the presence of organic substrates.

Biocides, such as chlorine, have been used to inhibit the development of biofilms that form as a result of warm surface waters flowing through tubes. Three chlorine-based water treatments are in common use: hypochlorination using liquid sodium hypochlorite; chlorination using elemental chlorine, either liquid or gas; and electrolytic hypochlorination. Electricity passed through seawater causes the following reaction to take place:

$$2\,Na^+ + 2\,Cl^- + 2\,H_2O \longrightarrow 2\,Na^+ + 2\,OH^- + H_2 + Cl_2.$$

The chlorine gas discharged at the anode hydrolyzes with water to produce hypochlorous acid and hypochlorite ion. The ratio of the reaction products depends on pH and temperature. The chemistry of chlorine in seawater has been studied by several investigators (*38*, *39*, *40*). Due to the presence of the bromide ion in seawater, the compounds produced by chlorination of seawater differ from those produced in freshwater. The bromide ion (approximately 65 mg/L) is oxidized by chlorine to produce a series of brominated compounds, analogous to the chlorinated compounds produced in freshwater.

Chlorine is effective as a biocide as a result of the following processes:

(a) Chlorine species entering the water react with chlorine-demanding components (viable cells and organic compounds) in the bulk water.

(b) Chlorine-containing compounds are transported through the bulk fluid to the liquid-biofilm interface.

(c) Chlorine species diffuse into and react with components within the biofilm.

The rate at which chlorine is transported to the biofilm depends on the concentration of chlorine in the water and the intensity of the turbulence. The concentration of chlorine in the bulk water is a function of not only the amount of chlorine added to the water, but also the chlorine demand of the water and the substratum (*22*). It appears that the chlorine requirement is essentially set by the chlorine demand of the water and that a chlorine dosage slightly in excess of this demand may be expected to control microfouling. The concentration of added chlorine has been shown to decrease with time in two phases. The rate and the mechanism for the reactions are determined by temperature, pressure, photooxidation, bromide concentration, and the amount of total organic matter. Although the mechanistic pathways for each phase have not been elucidated, the dissipation rate is more rapid in the first few hours, followed by a slow, continuous loss of residual chlorine (*41*).

Transport rates of chlorine in the bulk water, as well as the concentration at the interface, are increased by turbulence through increased flow rates. It has been reported that increased turbulence resulted in increased chlorine reaction rates (*14*). Temperature has also been shown to affect significantly both the rate and extent of chlorine consumption in seawater (*42*).

Once chlorine has reached the biofilm, it reacts with organic and reduced inorganic components. Hydrochlorous and hypobromous acids are extremely powerful oxidizing agents that easily diffuse through the cell walls of microorganisms, oxidizing the active sites of certain enzyme sulfhydryl groups and preventing adenosine triphosphate production (*14*).

In addition to this biocidal property, chlorine can also react with the extracellular polysaccharides within a well-developed biofilm to disrupt the physical integrity of the film. Characklis and Zelver (*22*) have estimated the process rate and process stoichiometry of the reaction of chlorine with immobilized biofilms. Biofilms rich in extracellular polymeric substances exhibit greater chlorine demands than those composed predominantly of living cells. This destructive action of chlorination differentiates chlorine from nonoxidizing biocides and indicates that chlorine can be used not only for disinfecting and controlling biological fouling, but also for removing established biofilms as well. The use of cationic flocculants with chlorine improves the efficiency of chlorine in deposit removal and eliminates the need for chemical dispersants and nonoxidizing biocides.

The rapid reformation of a microbial film immediately following chlorine treatment has been observed by many investigators. Explanations for this enhanced biofilm development compared with that on a clean surface are numerous. There is the possibility that some microorganisms remain viable in the film and continue to grow after chlorine treatment. The susceptibility of microorganisms to the action of chlorine certainly varies among species. There is also the possibility that chlorine may oxidize the extracellular polymeric secretions of the microorganisms without completely removing them from fouled surfaces, resulting in the nutrient-rich surface that is suitable for recolonization. Such a partially oxidized residue may be sticky and be capable of entrapping species from the water column. The surface may also be roughened as a result of chlorination enhancing transport of microbial cells and nutrients to the surface.

2. Mechanical Cleaning

Recirculating rubber and carborundum balls, flow-driven brushes, and diatomaceous earth slurries have been used with limited success to mechanically remove biofilms from the interiors of piping systems (*43*, *44*, *45*). Wear and contamination are the most often cited reasons for the failure of control systems based on mechanical techniques.

II. Marine Macrofouling

A. Principal Foulers

Macrofouling organisms are found at all depths and in all seas. The accumulation of macrofouling impacts the performance of all materials exposed to marine environments. They are most numerous along the margins of continents where they are sometimes organized in communities of up to 500 species. A few of the larger, more prolific taxa dictate the size and shape of fouling communities, and the ability to predict fouling conditions is based on knowledge of the habits and ranges of these dominants and on their rates of colonization in various parts of the world ocean.

At least 4000 different species of organisms are recorded as marine-fouling nuisances. However, only ten groups of macrofouling organisms are usually categorized as a major nuisance.

1. Barnacles

Barnacles are distributed worldwide and are the most common fouling organisms. In coastal waters the dominant fouling organisms tend to be acorn barnacles (Figure 5), which cement themselves directly to the substratum and produce a maximum hardshelled buildup of about 5 cm. This hardshelled layer is sometimes overgrown to greater heights by soft-bodied forms. In relatively turbid water (Secchi values of less than 20 m), barnacle communities will add up to 10 kg/m^2 to the weight-in-water of submerged structures; only about half as much weight is added in clear water. The horizontal limits of turbid and clear, warm coastal waters are shown in Figure 6.

Below the surface layer, fouling rates diminish progressively with distance from the continental shore lines. Figure 7 shows the quantitative vertical changes in fouling production in coastal regions where currents generally flow parallel with the shoreline. Where the current regimes and shorelines are more complicated, this regular pattern becomes somewhat less apparent.

Pelagic fouling rates are considerably less at all latitudes. Stalked barnacles are usually the dominants in the surface layers, and they can cover introduced

Fig. 5. Acorn barnacles attached to pipe interior.

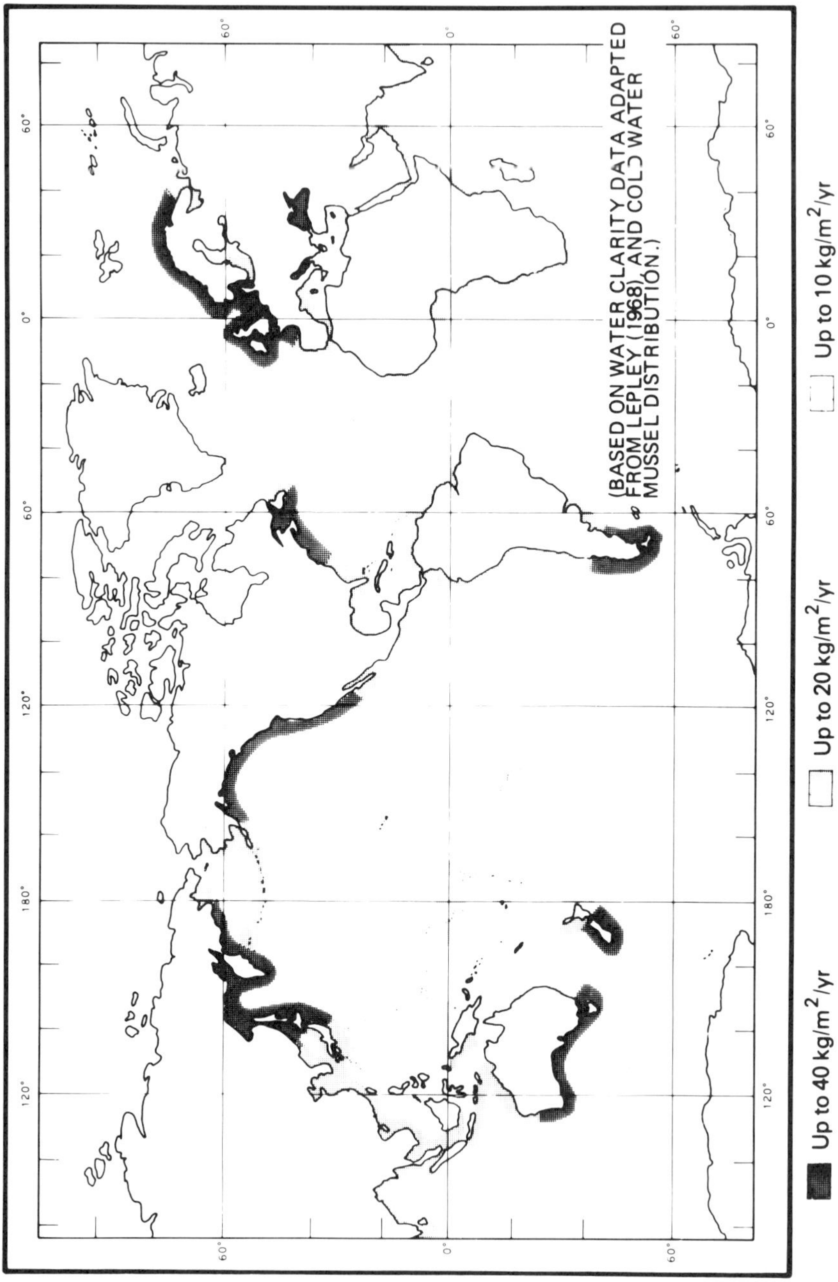

Fig. 6. Biofouling production in the world ocean.

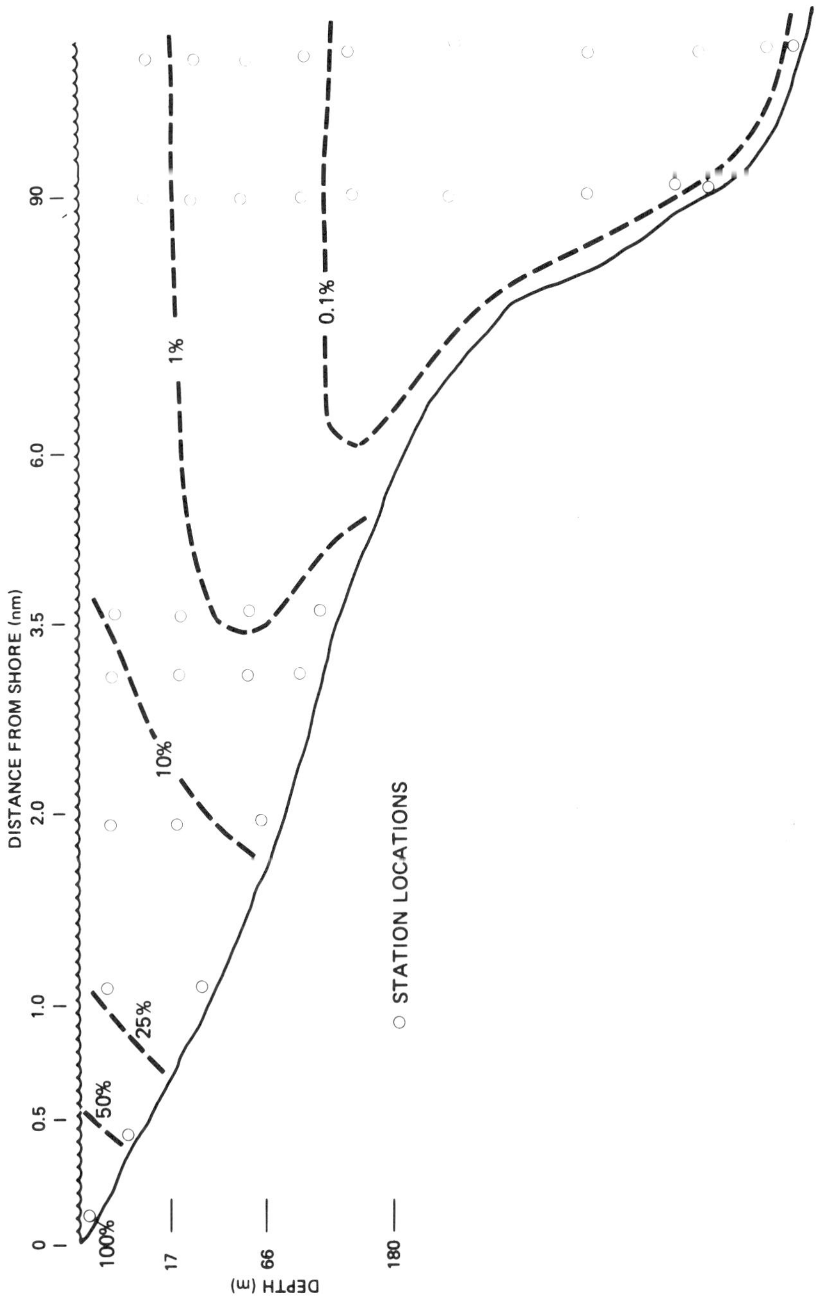

Fig. 7. Diminution of fouling rate with depth and distance from shore, South Florida.

substrata to a thickness of about 5 cm in just a few months in tropical regions. However, these mostly soft-tissued organisms contribute no more than an additional 2 kg/m^2 to a substratum weight-in-water.

Directly on the deep sea floor, exposed surfaces will be colonized by small bivalves, barnacles, or tubeworms, but no more than 0.3 kg/m^2 of dry weight fouling biomass has been collected, even after several years. In the deep ocean the slow accumulation of fouling organisms can usually be accommodated by conservative design and careful material selection.

2. Mussels

The most severe fouling conditions occur in the high and middle latitudes (Figure 6) where *Mytilus* mussels are one of the important fouling dominants. Mussels have a free-swimming larval stage and juvenile stage when the animal attaches. Young mussels may often move about on a surface before settling. The shell is composed of tanned protein and calcium carbonate. The shell can be closed when the environment becomes unfavourable, for example, when toxic chemicals are applied. In addition, byssal threads attach the organism firmly to the substratum so that mussels can resist physical cleaning methods (Figure 8). Cold-water mussel communities under ideal conditions can accumulate to a thickness of about 15 cm—often in one year. A 15 cm thickness of hardshelled foulers adds about 20 kg/m^2 additional weight-in-water to submerged structures.

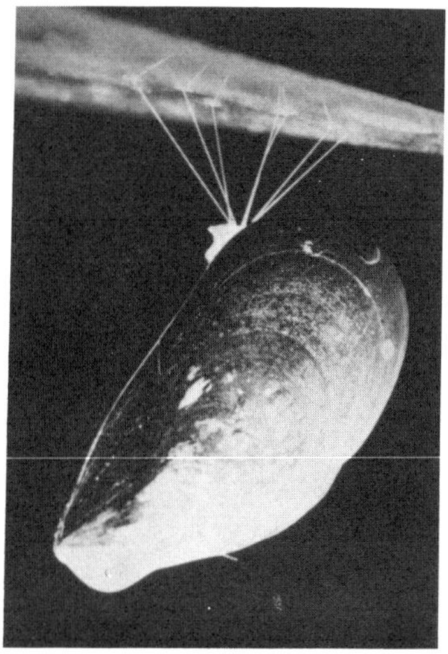

Fig. 8. *Mytilus* attached to glass slide with byssal threads.

3. Sponges

Sponges are composed of silica-containing spicules which form a skeleton and a slimy covering. Sponges of various colors colonize surfaces in warm quiet waters and are usually found in combination with other fouling forms such as bryozoans or tunicates.

4. Coelenterates

The least fouling productive ocean regions are in deep water below the mixed layer where plantlike coelenterates—the dominant (often exclusive) taxon—attach and grow very slowly. Sea anemones and hydroids belong to the coelenterate group. The fouling forms have polyps, tubelike animals with an opening at the top (Figure 9). These organisms are motile. When an environment becomes unfavorable, they can detach their pedal disc at the base of the colony and float to another site and attach to a new substratum.

5. Bryozoans

Bryozoan colonies resemble red or brown algae, sponges, hydroids, and even tunicates. A colony consists of gelatinous or calcareous boxlike compartments arranged in patterns that are characteristic of the species. Each

Fig. 9. Mat of hydroids and serpulids removed from copper/nickel screen.

compartment contains an individual animal. Fouling forms of bryozoans metamorphose and grow over the surface or grow upright like a plant. *Bugula* is a common upright bryozoan normally appearing in a brushy growth made up of branches in double rows. The color is usually straw or red. The larvae are nonfeeding and remain free-swimming only a few hours before settlement and metamorphosis. The behavior of larvae can vary widely among species, but generally it can be divided into free-swimming and substratum selection phases.

6. Serpulids

Serpulids are worms that live in calcareous tubules, usually an inch or more in length (Figure 9). Many of these organisms are filter feeders. Eggs of the tubeworms are liberated into the water and the larvae develop into free-swimming members. The free-swimming form settles on the substratum, secretes a tube, and undergoes metamorphosis into an adult worm.

The larvae may visit a large number of substrata, and when a favorable one is found, they attach. A milky fluid is secreted from an attachment gland, forming the initial tube. The process from initial fixation to the formation of the tube takes one to two minutes.

7. Tunicates

Tunicates are members of a group of organisms that have backbones and are closer to free-swimming fish than they are to the mollusks and crustaceans. The larval form is tadpole-like and has a primitive backbone. The adult forms are soft saclike creatures that grow either singly or in colonies. The simple tunicates have two body openings called syphons. A stream of water is drawn in one syphon and expelled through the other. Simple tunicates can be recognized by their production of jets of water when they are taken out of the sea, which gives them the name sea squirts. These fouling organisms are covered by a protective layer or tunic of tough cellulosic material. The colonial forms can be recognized by their appearance in large, often brightly colored purple, orange, or yellow groups on marine surfaces.

8. Amphipods

Fouling amphipods are relatively small—seldom more than 10 mm long. They resemble small shrimp and often produce membranous or muddy tubes. Ultimately, large masses of these tubes can completely cover a surface.

9. Algae

Microscopic diatoms and blue-green algae have been reported as microfouling organisms that contribute to the initially formed primary-film community, but the great majority of fouling algae belong to the macroalgae groups including Chlorophyta (green algae), Phaeophyta (brown algae), and Rhodophyta (red algae).

Algae dominate the unstable waterline regions of partially immersed structures (*46*). Slimes of algal spores form within a week or two. Eventually, filamentous forms grow on the surface and are often a major source of fouling. Both the sea lettuce, *Ulva*, and the complex branched green alga, *Enteromorpha*, are commonly found on fouled surfaces.

10. Marine Borers

The majority of marine borers penetrate for protection and support, but borers belonging to Teredinidae, Xylophagainae, Limnoridae, and Sphaeromidae use their borings as a source of nutrients (Figure 10) (*47*).

Some marine borers attack wood (Figure 11), plastic, lead sheathing, concrete, and rocks. The most successful and highly adapted stone borers, the

Fig. 10. X-ray of wood panel invaded by teredinids.

Fig. 11. Wooden piling riddled by *Bankia gouldi* (internal) and *Limnoria* (surface) taken from Oyster Creek, Barnegat Bay, New Jersey.

pholads, are bivalve mollusks. They have shells with sharply serrated edges and use these, possibly in combination with chemical mechanisms, to bore.

The teredine borers *Bankia*, *Lyrodus*, and *Teredo*, as well as the pholad, *Martesia*, have been shown to attack plastics (*48*). The polyvinyl chlorides are most susceptible to severe damage and deep penetration. *Martesia* has also been observed to attack polyvinylchloride, nylon, and some low density polyethylenes.

B. Factors Influencing Attachment and Adherence of Macrofouling Organisms

1. Relation to Local Environments

The chemical and environmental factors that control and influence the settlement of macrofouling organisms have not been elucidated. Predictions concerning rate of attachment and extent of growth cannot be related to the nutritional status of the water. However, it is recognized that water pollution affects settlement. Pollutants from domestic sewerage outfalls may be nutritionally beneficial and encourage settlement of fouling organisms, while oxidizable organic pollutants will adversely affect their settlement by causing an intolerable decrease in dissolved oxygen. Other pollutants occurring in industrial waste outfalls may contain toxic components that prevent, directly or indirectly, the settlement of foulers. Silt and other suspended matter, which can accumulate in landlocked or otherwise physically restricted waters, may affect development of filter feeders and photosynthetic plants.

2. Temperature

Temperature is an important environmental factor controlling both the rate of attachment and growth of macrofouling organisms, as well as the metabolic and reproductive activities of many fouling organisms. In colder waters, reproduction starts in the spring and continues throughout the summer; in tropic waters, a vigorously breeding fouling community can exist throughout the year. However, within a given environment all fouling organisms do not become reproductively active simultaneously, and the population of different species colonizing a surface will peak at different times over the growing season. Numerical dominance shifts rapidly and repeatedly as successive waves of planktonic larvae settle, grow, and die. Photosynthetic activity of aquatic plants, salinity changes, seasonal changes in turbidity, and other environmental factors affecting the water column will also influence the formation of a local, normal fouling community.

3. Water Velocity

The impact of water velocity on the attachment of macrofouling species to tube interiors and rotating discs has been evaluated by numerous investigators (*49*, *50*, *51*). The velocities that are cited for the prevention of settlement vary with species and conditions of flow. For example, it has been reported that the attachment of ascidians and bryozoans can be prevented by current speeds greater than 0.4 m/sec (*51*), whereas barnacle larvae can be prevented from settling on immersed plates by velocities in excess of 1.0 m/sec (*52*).

4. Substratum

The attachment of macrofouling organisms to submerged surfaces is a nonrandom process. Although the swimming ability of invertebrate larval forms is not sufficiently powerful (ca. 1 cm/sec) to overcome transport by currents, their selection ability—that is, the ability to delay or not delay the irreversible process of attachment—allows a small-scale response to environmental factors and permits them to make a choice. Virtually all benthic marine invertebrates that contribute to macrofouling are recruited as larvae from the plankton. This recruitment depends on detecting specific exogenous biochemical cues or triggers from the substratum (*53*, *54*). The natural inducers in larvae recruitment are, in several well characterized examples from different phyla, surface associated nondiffusing proteins or protein-conjugates (*53*). Additionally, most sedentary fouling organisms establish gregarious communities, a process that requires the settling larvae to recognize their own species. At release, most larvae are photophilic but become photophobic if their customary adult habitat is subsurface. Dissolved attractants and repellents, light, and gravity also influence swimming direction. Surface shape is also a factor for many species. Barnacle cyprids settle in depressions, and algal spores preferentially seek out surface cracks. Surface texture effects on larval settlement are difficult to generalize, except roughened surfaces do seem to foul faster than smooth surfaces (*55*, *56*, *57*).

C. *Methods for Quantifying Extent of Macrofouling*

Macrofouling rates are often determined by exposing standard 5-cm by 30-cm wood or asbestos test panels at stations in the water (Figures 12 and 13). The term "fouling rate" is defined as the dry weight of attached organisms per unit area of substratum per unit time of exposure. This dry-weight value provides a quantitative measure of fouling productivity. Panels are usually recovered monthly for exposure times of one to two years. The

Fig. 12. Wood panels on rack exposed in Oyster Creek, Barnegat Bay, New Jersey. Major fouling organisms: *Mytilus edulis*, barnacles, seaweed.

Fig. 13. Wood panels exposed in Admiralty Inlet, Washington. *Balanus* barnacles are the dominant fouling organism.

dry-weight values for a series of exposures are often plotted on a cross-sectional chart of the test area and contoured by percent of maximum accumulation using the highest value, the shoreline station, as 100%. These data provide predictive forecasts of fouling rates for stationary structures at specific locations.

D. Control

Control of macrofouling depends on a knowledge of the organisms and the processes that control their attachment. Methods of control can be divided into surface modifications, chemical techniques, and mechanical treatments.

1. Surface Modifications

One surface modification that has been investigated in recent years is the prevention of macrofouling by excluding the initial slime layer. The sticky bacterial slime found on all surfaces exposed to naturally occurring fluids was thought to encourage general adhesion and to promote macrofouling. However, there is no evidence that macrofouling can be controlled by regulating the initial events of microfouling film formation.

Surfaces can be designed to incorporate anti-adhesion agents so that macrofouling can be easily removed. The experimental development of surfaces with low surface energies, such as fluoropolymers, silicones, and silanes, has led to the concept of fouling-release surfaces. Tests evaluating barnacle adhesion have demonstrated that the fluoropolymer-barnacle attachment is more fragile than the attachment on surfaces of higher surface energy.

Copper alloys have a long history of use as antifoulants in the marine environment. It should be emphasized that this biocidal activity is limited to the macrofouling organisms. Copper and its alloys are readily colonized by microfouling organisms; however, copper does inhibit barnacle shell formation. The antifouling properties of copper and its alloys in seawater depend on the corrosion rate and the release of copper ions into the immediate environment. Therefore, the antifouling properties of copper depend on a number of such environmental parameters as water temperature, velocity, oxygen content, and salinity, as well as alloy composition.

The most successful antifouling methods depend on the surface modifications through such coatings as polyvinyls, polyacrylates, epoxy compounds, and polyurethanes that are impregnated with toxicants. The advantages of coatings are easy manufacture, high speed, low cost application, durability, and applicability to a variety of structural forms and compositions. The delivery of the toxic control agents to the desired location in sufficient quantity to retard or eliminate macrofouling depends on the structural design and composition. Despite their limited lifetimes, it is generally recognized that coatings provide the most cost effective means of macrofouling control.

Commercial cuprous oxide formulations must provide a leach rate of 10 g Cu/cm^2/day to be effective. The lifetime of present copper-containing coatings varies from 12 to 20 months. During normal exposures, these coatings form a surface layer of insoluble copper salts that tends to inhibit further leaching. Because of the limited lifetime and because they promote corrosion of aluminum surfaces, copper-containing coatings have limited applicability.

Organotin compounds tend to have a higher level of toxicity than copper-based compounds with required leach rates of 1–5 g Sn/cm^2/day. Coatings containing both copper and tin have been formulated with resulting synergistic toxicity.

Controlled release of antifouling compounds can be achieved by several means. Hydrophilic polymers that absorb water to about 75% of the original weight prevent excessive release of the toxin. Another system incorporates microsphere-encapsulated toxicants into a resin matrix. Ablative plastic films have been used as overlays for conventional painted coatings. These films provide drag-reducing properties and a controlled release of toxin under turbulent conditions. These films are made of hydrophilic materials that provide a stable laminar sublayer while a ship is in motion. In the absence of such a film, turbulence induced by microroughness normally reduces the thickness of the laminar sublayer. The outer layer of the coating softens with time in service, exfoliates, and is smoothed by flowing water.

Elastomeric products that contain toxicants, such as neoprene and nitrite rubber, can be made in a 10 mm thickness and can thus have a toxicant reservoir much greater than that found in paints. The advantages are resistance to abrasion, erosion, and cavitation. The elastomers can be formed by curing them on the object to be protected or formed as sheets that must be glued to the substratum.

In addition to these well-known toxicants, several substances, such as tannic acid, benzoic acid, acrylamides, and extracts from sargassum and gorgonian corals, have been shown to be effective in repulsing macrofouling species. Quinones and tannins are effective in blocking shell formation of teredinids, as demonstrated in laboratory studies. Current research on compounds used for macrofouling control has been tabulated in Fischer et al. (*58*).

2. Chemical Techniques

Creosote has traditionally been used to impregnate wood and to protect wood against marine borers. Vapam, Vorlex, and chloropicrin have also been used to control marine-borer infestations, as well as fungi-initiated decay. Polyvinyl chloride and polyethylene film wraps, as well as concrete encasements, have been used to limit the oxygen supply and thereby control marine borer attack. A recent approach to wood preservation has been to adopt the new polymer-based control release technology.

There is no consensus about the concentration of chlorine needed to control macrofouling. Similarly, no agreement has been reached about the relative advantages of low-level continuous chlorination compared to intermittent chlorination at higher concentrations. Experience has shown that the level of chlorination and the application rates depend on a variety of factors, including the predominant organisms, growth rates, location, season, and water temperature. In general, the soft macrofouling organisms can be controlled by intermittent chlorination at a level of 1.0 mg/L residual chlorine for one hour out of every eight hours. Hardshelled foulers, including barnacles and mussels, require continuous discharge of low-level chlorination—0.25–0.5 mg/L of free residual chlorine (*59*).

3. Mechanical Treatments

Underwater macrofouling can be removed by scrubbing with rotary brushes, hydraulic jets, and scraping devices to remove growths from large areas. These devices are effective in removing macrofouling from intact antifouling coatings on flat or slightly curved surfaces. Complex surfaces can be cleaned by scraping; however, shell bases are routinely left behind. The use of a flat wire brush or a cavitating water jet tool to remove fouling, followed by polishing with an abrasive pad or a carbide-impregnated bristle brush, can effectively remove hard fouling deposits.

Electric currents have been used to provide fouling control for short periods (*60, 61*). However, because of the high ionic strength and conductivity, these methods are of limited usefulness. A direct current density of 1 amp/ft^2 has been used to control macrofouling for 40 days (*62*). Alternating currents have also been used (*63*). Electric currents produce hydrogen peroxide and chlorine, which may be responsible for the biocidal impact (*64*).

III. Summary

Microfouling and macrofouling can be expected to occur on all unprotected surfaces in contact with seawater. Water chemistry, water temperature, pressure, shear stress, substratum composition, and design features can affect the settlement and attachment of organisms. The formation of a microfouling film results in increased fluid frictional resistance, decreased heat transfer efficiency, and biodeterioration. The buildup of macrofouling increases the weight-in-water of submerged structures and contributes to deterioration via numerous mechanisms. No single countermeasure is effective in the prevention of microfouling and macrofouling. However, microfouling can be controlled by chemical and mechanical treatments. Control measures for macrofouling depend on surface modifications that prevent the settlement of the macrofouling larvae.

Acknowledgements

This work was supported, in part, by NORDA Program Element 61154N, Herbert Eppert, program manager. Special thanks to Dr. Ruth D. Turner,

Museum of Comparative Zoology, Harvard University, Cambridge, MA for providing photographs of macrofouling organisms.

References

1. Little, B. J. (1983). In: *Marine Biodeterioration. An Interdisciplinary Study* (J. D. Costlow and R. C. Tipper, eds). Naval Institute Press, Annapolis, pp. 63–67.
2. Marszalek, D. S., Gerchakov, S. M., and Udey, L. R. (1979). *Appl. Environ. Microbiol.* **38**, 987–995.
3. Mitchell, R., and Benson, P. H. (1980). *Micro- and Macrofouling in the OTEC Program: An Overview*. Prepared for Argonne National Laboratory, ANL/OTEC-BCM-011.
4. Corpe, W. A. (1977). *Proc. OTEC Biofouling and Corrosion Symposium*, Seattle, WA, 31–44.
5. Baier, R. E., Shafrin, E. G., and Zisman, W. A. (1968). *Science* **162**, 1360.
6. Marshall, K. C. (1976). In: *Interfaces in Microbial Ecology*. Harvard University Press, Cambridge, MA.
7. Di Salvo, L. (1973). *Limnol. and Oceanog.* **18**, 165–168.
8. Friedman, B. A., Duggan, R. R., Pfuster, R. M., and Remson, C. C. (1969). *J. Bacterial.* **98**, 1328–1334.
9. Hendricks, S. W. (1974). *Appl. Microbiol.* **28**, 572–578.
10. Costerton, J. E., Geesy, G. C., and Cheng, K. J. (1978). *Ind. Eng. Chem.* **43**, 901–904.
11. Bryers, J. D. (1980). "Primary Biofilm Formation." Ph.D. Thesis, Rice Univ., Houston, TX.
12. Characklis, W. G. (1973). *Water Res.* **7**, 1113–1128.
13. Young, L. Y. (1972). "The Ecology of Bacterial Populations on Surfaces in Seawater." Ph.D. Dissertation, Harvard University, Cambridge, MA.
14. Novak, L. (1981). "Comparison of the Rhine River and the Aresund Sea Water Fouling and Its Removal by Chlorination." Presented at 20th National Heat Transfer Conference, Milwaukee, WI.
15. Jannasch, H. W., Eimjellen, K., Wirsen, C. O., and Farmanfarmian, A. (1971). *Science* **171**, 672–675.
16. Berger, L. R. (1974). *Rev. Phys. Chem. Japan* Special Issue, 639–642.
17. Characklis, W. G. (1977). *Proc. of Symp. on the Microbiology of Power Plant Thermal Effluents.*
18. Gerchakov, S. M., and Sallman, B. (1978). In: *Microbiology of Power Plant Thermal Effluents* (R. M. Gerkala, ed). University of Iowa Press, Iowa City, pp. 67—72.
19. Loeb, G. I., and Niehof, R. A. (1975). In: *Applied Chemistry at Interfaces*, Advances in Chemistry Series no. 145. American Chemical Society, Washington, DC, pp. 319–335.
20. Characklis, W. G. (1981). *Biotechnol. Bioeng.* **23**, 1923–1960.
21. Loeb, G. I., Laster, D., and Gracek, T. (1984). In: *Marine Biodeterioration: An Interdisciplinary Study* (J. D. Costlow and R. C. Tipper, eds.). Naval Institute Press, Annapolis, MD, pp. 88–91.
22. Characklis, W. G., and Zelver, N. (1983). "Biofilm Development and Control in Simulated Shipboard Condenser Tubes." Final Report to David Taylor Ship Research and Development Center, 203 pp.
23. Staffeldt, E. E., and Calderon, O. T. (1967). *Div. Ind. Microbiol.* **8**, 321–326.
24. Burnes, J. M., Staffeldt, E. E., and Calderon, O. H. (1967). *Develop. Ind. Microbol.* **8**, 327–334.
25. Ashton, S. A., Miller, J. D. and King, R. A. (1973). *Br. Corr. J.* **8**, 185–189.

26. Webb, C. A. (1975). "A Polarization Resistance Study of Copper Corrosion Kinetics in Aqueous Chloride and Amino Acid Solutions." Ph.D. dissertation, Univ. of Miami, FL.
27. Kramer, D. A. (1976). "The Corrosion Kinetics of Copper in Aqueous Amino Acids and Carboxylic Acid Solutions." Ph.D. Dissertation, University of Miami, FL.
28. Miller, J. D. and Tiller, A. K. (1970). In: *Microbial Aspects of Metallurgy* (J. D. Miller, ed). Elsevier, New York.
29. Mara, D. D., and Williams, D. J. (1971). *Chem. Ind.* **21**, 566-567.
30. Salvarezza, R. C., de Mele, M. F. L., and Videla, H. A. (1983). *Corrosion* **39**, 26-32.
31. Karl, D. M., and La Rock, P. A. (1975). *J. Fish. Res. Bd. Can.* **32**, 599—607.
32. Karl, D. M., and Holm-Hansen, O. (1976). *Analyt. Biochem.* **75**, 100-112.
33. Karl, D. M. (1978). *Microbiol.* **36**, 349-355.
34. King, J. D., and White, D. C. (1977). *Appl. Environ. Microbiol.* **33**, 777-783.
35. Moriarty, D. J. (1977). *Occologia (Berl.)* **26**, 317-323.
36. White, D. C., Davis, W. M., Nickels, J. S., King, J. D., and Bobbie, R. J. (1979). *Occologia (Berl.)* **49**, 51-62.
37. Hobbie, J. E., Daley, R. J., and Jasper, S. (1977). *Appl. Environ. Microbiol.* **33**, 1225-1228.
38. Macalady, D., Carpenter, J., and Moore, C. (1977). *Science* **195**, 1335-1337.
39. Sugam, R., and Helz, G. R. (1977). *Chesapeake Science* **18**, 113-116.
40. Eppley, R. W., Renger, E. H., and William, P. M. (1976). *Est. and Coast. Mar. Sci.* **4**, 147-161.
41. Wong, G. T., and Oatts, T. J. (1982). *Water Res.* **17**, 1533—1535.
42. Wong, G. T. (1980) In: *Water Chlorination: Environmental Impact and Health Effects*, Vol. III (R. L. Jolly, W. A. Brungs, and R. B. Cumming, eds). Ann Arbor Science Publishers, Inc., Ann Arbor, pp. 395-406.
43. Lott, D. F., and Tuovila, S. M. (1979). *Proc. 6th OTEC Conference*, 12.12-1-12.12-5. Washington, DC.
44. Nubel, E. D. (1977). *Proc. 4th Ann. Meet. of the OTEC Conference, VII*, 61-63. New Orleans, Louisiana.
45. Kern, W. I. (1977). *Proc. 4th Ann. Conf. on Ocean Thermal Energy Conversion, VII*, 64-78. New Orleans, LA.
46. Fletcher, R. S. (1980). In: *The Shore Environment*, Vol. 2 (J. H. Price, D. E. G. Irvine, and W. F. Farnham, eds). Academic Press, London, pp. 843-874.
47. Turner, R. D. (1984). In: *Marine Biodeterioration: An Interdisciplinary Study* (J. D. Costlow and R. C. Tipper, eds). Naval Institute Press, Annapolis, MD, pp. 3-16.
48. Muraoka, J. S. (1971). *Ocean Industry.* (Feb.), **6**, 21-23.
49. Smith, F. G. W. (1946). *Biol. Bull.* **90**, 51-70.
50. Wood, E. J. (1955). *Austral. J. Sci.* (Aug.), **10**, 34-37.
51. Relini, G., and Dabini-Oliva, G. (1972). *Proc. 3rd Int. Cong. on Marine Corrosion and Fouling*, 757-766.
52. Huve, P. (1953). *C. R. Acad. Sci., Paris.* **236**, 418.
53. Morse, D. E. (1984). In: *Marine Biodeterioration: An Interdisciplinary Study* (J. D. Costlow and R. C. Tipper, eds). Naval Institute Press, Annapolis, MD, pp. 134-140.
54. Chia, F. S., and Rice, M. E., editors (1978). *Settlement and Metamorphosis of Marine Invertebrate Larvae.* Elsevier, New York.
55. Wiseley, B. (1959). *Austral. J. Mar. and Freshwater Res.* **10**, 30-44.
56. Corlett, J. (1948). *Estuary. Proc. Trans. Liverpool Biol. Soc.* **56**, 3-28.
57. De Block, J. W., Hannie, J., and Geelen, F. M. (1959). *Arch. Neerl. Zool., Suppl* **1-13**, 446-460.
58. Fischer, E. C., Castelli, V. J., Rodgers, S. D., and Bleile, H. R. (1984). In: *Marine Biodeterioration, An Interdisciplinary Study* (J. D. Costlow and R. C. Tipper, eds). Naval Institute Press, Annapolis, MD, pp. 261-290.
59. James, W. G. (1967). *Chem. and Ind.* **Jun**, 994-996.

60. Paul, S. K., Chari, K. V., and Bhattacharyya, B. (1975). *J. Am. Wat. Wks. Assoc.* **67**(3), 140–141.

61. Czenski, P. (1975). *Proc. IEEE*, **63**(11), 1540–1544.

62. Castle, E. S. (1951). *Ind. Eng. Chem.* **43**, 901–904.

63. Roe, T. (1967). Report NCEL No. N-82.

64. Dhar, J. P., Bockris, J. O'M., and Lewis, P. H. (1981). *J. Electrochem. Soc.* **128**(1), 229–231.

65. Costerton, J. M., Geesey, G. C., and Ching, K. J. (1978). *Sci. Amer.* **238**, 86–95.

66. Lepley, L. K. (1968). *Photogrammatric Engineering* **34**, 667–674.

4

Marine Corrosion

G. H. KOCH, J. A. BEAVERS, and W. E. BERRY

BATTELLE
Columbus Division
Columbus, Ohio

ISBN 0-12-341828-3

I. Marine Corrosion

This chapter discusses the performance of construction materials in marine environments. The first section provides a general description of the different forms of corrosion that can be experienced by the various alloys. Also, different marine environments (viz., atmosphere, splash and tide, immersion, and mud) and the factors in these environments that affect the corrosion of alloys are discussed. These factors include oxygen, biological activity, temperature, velocity, salinity, and pH.

Following this general discussion, the marine corrosion behavior of specific commonly used alloys is described. These alloys include carbon steel, stainless steels, nickel alloys, aluminum alloys, titanium alloys, and copper alloys. Although these sections provide a thorough description of the corrosion performance of these alloys in marine environments, little actual numerical information is presented. For a more extensive and quantitative discussion of marine corrosion of the alloys discussed in this chapter, reference is made to a Department of Defense publication (*1*).

II. Classification of Metals and Alloys

The common metallic materials can be divided into two classifications according to their corrosion behavior in marine environments: corrosion resistant materials and corrosion allowance materials. Most of the corrosion resistant materials depend upon the presence of a thin, tightly adherent, passive oxide film to limit the rate of the corrosion (oxidation) reaction. Examples of these materials are aluminum, stainless steels, certain nickel-base alloys, and titanium.

Corrosion allowance materials corrode at relatively high rates in marine environments and rates of attack are generally limited by the rate of the cathodic (reduction) reaction. Examples of these materials are steel and zinc. Copper-base alloys possess characteristics of both corrosion resistant and corrosion allowance materials. Rates of attack of virgin specimens in seawater are generally limited by the rate of oxygen reduction, but semiprotective passive films form on copper base alloys with time.

Discussion of the forms of corrosion for these materials is given below.

III. Forms of Corrosion

Aqueous corrosion generally is divided into eight common forms: uniform attack, pitting, crevice corrosion, galvanic corrosion, erosion-corrosion, environmentally assisted cracking, dealloying (selective leaching), and intergranular corrosion.

A. *Uniform Attack*

Uniform attack, as its name implies, is the general wastage of a metal. In aqueous environments, it is generally electrochemical in nature, with two or more half cell reactions occurring simultaneously and randomly over the metal surface. The anodic reaction is metal oxidation, while the cathodic reaction may be one or more of several reactions, such as oxygen reduction, proton reduction, water reduction, metal ion reduction, or reduction of an oxidant such as nitrate or H_2O_2. In marine environments, the dominant reduction reaction generally is oxygen reduction. Although uniform attack represents the greatest destruction of metal on a tonnage basis, it is by far the least insiduous form since rates of attack can be conservatively estimated and there are a variety of mitigation procedures.

B. *Pitting*

In the corrosion literature, there are two common definitions of the term pitting: a morphological one and a mechanistic one. In the former, the mechanism by which the 'pits' propagate may be anything from erosion-corrosion to differential cell corrosion. The most generally accepted mechanistic definition of pitting is localized corrosion that occurs on boldly exposed surfaces (no macrocrevices) of passive metals under relatively stagnant conditions. Pitting may also occur on relatively film-free metals at local metallurgical imperfections such as at inclusions. Although the mechanism of initiation is not well understood for either type of surface and may vary from one alloy system or environment to the next, it is generally considered that the mechanism of propagation of pitting is common to most systems. The propagation stage consists of metal oxidation within the pit, migration of anions into the pit to maintain charge neutrality, acidification in the region as a result of hydrolysis of the metal ion, and oxidizer (oxygen) depletion once an occluded cell has formed. All of these processes tend to further destabilize the system; thus, pitting corrosion is autocatalytic in nature.

Metallurgical factors which favor pitting include any inhomogeneity in the material that can act as an initiation site, such as inclusions, segregation, welds, scrapes, or heat-tinted areas. Under submerged exposure, environmental factors which favor pitting include stagnant conditions, the presence of specific anions such as chloride and heavy metal ions such as copper (in aluminum alloys), the presence of highly oxidizing conditions, or local variation in the oxidizing strength of the environment.

C. *Crevice Corrosion*

Like pitting, crevice corrosion occurs in both passive metals and in relatively film-free materials. The presence of a crevice or occluded cell is essential for initiation of crevice corrosion. The first step involves the depletion of an oxidizer (oxygen) within the crevice, resulting in the formation of a differential aeration cell and, on passive metals, a reduction in the stability of the passive films within the crevice. Once initiation has occurred, the processes associated with propagation of attack in crevices and pits are thought to be essentially identical in most systems.

The common materials of construction are not uniformly susceptible to crevice corrosion in seawater. The stainless steels and some aluminum alloys are highly sensitive, whereas titanium and the high molybdenum nickel base alloys are resistant.

The conditions that promote crevice corrosion include the presence of crevices, deposits, and fouling organisms. Common design features which promote crevice corrosion are the presence of gaskets, washers, and rivets. Crevice corrosion is generally more prevalent in the splash and tide zones, where there is a plentiful supply of oxygen, than under submerged conditions.

D. *Galvanic Corrosion*

Galvanic corrosion is the accelerated attack of a metal when electrically coupled to a dissimilar metal and contacted with an electrolyte. Electrochemical potential differences between metallic materials that are immersed in the electrolyte provide the driving force for galvanic corrosion. A listing of these potentials in order of ascending or descending nobility, which is referred to as a galvanic series, is environment specific; a series in flowing seawater at 25°C is given in Figure 1 (*3*). When dissimilar metals are placed in electrical contact in the electrolyte, the corrosion rate of the more noble (electropositive) material will be decreased and the corrosion rate of the more active (electronegative) material will be increased. For this galvanic effect to occur, the dissimilar metals must be in electrical contact such that electron flow

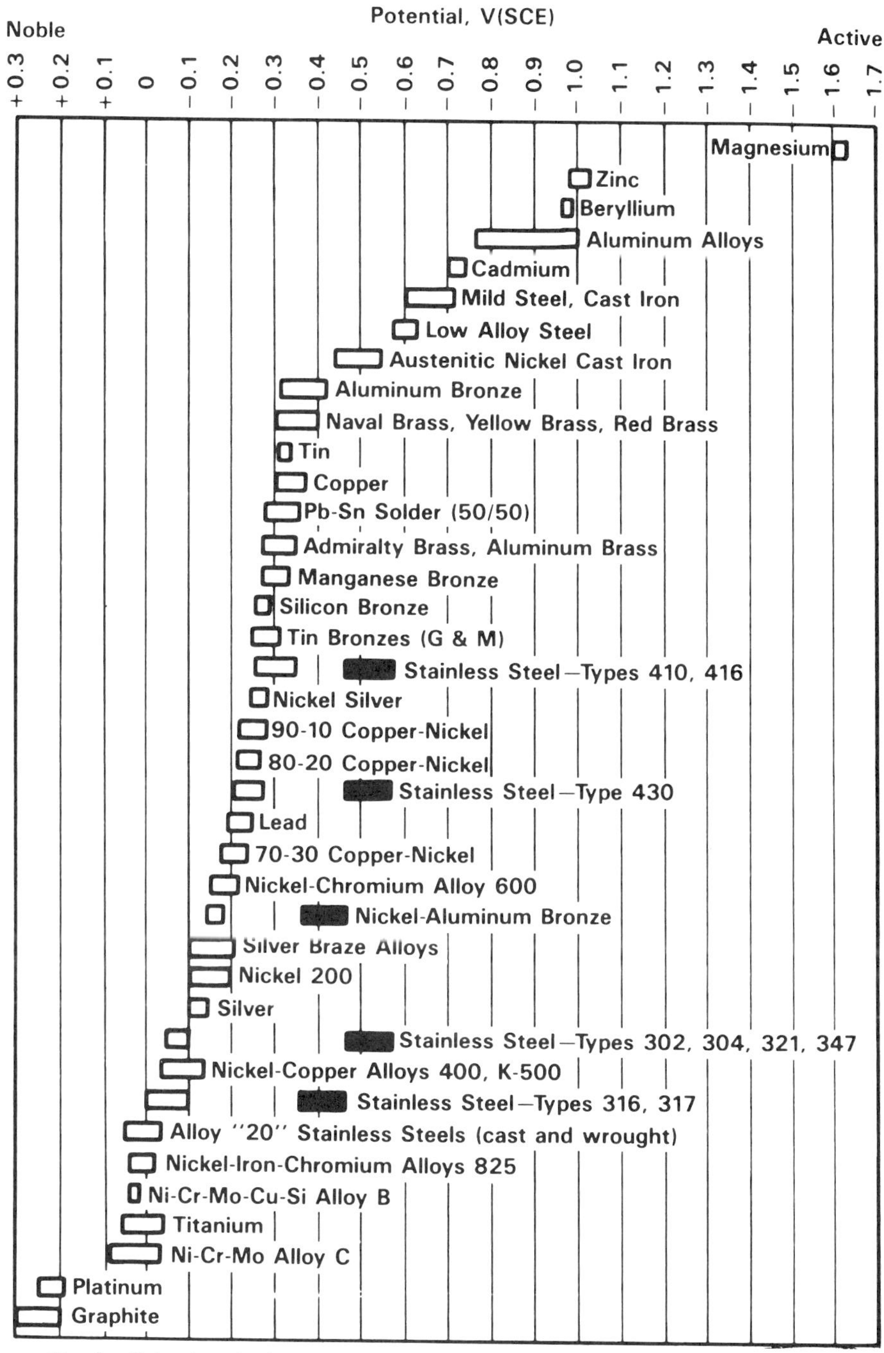

Fig. 1. Galvanic series in seawater (*3*). Alloys are listed in the order of the potential they exhibit in flowing seawater. Certain alloys indicated by the symbol: ■ in low-velocity or poorly aerated water, and at shielded areas, may become active and exhibit a potential near −0.5 volts.

between the two metals can occur; close proximity in an ionic conducting media is not sufficient. However, in a complex structure or component, the path or paths of electrical conduction are not always readily apparent.

The severity of galvanic attack is dependent upon many variables, among them are the electrolyte conductance, the electrical contact conductance, the relative surface areas of the metals being coupled, the relative position of the metals in the galvanic series, and the ease with which the metals are polarized. Thus, copper will promote more severe galvanic attack of carbon steel than stainless steel or titanium, because the latter two materials are more readily polarized than the copper.

It is found that, for a metal like carbon steel whose corrosion rate is usually controlled by the total cathodic area available, the ratio between the area of the cathode and the area of the anode is important. A small anode, e.g., steel, coupled to a large cathode (e.g., copper) and immersed in seawater, will result in a greatly increased rate of attack of the anode. The reverse situation of a small cathode coupled to a large anode will have only a minor influence on the rate of attack of the anode.

To control or prevent the accelerated attack as a result of galvanic coupling, certain principles should be observed. First, the possibility of breaking the electrical circuit by providing an insulating barrier at the junction between the two metals should be considered. Second, if dependable isolation is not feasible, the cathodic member of the couple should be covered with a nonconductive protective coating. By reducing this area, or eliminating the cathode entirely, enhanced corrosion of the anode material due to the couple is prevented in a safe manner. Under no circumstances should only the anode be painted. Any defect (holiday) in the coating will result in the entire cathode area being coupled to a small area of anode material, and extremely high rates of penetration can be expected.

In the atmosphere, galvanic corrosion is confined to a short distance, usually of the order of several millimeters, from the joint between the two metals because of the highly resistive nature of the condensed layer of moisture. On the other hand, galvanic corrosion may be significant over large distances in seawater because of its high conductivity. Some examples of couples and their corrosion behavior are given in Table I (*2*).

E. Erosion-Corrosion

Erosion-corrosion is an accelerated form of attack and is promoted by the relative movement between the corrosive media and the metal surface. Erosion-corrosion conditions can vary from relatively low velocity, lamellar flow to more severe impingement type attack where highly turbulent condi-

TABLE I

EXAMPLES OF GALVANIC COUPLES IN SEAWATER (*2*)

Metal A	Metal B	Comments
		Couples that usually give rise to undesirable results on one or both metals
Magnesium	Low-alloy steel	Accelerated attack on *A*, danger of hydrogen damage on *B*.
Zinc	Low-alloy steel	Accelerated corrosion of metal *A*, protection of metal *B*.
Bronze	Low-alloy steel	Protection of metal *A*, possible fouling of metal *A*, accelerated corrosion of metal *B*.
Aluminum	Low-alloy steel	Accelerated corrision of metal *A*, protection of metal *B*.
Stainless steel	Low-alloy steel	Reduction in pitting of metal *A*, accelerated corrosion of metal *B*.
Aluminum	Copper	Accelerated pitting on *A*; ions from *B* attack *A*. Reduced corrosion on *B* may result in biofouling on *B*.
Bronze	Stainless steel	Increased pitting on *A*.
		Borderline, may work, but uncertain
Copper	Solder	Soldered joint may be attacked but may have useful life.
Graphite	Titanium or Hastelloy C	
Monel-400	Type 316 SS	Both metals may pit.
		Generally compatible
Titanium	Inconel 625	
Lead	Cupronickel	

tions exist. In most cases, the mechanisms are related to interaction of the corrosive media with a protective surface film. The relative movement of the solution may result in mechanical damage to the film or may replenish the electrolyte of aggressive species at the metal surface that, in turn, affects the protective nature of the passive film.

Erosion-corrosion is characterized by the presence of holes, grooves, or valleys which usually exhibit a directional pattern. A variety of adjectives has been used to describe erosion-corrosion, such as horseshoe, star, crescent, and slot. Each is generally associated with a specific geometry. For example, horseshoe attack is associated with the erosion pattern around lodged obstructions in heat exchanger tubes.

Cavitation is a special form of erosion-corrosion which is caused by the formation and collapse of vapor bubbles in a liquid near the metal surface. Cavitation occurs most commonly in rotating machinery, such as pump impellers and ship propellers, where high pressure drops are encountered. The pressure drops, if sufficiently large, can promote local boiling. The vapor bubbles which form subsequently collapse, producing shock waves with very high pressures. These high pressures can produce plastic deformation of many metals. Thus cavitation damage is attributed to both corrosion and mechanical effects. The appearance of cavitation damage is similar to pitting except that the attacked areas are closely spaced, resulting in a roughened surface appearance.

Environmental factors that control erosion-corrosion include water velocity, oxygen content, water pH, temperature, flow geometry, and the presence of particulates in the water. Both the mechanical properties of the metal and the nature of the passive film on the metal affect the erosion-corrosion susceptibility. Stainless steels, nickel-base alloys, and titanium are generally resistant to the various forms of erosion-corrosion, whereas the copper-base alloys are usually more susceptible.

F. Environmentally Assisted Cracking

Environmentally assisted cracking (EAS) refers to the fracture or cracking of material by the conjoint action of an applied tensile stress and a specific corrosive medium. During EAS, the metal is unattacked over most of its surface but fine cracks progress through it. Depending upon the alloy-environment system and the specific form of the EAS, the cracking may be intergranular, transgranular, or a mixed mode. Cracks may follow straight paths or branch in several directions.

EAS can be subdivided into three general categories, anodic stress corrosion cracking, hydrogen stress cracking, and corrosion fatigue. Anodic stress corrosion cracking generally occurs where there is an appropriate balance between electrochemical activity of the advancing crack tip and passivity of the crack walls and the free surfaces. Cathodic protection is generally effective in preventing this form of cracking. On the other hand, cathodic protection may accelerate hydrogen stress cracking by facilitating hydrogen entry. The effect of cathodic protection on corrosion fatigue is dependent upon whether crack propagation is being assisted by anodic dissolution or hydrogen. Other important variables that affect environmentally assisted cracking include metal composition, heat treatment, solution composition, temperature, and stress-related parameters such as the maximum stress or stress intensity and the presence of load cycling.

All engineering materials are susceptible to one or more forms of EAS under certain environmental and applied stress conditions. However, the environment that promotes cracking in one material may be completely harmless to another. For example, practically all copper base alloys are susceptible to stress corrosion cracking in ammoniacal environments, whereas the austenitic stainless steels are immune in the same environment. On the other hand, copper-base alloys are generally considered to be immune to stress corrosion cracking in seawater environments, but many stainless steels are extremely susceptible. Similarly, the high strength steels are susceptible to hydrogen stress cracking in chloride containing environments.

G. *Dealloying*

Dealloying is the selective removal of the more active element form a solid solution alloy by electrochemical processes. This phenomenon is also referred to as selective leaching. The most common example is the selective removal of zinc in brass alloys (dezincification). Similar processes occur in copper-nickel and copper-aluminum alloys. Graphitic corrosion, a process in which iron is selectively removed from cast iron, leaving a graphite layer, is also considered to be a form of dealloying.

Dealloying failures in seawater applications are most common in copper-base alloys and generally are not observed in stainless steels or titanium. Two types of dealloying are observed in copper-base alloys, namely, uniform or layer type and plug type. The latter type generally initiates at sites of deposits or foreign bodies.

Two mechanisms for dealloying have been proposed. One involves the codissolution of the two alloying elements and the deposition of the more noble material onto the metal surface. The other involves the surface or bulk diffusion and the selective dissolution of the active element.

The occurrence of dealloying is affected by the active elements present in the alloy as well as minor elemental additions such as tin and arsenic (in the case of dezincification). Other factors that affect the failure include solution composition, temperature, flow velocity, pH, and electrochemical potential.

H. *Intergranular Corrosion*

Intergranular corrosion refers to localized attack of grain boundaries in certain aqueous environments. It can be caused by impurities at the grain boundaries, enrichment of one of the alloying elements, or depletion of one of these elements in the grain boundary areas. Intergranular corrosion is most commonly observed in austenitic stainless steels where the attack is related to heat treatment in the temperature range of 500°C to 800°C. Heating the alloy in this temperature range promotes the precipitation of chromium carbides thereby removing the chromium from solid solution and resulting in a metal with lower chromium content than the adjacent areas.

Intergranular corrosion is also observed in susceptible aluminum alloys. Due to the strong directionality of the grains in plate materials, the intergranular corrosion is recognized by the formation of surface flakes and is therefore referred to as exfoliation corrosion.

IV. Ocean Environments

Seawater covers more than 70% of the earth's surface and is the most abundant naturally occurring electrolyte. Most of the common metals and alloys of construction are attacked by seawater or moisture-laden sea air. Since the behavior of materials may vary widely, depending on the exposure conditions, their performance is discussed according to the specific environmental zone involved, namely atmosphere, splash and tide, immersed, and mud. Characteristics of these zones and their influences on corrosion are briefly discussed below.

A. Atmosphere

Deposition of sea mist and sea salt occurs on materials exposed in marine atmospheres. The salts are deleterious to most common alloys, promoting localized attack; their hygroscopic nature and sea mist provide an electrolyte that is necessary for the attack to occur, while the anions in the salt promote film breakdown on passive metal surfaces. The mere presence of the deposits also may promote the formation of differential aeration cells.

The salt deposition process is a function of many variables, including wind and wave conditions, height above the sea, distance from the shore, panel orientation, degree of sheltering, and the amount of rain and distribution during a given time period. In general, those factors which promote salt buildup and prevent rainfall from removing the salt promote the most severe corrosion. Thus, the bottom sides of panels or those panels which are sheltered from the rain generally are more severely attacked.

Fungi and mold also may deposit on metal surfaces in marine atmospheres. These organisms may promote corrosion by the formation of differential aeration cells and by holding moisture on the metal surface.

B. Splash Zone

Materials in the splash zone are almost continuously wetted with well aerated seawater and biofouling generally does not occur. For materials such as carbon and low alloy steels, which do not form thin, tenacious, passive films, the splash zone is the most aggressive of the marine zones. In this region, there is an ample supply of oxygen, and the entrained bubbles in the seawater are highly destructive toward the thick semiprotective films that form on some of these materials.

On the other hand, passive metals such as the stainless steels and titanium generally perform well in the splash zone because the well aerated conditions promote passivity.

C. Tide Zone

In the tide zone, a material alternates between fully immersed exposure and splash zone exposure; the relative amount of time under each exposure condition varies depending upon location within the zone. Thus, the tide zone is a transition region with a range of exposure conditions. For isolated test panels, upper tide zone exposure is similar to splash zone exposure, whereas lower tide zone exposure is similar to immersed conditions. In general, the aerated conditions within the tide zone promote rapid attack of metals such as steel, whereas passive metals such as stainless steels and titanium are protected. However, fouling does occur in the tide zone, and this may result in partial protection of steels or may promote localized attack of stainless steels.

One also has to differentiate between the behavior of isolated panels in the tide zone and typical case occurring in practice, where a structure such as a pile extends from the atmosphere through the splash and tide zones into submerged and mud zones. For a continuous steel pile, rapid attack takes place on that part of the surface just below the water line, with the portion at the water line being the cathode, which receives protection at the expense of the metal below.

D. Immersion Zone

The major factors that affect the corrosivity of the immersion zone are: oxygen, biological activity, pollution, temperature, and velocity. The seawater pH and salinity also may influence corrosivity. These factors are, in turn, a function of location, the depth, and the season.

Oceanographic data showing the variation in oxygen, temperature, pH, and salinity with depth for a site in the Pacific Ocean are shown in Figure 2 (*4*). These data show that the oxygen concentration is the highest near the surface (approximately 6 ppm), reaches a minimum of about 0.2 ppm at about 2000 feet in depth, and increases slightly to about 1 ppm at about 6000 feet in depth. Near the surface, photosynthesis and wave action oxygenate the seawater, whereas at intermediate depths, decomposition of organisms consumes the oxygen. The oxygen found at greater depths in the Pacific is attributed to supply of oxygenated arctic water by bottom currents.

On the other hand, analyses of seawater from various depths in the Atlantic, at the Navy's TOTO deep sea test site, indicate that the oxygen concentration does not vary appreciably with depth. A higher oxygen concentration at depths in the Atlantic than in the Pacific has been attributed to the lower biological oxygen demand in the Atlantic. In other words, decaying organisms in the Atlantic do not exhaust the available oxygen supply.

The lower pH found in waters at great depth may be partially explained on thermodynamic grounds, since the pH is reduced by extreme pressures alone. However, it is also known that the carbon dioxide-calcium bicarbonate-calcium carbonate equilibrium tends to be shifted; that is, the dissolved calcium carbonate in the deep ocean tends to be slightly below saturation; while, conversely, it is frequently supersaturated at the surface (*1*). One would therefore expect that, in deep waters, there would be less tendency to form a protective mineral scale, as compared with surface waters. It has been observed, for example, that the consumption of sacrificial anodes for providing cathodic protection is significantly greater at the extreme depths, e.g., 5000 feet, than in surface waters.

E. Mud Zone

The environmental conditions within the mud zone are highly dependent on geographic location and have been well characterized in only a few instances. In general, there are opposing influences of the environment on corrosivity within the mud zone. The oxygen concentration is low, which is beneficial for materials such as steel and copper. On the other hand, sulfate-reducing bacteria, which are found in the mud zone, produce sulfides that are highly detrimental to these materials. Practical experience from examination of offshore structures and with experimental piles indicates that steel is often attacked more slowly in the mud zone than in the seawater above it.

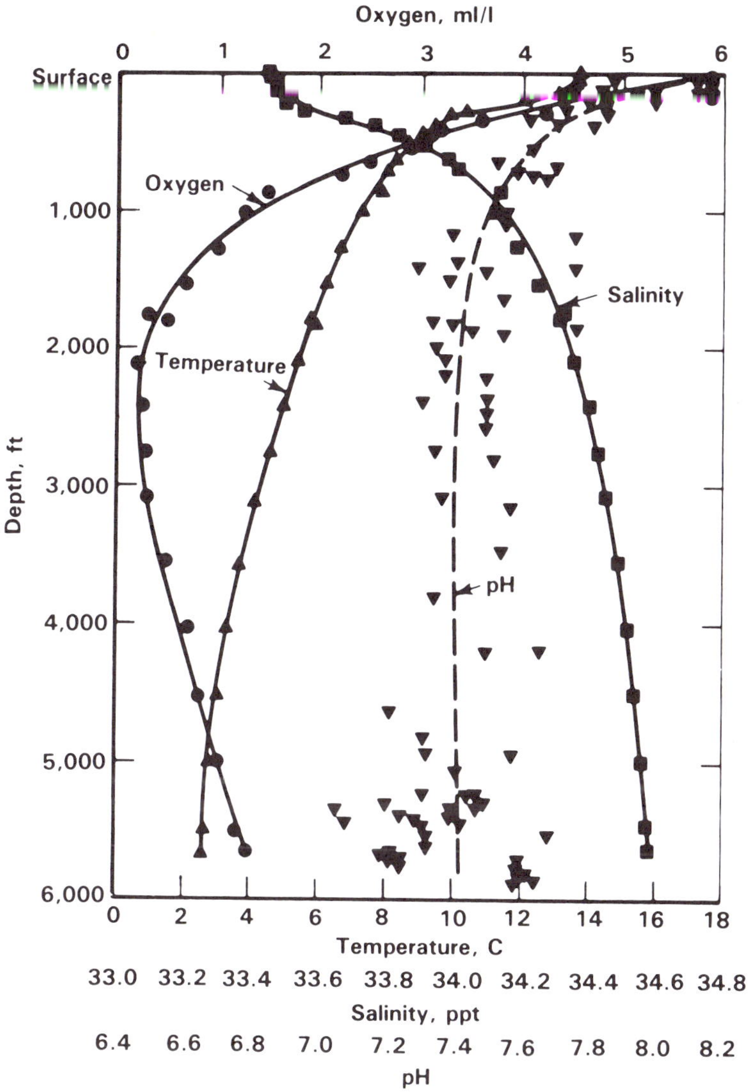

Fig. 2. Oceanographic data taken in the Pacific Ocean at a site west of Port Hueneme, California (*4*).

V Factors Affecting the Corrosivity of Seawater

The role of oxygen, biological activity, temperature, velocity, salinity, and pH on the corrosivity of seawater are discussed below.

A. *Oxygen*

The oxygen concentration in seawater may range from as low as 0.2 ppm at great depths to up to 12 ppm near the surface. Photosynthesis of green plants, wave action, etc. increase the oxygen level, whereas the biological oxygen demand of decomposing organisms will reduce it. The solubility of oxygen in seawater also increases with decreasing temperature. For a given location, seasonal variations in oxygen level resulting from the above actions will, in turn, influence corrosion behavior.

Oxygen affects the corrosion rate of metallic materials by providing a cathodic reaction ($O_2 + 2\,H_2O + 4\,e^- \rightarrow 4OH^-$), having a relatively noble equilibrium potential and rapid kinetics. For corrosion allowance materials such as steels, the presence of oxygen greatly accelerates rates of attack since the corrosion kinetics are limited in many cases by the rate of the reduction reaction. On the other hand, the presence of oxygen generally reduces rates of attack of corrosion resistant materials, such as stainless steels and titanium, by promoting passivation. However, where the oxygen supply to a passive metal surface is limited locally, such as at crevices, the presence of oxygen on the boldly exposed surfaces may promote localized attack.

For the corrosion allowance materials, the rate of the oxygen reduction reaction generally is limited by the rate of transport of oxygen to the metal surface. Thus, factors which influence the rate of transport, such as the oxygen concentration of the seawater and velocity of the seawater, also affect the corrosion rate.

B. *Biological Activity*

When a metal or other surface is first immersed in seater, a biological slime tends to develop within a matter of hours. Subsequent to the formation of the biological slime film, embryonic sessile organisms become firmly attached. Once attached, they rapidly transform to the mature form. Clapp (*5*) lists the following most common forms of sessile fouling organisms:

(1) Organisms Which Build Hard Shells
  (a) Annelids
  (b) Barnacles

- (c) Encrusting bryzoa
- (d) Mollusks
- (e) Corals

(2) Organisms Without Hard Shells

- (a) Marine algae
- (b) Filamentous bryzoa
- (c) Coelenterates or hydroids
- (d) Tunicates
- (e) Calcareous and silaceous sponges

The growth of sessile organisms, referred to as biofouling, is a major factor in the performance of metals in marine environments. For corrosion allowance materials, fouling may actually reduce rates of attack by limiting the supply of oxygen to the metal surface. However, the decay and by-products of marine organisms can be aggressive and promote localized attack of both corrosion resistant and corrosion allowance materials. Moreover, the simple presence of the marine organisms can aid in the development of differential aeration cells. For heat exchanger applications, lodgement of organisms also promotes erosion-corrosion of susceptible materials, such as copper base alloys, by increasing the seawater velocity in localized areas.

C. *Temperature*

The effect of temperature on the corrosion of metals is complicated by the interdependence of other relevant parameters and temperature. The kinetics of activation-controlled oxidation reactions and diffusion-controlled reduction reactions increase with increasing temperature. On the other hand, the passivation phenomenon is a complex function of temperature, since the stability of the films and the rate of formation are affected. In addition, oxygen solubility in seawater decreases with increasing temperature, and calcareous scales exhibit retrograde solubility, increasing the likelihood of precipitation of protective scales at elevated temperatures. Biological activity increases with increasing temperature over the temperature range near ambient, which can be either beneficial or deleterious. At higher temperatures, the marine organisms are actually killed, and thus biofouling does not occur. Thus, the overall influence of temperature on corrosion must be established for each unique set of conditions.

D. *Velocity*

The corrosion rate of corrosion allowance materials such as steels and copper-base alloys normally increases with increasing seawater velocity, since rates of attack of these metals are frequently limited by the availability of oxygen at the metal surface. In addition, high velocities tend to erode barrier films which form on alloys such as the copper based alloys. Passive metals such as titanium and stainless steels are highly resistant to flowing seawater. Indeed, pitting of stainless steels is less likely in flowing seawater than under stagnant conditions.

E. Salinity

The major oceans of the world are completely connected in the southern hemisphere and mixing is continuous. Not surprisingly, the relative proportion of salts does not vary appreciably in these interconnected seas; variations in salinity in open ocean surface waters typically range from 3.2% to 3.75%.

On the other hand, large variations in salinity are found in more isolated seas, see Table II (*6*). These salinity variations are accompanied by other chemical changes in the seawater, and thus the overall effect on corrosion behavior is difficult to predict. The concentration of aggressive ions such as chloride is lower in diluted seawater and in seawater having lower salinity, but the solubility of oxygen increases with decreasing salinity. Moreover, diluted seawater, as is found in the mouths of rivers, is usually unsaturated with respect to carbonate and thus is less likely to scale. Biofouling also is less likely in diluted seawater due to the reduced activity of marine organisms.

F. pH

The pH of seawater may vary slightly depending on the photosynthetic activity. Plant matter consumes carbon dioxide and affects the pH during daylight hours. Carbon dioxide content in seawater is influenced close to the

TABLE II

TOTAL SOLIDS IN OCEAN WATERS (*6*)

Body of Water	Total Dissolved Solids, ppm
Baltic Sea	8 000
Black Sea	22 000
Atlantic Ocean	37 000
Mediterranean Sea	41 000
Caspian Sea	13 000
Irish Sea	32 500

surface by the exchange with carbon dioxide in the atmosphere. The daily shift in pH from, say, 8.0 to 8.2, has little direct effect on the corrosion behavior; however, it can be factor in calcareous scale deposition, which, in turn, influences corrosivity.

As pressure is increased, pH is reduced according to thermodynamic considerations. Thus, at great depths, there is some evidence of less tendency for protective carbonate type scales to form.

VI Corrosion Behavior of Common Metals and Alloys

In the following section, the corrosion behavior of alloys that are commonly used in marine environments is discussed. These include carbon steels, stainless steels, nickel base alloys, titanium alloys, and copper alloys.

A. Carbon and Low Alloy Steels

Plain-carbon steel is the most commonly used metal in marine service. More recently, low-alloy steels of higher strength have been finding increasing application. Steels are selected for marine service because of such factors as availability, cost, ease of fabrication, design experience, and physical and mechanical properties.

The behavior of steel and other ferrous materials tends to show striking differences in the various exposure zones. Thus, one cannot apply the results from an atmospheric exposure to an application involving fully immersed conditions. The sections which follow give selected details of the corrosion of steel in each type of marine environment.

1. Atmosphere

General and Pitting Corrosion

Carbon and low alloy steels undergo general, pitting, and crevice corrosion in marine atmospheres. General corrosion rates, based on weight or average thickness losses, vary widely, between about 1 and 30 mils per year (*6*). Similarly, pitting factors[1] vary widely, between about 3 and 15 (*8*). Both

[1] Ratio of deepest measured penetration divided by the average penetration calculated from weight loss.

general and localized corrosion of carbon and low alloy steels are greatly influenced by a number of parameters, which include geometrical factors, climate (both local and regional), and alloy composition. Discussions of these factors are given below.

Effect of Alloy Composition. It is well established that alloy additions to carbon steel are beneficial in improving corrosion resistance in marine atmospheric exposure. Studies by Larrabee (*9*) and Copson (*7*) indicate that a wide range of compositions give improved resistance to marine atmospheric corrosion. When selected corrosion data from these studies are plotted as a function of the total concentration of elements added, the results suggest that a total alloy content of about 2% provides optimum performance, see Figure 3 (*7*).

The effects of addition of individual alloying elements to carbon steel on corrosion in marine atmospheres also have been studied in some detail. Results of a study by Copson (*7*) on the effects of individual additions of Cu, Mn, Ni, and Cr on atmospheric corrosion at Kure Beach, North Carolina, show that additions of up to about 1% Cu are beneficial for high P (0.1% P) steels containing 0.030–0.06% S, but the greatest benefit occurs with about the first 0.1% Cu addition. On the other hand, Fyfe et al. (*10*) showed that the effect of Cu on corrosion of S free iron was negligible in marine exposure at Pilsey Island, Sussex, England. Thus, Fyfe (*10*) speculated that the beneficial effect of Cu in steels is related to the formation of insoluble copper sulfides. Much earlier, Buck (*11*) had speculated that ferrous sulfides are much more corrosive toward steels than copper sulfides.

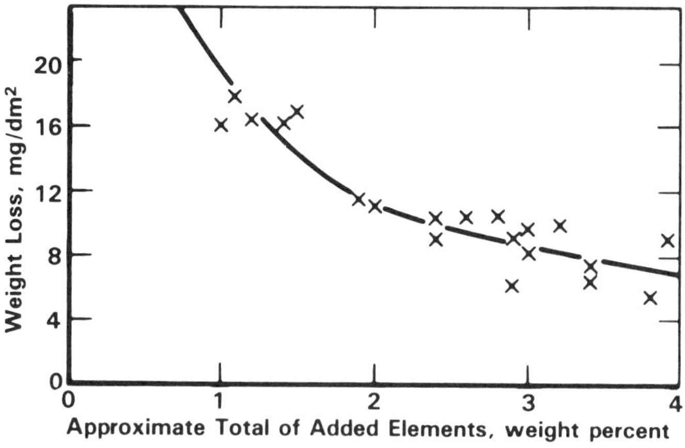

Fig. 3. Corrosion of selected low-alloy steel panels exposed in marine atmosphere for 15.5 years at Kure Beach, NC (800-foot lot) (*7*).

The data by Copson (*7*) also show that Mn, Ni, and Cr additions are beneficial, but larger percentage additions of Ni or Mn are required than Cu or Cr to impart the same degree of resistance. Copson (*12*) noted in an earlier study that Ni was beneficial in reducing pitting. On the other hand, the Cr bearing steels tended to have fewer but deeper pits than the other low alloy steels (*12*).

Phosphorous also is highly beneficial with regard to atmospheric corrosion resistance when added to Cu bearing steels. Larrabee and Coburn (*13*) studied the effect of Cu and P on the corrosion rates of low alloy steels in 15.5 year exposures at Kure Beach, North Carolina. Results of the study showed that the corrosion rates of the high P alloys (approximately 0.06 P) were consistently lower than those of the low P (< 0.01 P) steels for all Cu contents studied; for a 0.5 Cu alloy, corrosion rates were 0.68 mpy for the low P alloy and 0.45 mpy for the high P alloy.

Effect of Surface Condition. Southwell et al. (*8*) studied the effect of surface preparation on the general and pitting corrosion of carbon steel. Specimens having pickled and mill-scaled surfaces were exposed for up to eight years in marine and inland sites in the Panama Canal Zone. Results of this study showed that there was very little difference in the performance of the mill-scaled and pickled surfaces. This behavior is contrasted to that of submerged specimens, where the presence of mill-scale was found to be highly detrimental (see section on immersed exposure).

Geometrical Factors. Larrabee (*9*) studied the effect of sheltering and facing direction (north, south, east, or west) on the corrosion of carbon steel on a lot 800 feet from the ocean at Kure Beach, North Carolina. Specimens were mounted vertically in five horizontal rows on a sheltered test rack such that the upper two rows were completely sheltered from the rain. Results showed that the east exposure was less severe than the west exposure and that the higher, more sheltered specimens generally exhibited the highest weight losses. This behavior was attributed to the deposition of sea salt on the specimens facing the ocean and the washing action of rain in removing the sea salt from the lower specimens. Actual corrosion rates ranged from 3.3 mpy for the sheltered east exposure to 1.3 mpy for the unsheltered west exposure.

Larrabee (*9*) also studied the corrosion of carbon steel specimens exposed at an angle of 30° to the horizontal, one above the other, in an unsheltered south facing rack. Results indicated that attack generally was greater in these exposures than in the sheltered exposures and was highest for the uppermost specimens. Corrosion of the skyward faces was mild and uniform, whereas the downward faces underwent severe attack. Thus, it was concluded that rain washed the deposited salts from the top sides of the specimens and that the amount of salt deposited on the bottom sides of the specimens increased with increasing height above the ground.

Ambler and Bain (*14*) also studied the effect of orientation on the marine atmospheric corrosion of carbon steel. Results of these studies showed that the seaward exposed faces experienced the highest corrosion rates. Analysis of the effect of skyward versus downward facing of the specimens on corrosion was confounded in this study by the fact that the specimens were oriented at a 45° angle to the horizontal and the upper faces also were the seaward faces. Ambler and Bain (*14*) did observe that exposures at 20 feet above the ground were more severe than exposures at 5 feet above the ground, which is qualitatively in agreement with the results of Larrabee (*9*). However, Ambler and Bain (*14*) reported that the exposure height effect on corrosion rates was not observed away from the salt spray.

Ambler and Bain (*14*) further studied the effect of screening specimens from the prevailing wind on the corrosion rate of mild steel and other alloys. Results indicated that specimens which were sheltered from the wind exhibited lower rates of attack than those which were not.

2. Splash and Tide

The corrosion behavior of carbon and low alloy steels in the splash and tidal zone has been investigated in some detail as a result of corrosion problems that have been encountered with pilings in seawater applications. A typical corrosion profile of a steel piling is given in Figure 4 (*15*) where it can be seen that rates of attack in the splash zone (zone 2) and in the continually immersed zone which is adjacent to the tidal zone (zone 4) are much higher than corrosion rates elsewhere (i.e., atmosphere (zone 1), mid-tidal region (zone 3), and deep submerged and mud (zone 5)). Humble (*15*) has presented explanations to account for these wide variations in corrosion rates. He suggests that zone 1 is wetted only infrequently and thus the corrosion rates in zone 1 are similar to those observed for marine atmospheres. In zone 2, the splash zone, materials are continually exposed to, and there is a plentiful supply of, oxygen in the thin aqueous layer on the metal surface, resulting in high corrosion rates. The high corrosion rates in the upper tidal zone (zone 3) also are attributed to this splash zone effect. The tidal zone, zone 3 and the upper portion of zone 4, form a galvanic cell which is driven by differential aeration; the tidal zone, which is fully aerated, forms the cathode in the cell and promotes accelerated attack of the underlying region. It should be pointed out that this galvanic protection of the tidal zone is not available to the splash zone, since the electrical resistance of the moisture layer on the material in the latter zone is high. Finally, rates of attack are low in zone 5 and the lower portion of zone 4 since the seawater is relatively deaerated in

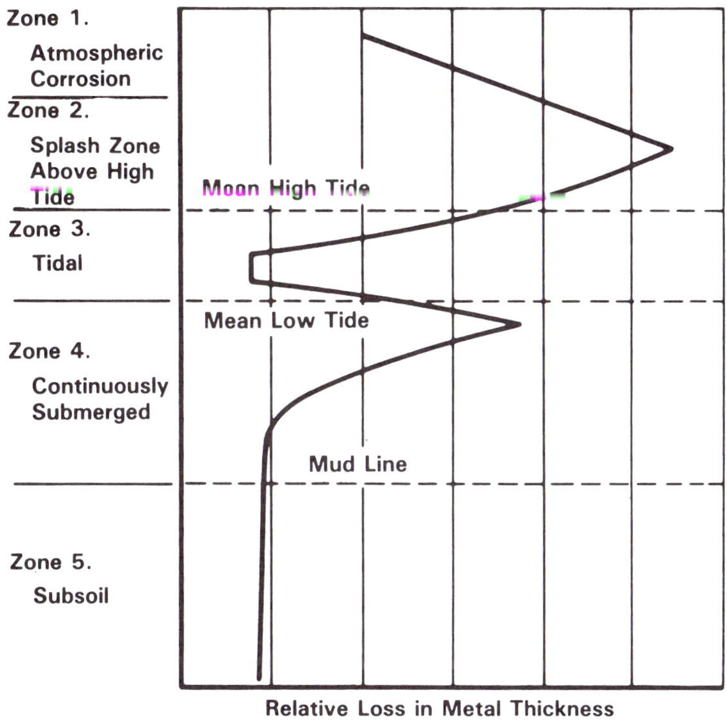

Fig. 4. Corrosion profile of steel piling in 5 year exposures in seawater at Kure Beach (*15*).

this region and the region is sufficiently far away from zone 2 to avoid deleterious galvanic effects. Actual corrosion rate data for steel pilings for these zones are given in Figure 5 (*16*).

Effect of Alloy Composition. Several researchers have investigated the effect of alloy composition on corrosion performance of carbon and low alloy steels in the splash and tide zone. Larrabee (*16*) studied the behavior of six steels having various combinations of Ni, Cu, and P at Wrightsville Beach, North Carolina. Results of the study indicated that the Ni-Cu-P alloyed steels are much more resistant than carbon steel to attack in the splash zone. Results of studies by Nose et al. (*17*) reached similar conclusions.

Schultze and van der Wekken (*18*) statistically analyzed published corrosion data for carbon and low alloy steels under tidal exposure. Results of their study indicated that the excellent performance of the alloy steels described by Larrabee (*16*) and Nose (*17*) were solely the result of the presence of phosphorus. However, the statistical analysis did not address possible synergistic interactions of the alloying elements.

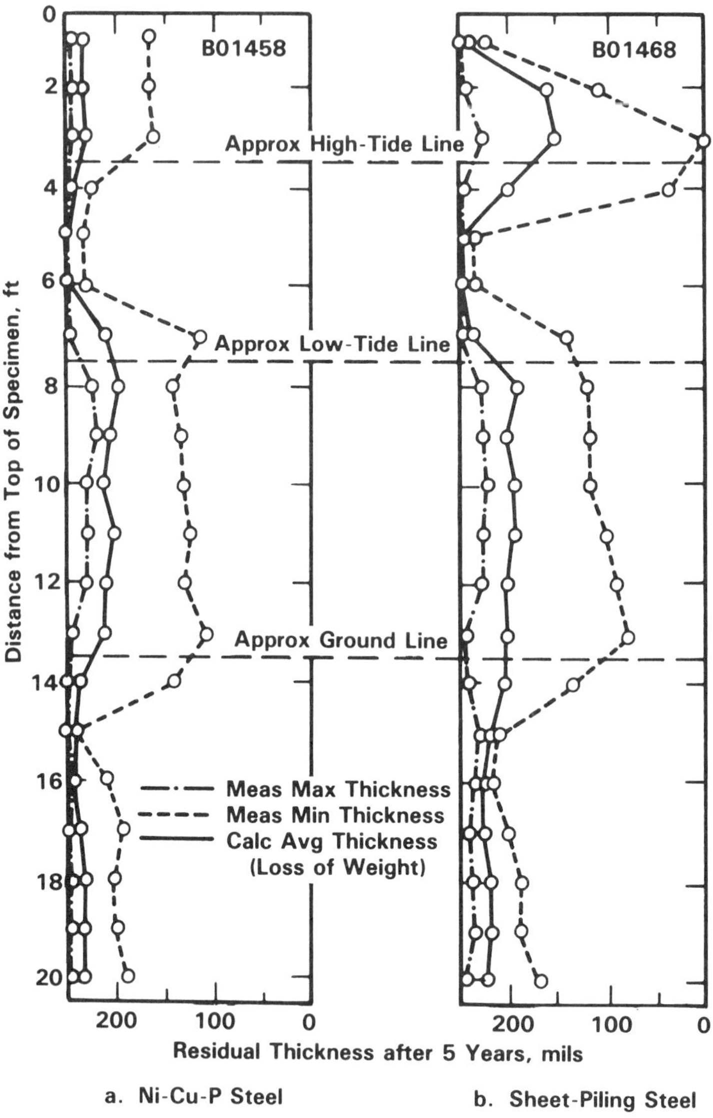

Fig. 5. Comparative corrosion of two carbon steels (Ni-Cu-P and B014G8) in marine environments (*16*).

3. Immersion

General, Pitting, and Crevice Corrosion

Corrosion rates for carbon and low alloy steels in quiescent seawater at ambient temperature, as determined from weight loss, generally are less than about 5 mpy after several years of exposure. These alloys are susceptible to pitting and crevice corrosion with pitting factors as large as about 30. Rates of attack are not particularly sensitive to alloy composition or surface properties, although the presence of mill scale is deleterious.

Effect of Exposure Time. Reinhart (*19*) reported on the results of a number of studies of the effect of exposure time on the corrosion rates of low carbon steels at various geographical locations and depths. Results indicated that rates decreased rapidly with exposure time over the first one to two years of exposure. Results of longer term exposures performed by Larrabee (9) and Southwell (*20*) are qualitatively in agreement with those reported by Reinhart (*19*). Larrabee (*9*) found that the corrosion rates of five low alloy steels decreased with exposure time over 4.5 years exposure at Kure Beach, North Carolina, but the decrease over the longer periods was much more gradual than that which was observed over the first year of exposure, see Table III. (*9*). Similarly, Southwell (*20*) found a logarithmic relationship between the average depth of penetration and exposure time for carbon steel exposed in seawater in the Panama Canal Zone. The one year average rate of attack was 5.8 mpy, whereas the rate calculated for the 16-year exposure was less than 3 mpy.

TABLE III

Corrosion Rates as a Function of Exposure Time for Steels Submersed in Seawater at Kure Beach, NC (*9*)

	Average Penetration, mpy		
Steel	1.5 Yr	2.5 Yr	4.5 Yr
Cor-Ten brand	4.2	4.3	3.8
Tri-Ten brand	4.4	3.8	3.0
Ni–Cu	5.3	4.5	3.5
Cr–Mo	1.4	1.6	1.6
Structural carbon	4.8	4.1	3.3

Rates of pitting of carbon and low alloy steels also decrease with exposure time, although pitting attack of these steels is severe even for short exposures. Data by Southwell (*20*) and Bultman et al. (*21*) show that depths of attack of as much as 100 mils were observed for carbon and low alloy steels within one year's exposure. The data also show that rates of pitting actually decreased more rapidly than rates of general attack; the values for the pitting factor generally decreased with exposure time.

Effect of Alloy Composition. General corrosion rates of low alloy steels fall between about 2.5 and 5 mpy for fully submersed conditions in seawater and are not greatly affected by alloying. Thus, the use of most alloy steels offers little corrosion advantage over carbon steel in submersed applications.

Moreover, results of long-term exposures of carbon and low alloy steels performed by Southwell and Alexander (*20*) do not indicate any beneficial effect of alloy addition in improving pitting and crevice corrosion resistance. In fact, their data indicate that pitting factors for some alloy steels, such as Ni and Cr steels, frequently are higher than those for carbon steel. Thus, some alloying additions appear to be detrimental to the pitting performance.

Effect of Depth and Oxygen. Reinhart and Jenkins (*19*, *22*, *23*) have performed extensive studies of the corrosion of structural materials in deep ocean environments. Included in the studies were the carbon and low alloy steels and cast iron. These studies were performed at sites in the Pacific Ocean near Port Hueneme, California, at depths of 1830 meters, 760 meters, and near the surface. The environments at these sites were characterized, and the O_2, temperature, salinity, and pH were found varying as described in Figure 2 (*4*).

It was found that the various steels exhibited similar corrosion rates at a given site over one year's exposure, and thus the values were averaged for subsequent analysis. Data showing the relationship between corrosion rate and depth are given in Figure 6 (*23*); superimposed upon the graph is the O_2 concentration as a function of depth. These data show that the corrosion rate appears to be a strong function of the O_2 concentration. A relationship between oxygen concentration, temperature, and corrosion rate was obtained using a linear regression analysis of the data. The derived formula:

$$\text{Corrosion Rate}\left(\frac{\mu\text{m}}{\text{y}}\right) = 21.3 + 25.4\left[\text{O}_2\left(\frac{\text{ml}}{\text{l}}\right)\right] + 0.356[\text{T}(^\circ\text{C})]$$

shows that the corrosion rate was much more sensitive to O_2 concentration of the seawater than to temperature and varied linearly with respect to either parameter.

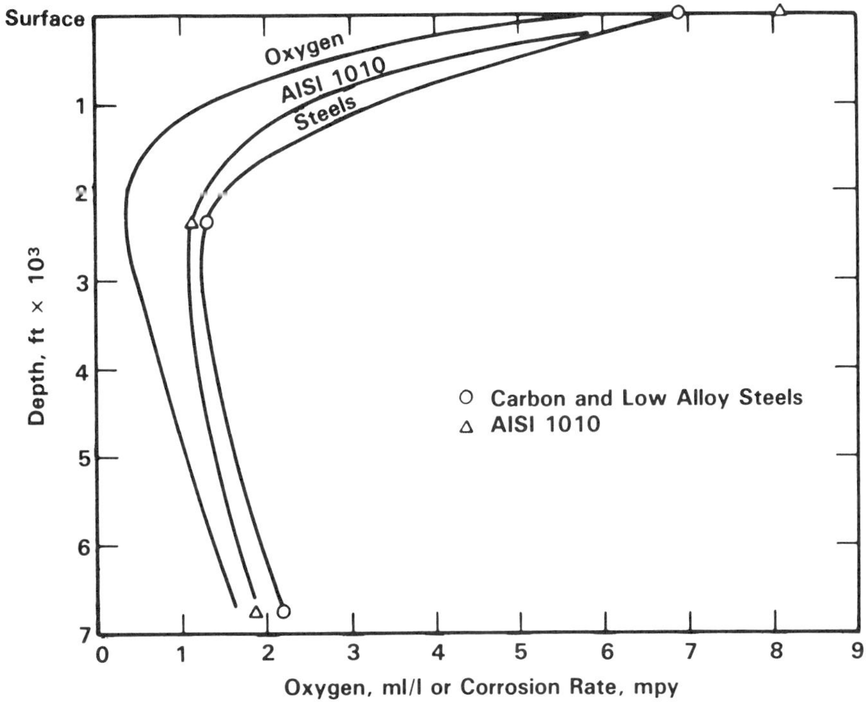

Fig. 6. Corrosion of steels versus depth in seawater after 1 year of exposure (*23*).

The specimens in these tests also were examined for evidence of pitting and crevice corrosion, and it was found that these forms of attack only occurred under near-surface exposure where oxygen concentrations were highest.

Effect of Temperature. General corrosion rates of carbon steel have been shown to increase with increasing temperature in laboratory studies in which deaerated aqueous chloride solutions have been used. (*24*, *25*). Typical data for concentrated synthetic or natural seawater indicate that weight loss followed Arrhenius behavior. In natural environments, the effect of temperature on general corrosion rates of carbon steel is complicated by the changes in oxygen solubility and biological activities that accompany changes in temperature. Data showing the seasonal variation in these parameters and in the corrosion rate for carbon steel in polluted ad unpolluted seawater are given in Figure 7 (*25*). These data show that in unpolluted seawater the corrosion rate was proportional to the oxygen concentration of the water, which was highest during the winter when temperature and biological activity were a minimum. On the other hand, fluctuation in corrosion rates

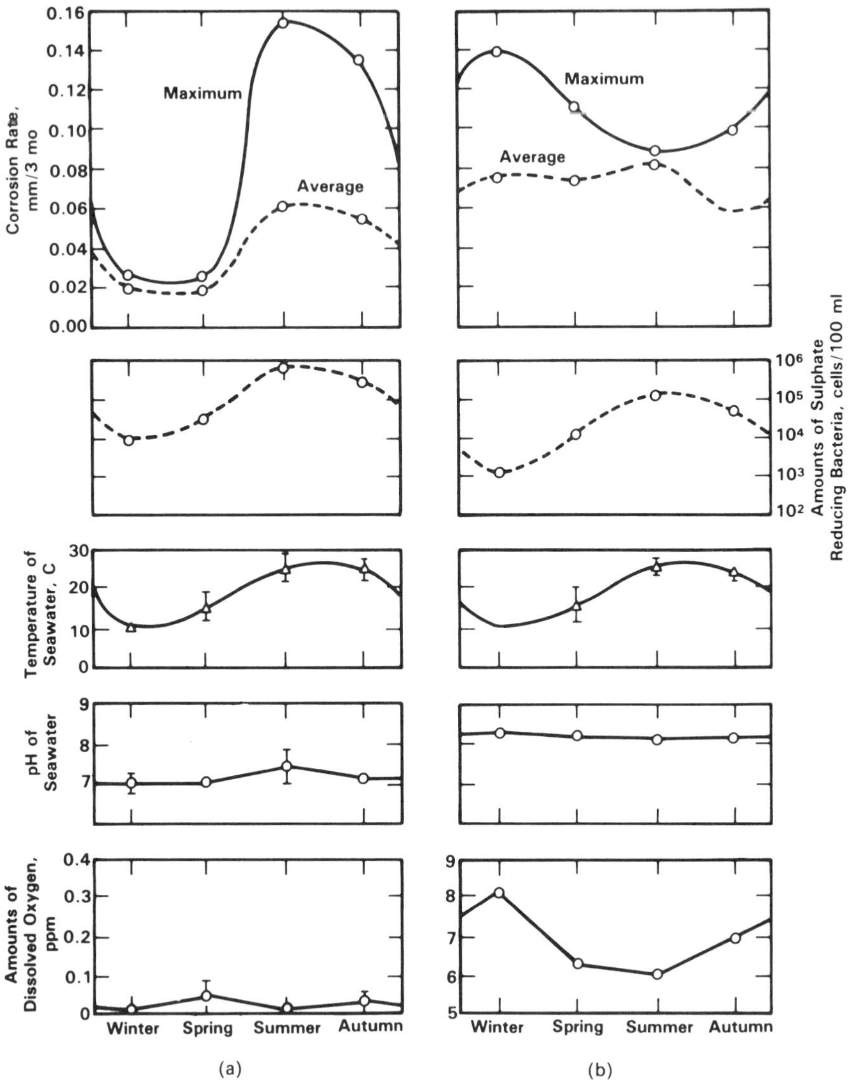

Fig. 7. Seasonal factors influencing the corrosion behavior of normal carbon steel in two sea areas (*26*): (a) the polluted sea area "Okudokai Harbor," (b) the pollution free sea area "Kammon Strait."

were proportional to temperature fluctuations in polluted seawater, where the oxygen concentration was low.

As discussed above, Reinhart and Jenkins (*22*) examined the relationship between the corrosion rate of steels in seawater and the temperature and oxygen concentration of seawater in their studies of the effect of depth on materials performances and found that O_2 concentration had much larger effect on corrosion of steels in unpolluted seawater than did temperature.

Effect of Biofouling. Efird (*27*) and Chandler (*28*) described the fouling characteristics of a variety of materials in ambient temperature seawater. For carbon and low alloy steels, classified as highly corrodible alloys, the fouling was characterized by the periodic growth and sloughing of marine growth.

Boyd and Fink (*2*) speculated that biofouling layers which develop on steels in ambient temperature seawater play a role in the decreasing corrosion rate with time observed for carbon and low alloy steels. However, the limited experimental data available do not support this contention. Traverso *et al.* (*29*) performed immersion tests of three years duration on carbon steel in filtered natural seawater from Genoa Harbor, Italy. Although macrofouling did not occur on the specimens, the typical decrease in corrosion rate with increasing time was observed, and rates of attack after three years were comparable with those observed in unfiltered seawater, 2.5 to 3 mpy. Traverso et al. (*29*) also reported pit depths in carbon steel for filtered seawater (–50 mils in three years) that were similar to those reported by Southwell (*20*) for unfiltered seawater. Microfouling may have played a role in the decreasing corrosion with time rate observed by Traverso (*29*).

Effect of Pollution. The primary effect of pollution on the seawater chemistry is to decrease the oxygen content and increase the sulfide content. pH values also are lower in polluted seawater. It is well known that the presence of sulfide is detrimental to the corrosion behavior of steels and its effect has been studied in some detail (*30, 31*). On the other hand, oxygen also is detrimental to the corrosion performance of carbon and low alloy steels in seawater, and thus one might speculate that O_2 and $S^=$ would have counterbalancing effects in polluted seawater. Results of long term exposures seem to support this contention. Results from Japanese (*26*) (Figure 7) and American studies (*32, 33*) show that general corrosion rates for polluted and unpolluted seawater are comparable, although rates of attack quoted for the Japanese study were somewhat higher than those in the American study.

Although average general corrosion rates for carbon steel in polluted and unpolluted seawater appear to be comparable for long exposures, the seasonal variation in the corrosion rate for polluted and unpolluted seawater is quite different. Shimada et al. (*26*) have found that the corrosion rate of carbon steel in polluted seawater is proportional to the temperature which, in turn, is proportional to the concentration of sulfate-reducing bacteria in the seawater. On the other hand, the general corrosion rate of carbon steel in unpolluted seawater is proportional to the oxygen concentration of the seawater and inversely proportional to the temperature; corrosion rates were highest in the winter months when the temperature was the lowest and the oxygen concentration was the highest, see Figure 7 (*26*). Of course, these seasonal effects on corrosion of carbon steel are probably location dependent and may be significantly different at other latitudes.

The relationship between pollution and localized corrosion of carbon steel appears to be qualitatively similar to that between pollution and general corrosion. Thus, Shimada et al. (*26*) observed similar maximum rates of localized attack in polluted and unpolluted seawater. It can be speculated, however, that polluted seawater is potentially very aggressive where oxygen is introduced periodically.

Stress Corrosion Cracking and Hydrogen Embrittlement

Sandoz (*34, 35*) compiled stress corrosion cracking (SCC) data for some precracked high strength steels in actual marine environments and found a trend of increasing susceptibility to SCC with increase in alloy yield strength. Thus, steels having yield strengths less than about 100 ksi are not normally considered to be susceptible to SCC in marine environments in the absence of sulfides.

Effect of Alloy Composition. The effect of alloying elements on SCC of low alloy steels are confounded, in most instances, by the influence these elements have on the yield strength of the materials. Nevertheless, a few experimental studies have been performed in which the SCC susceptibility of alloys of similar yield strengths, but varying compositions, were assessed. Sandoz (*34, 35*) studied the effects of Mn and C in a base 4340 steel alloy on K_{ISCC} in seawater and in 3.5% NaCl at two yield strength levels. Results showed that Mn is detrimental, up to a 3.0% level, as is C up to approximately 0.40% C; a slight improvement in K_{ISCC} values was observed at 0.5% C.

Results of studies of the effect of other alloying elements on SCC of low alloy steels showed that, with the exception of C and Mn, the alloy additions did not significantly alter K_{ISCC} values (*36*). There are a number of caveats however; first of all, synergistic interactions of the alloying elements were not considered in the studies reported above. Moreover, some of the alloying elements which had a negligible effect on K_{ISCC}, such as Si, altered cracking velocities above K_{ISCC} (*37*). Finally, the alloying elements may affect cracking where local (crevice or pit) chemistry conditions control susceptibility. Such conditions are common in the stainless steels (*37*, *38*).

Melting practice has been shown to influence SCC susceptibility of steels having yield strengths in the ranges of 180 to 200 ksi (*37*, *39*, *40*). Vacuum melted steels in this strength range apparently have higher resistance to cracking (higher K_{ISCC} values) in aqueous environments than similar air melted steels. Carter (*37*) attributed the improved performance of vacuum melted 4340 to lower impurity levels of sulfur and phosphorus, a conclusion which appears to contradict the previously reported studies of the effect of alloying additions on cracking susceptibility. The beneficial effect of melting practice on K_{ISCC} diminishes with increasing yield strength; steels above the 200 ksi yield strength range all tend to show low values of K_{ISCC} (*34*).

Effect of Heat Treatment. The overriding effect of heat treatment on SCC susceptibility of low alloy steels is through its influence on yield strength (*34*). Thus, the general trend is for SCC resistance to increase with increasing tempering or aging temperature (*41–44*). Some intermediate temperatures do produce maximum susceptibility. For example, Davis et al. (*45*) found that the maximum SCC of H11 and 4340 steels in 3.5% NaCl occurred at the "500 F embrittlement condition" where the impact energy was a minimum.

Effect of Environmental Variables. With the exception of a few deleterious species such as H_2S, the SCC of high strength low alloy steels in neutral aqueous environments is not particularly sensitive to the composition of that environment. Thus, K_{ISCC} values of 4340 steels in distilled water, 2% to 4% NaCl or seawater are similar at a given strength level (*34*, *46–48*).

Cracking susceptibility of these steels is influenced by solution pH, but the effect is not pronounced. Highly acidic conditions promote cracking, whereas highly basic conditions retard cracking, but there is little change in SCC susceptibility over the pH range of 3 to 10 (*34*, *49–53*). Sandoz (*34*) speculated that the absence of a large bulk solution pH effect is the result of the fact that the pH within a growing crack is invariably about 3.7 regardless of bulk solution pH.

Effect of Sulfur Species. The SCC susceptibility of high strength steels in aqueous environments is greatly increased by the presence of sulfides or other reduced sulfur species (*49*).

Thus, Uhlig (*54*) studied the effect of H_2S and $Na_2S_2O_3$ concentration on time to failure of a high strength (210 ksi tensile strength) carbon steel in acidified seawater and 3% NaCl and found that failure times dropped dramatically with the addition of 2 ppm S added as $S^=$ or $S_2O_3^=$. Cracking in the presence of H_2S is generally considered to occur by a hydrogen embrittlement mechanism (*49*).

The presence of H_2S also promotes hydrogen related failures of carbon and low alloy steels with low yield strengths (*55*, *56*) where both internal cracking and blistering are reported. The internal cracks may be either parallel to the rolled surface of the material or pependicular to the applied stress (*55*).

Galvanic Corrosion

Carbon and low alloy steels are quite active in seawater and thus are susceptible to galvanic attack when coupled with more noble materials. A galvanic series for metals in flowing seawater is given in Figure 1 (*3*). These data show that, of the common metals, only Al, Zn, Cd, and Mg are more active in seawater than carbon steel. In a comprehensive study of the galvanic corrosion of metals in tropical seawater, Southwell et al. (*57*) demonstrated the detrimental effect on carbon steel of coupling to most other common metals. In fact, Southwell et al. (*57*) stated that carbon steel was a highly effective sacrificial anode for protecting the more noble materials such as the stainless steels and copper-base alloys.

Most of the couples in the work of Southwell et al. (*57*) were made of metals with large potential differences, but even where the free-corrosion potentials were similar, large galvanic effects were observed in some instances. For example, the 2% Ni steel increased the corrosion rate of carbon steel by about 25% when coupled to carbon steel at a surface area ratio of about 4:1. Similarly, other researchers have shown that the carbon steel-titanium couple is not as severe as the carbon steel-copper couple, even though steel and titanium are much further apart on the galvanic series. Thus, the relative position of the members of the couple in the galvanic series is not always a good indication of the potential for galvanic attack, although metals which are close together in the series are more likely to be compatible. This is because the corrosion rates of the members of the couple are a function of their polarizability as well as the values of their free-corrosion potentials.

Geometrical Effects. The relative surface areas of the two coupled materials can have a significant influence on the corrosion rate of the more active member of the couple. In an early study, Wesley (*58*) examined the influence of surface area ratio between copper and iron on the corrosion rate of iron in an aerated 3% NaCl solution. Results showed that the corrosion rate increased in a linear fashion with increase in surface area of the copper. This effect is sometimes referred to as the catchment area principle, after Evans

(*59*), and appears to be valid where oxygen reduction at the cathode (noble metal) limits the rate of attack of the anode (active metal).

These area ratio effects were demonstrated recently in studies of the corrosion of fasteners of various structural materials in seawater. Hack (*60*) exposed $\frac{1}{4}'' \times 4'' \times 12''$ panels for six months in quiescent seawater at Wrightsville Beach, North Carolina. Each panel contained a single $\frac{1}{4}''$ bolt. Results showed that intense galvanic corrosion of steel bolts occurred where attached to more noble metal plates. On the other hand, the low alloy steel plate underwent only mild attack when coupled to a more noble alloy bolt.

The electrical resistance of the galvanic circuit can have a significant influence on the corrosion rate of the anode. The resistance may be either the solution resistance or the electrical resistance in the metallic portion of the circuit. Thus, Wesley (*58*) studied the effect of the resistance between Cu and Fe on galvanic corrosion of Fe in aqueous 3% NaCl by measuring the current flow between metal electrodes as a function of the value of an external resistor. It was found that the galvanic current drops rapidly with increasing resistance. Thus, even relatively poor insulators installed between carbon steel and noble metals can reduce galvanic effects.

Environmental Effects. The severity of galvanic attack of carbon steel generally increases with increasing solution velocity in seawater. Data by Perkins et al. (*61*) for carbon steel-copper 30 nickel (1 : 1 area ratios) coupled in synthetic seawater indicate that the corrosion rate of carbon steel increased rapidly with increasing velocity (from 10 mpy at 0 fps to 40 mpy at 5 fps) but increased only gradually with further increase in velocity up to 10 ft/s.

Electrolyte composition also influences galvanic attack of carbon steel. Results of the study by Southwell et al. (*57*) showed that galvanic attack of carbon steel was lower in fresh water than in seawater over 16 years of exposure. Nevertheless, corrosion of the carbon steel was significantly accelerated by galvanic coupling in fresh water.

4. MUD

Several researchers (*16*, *62*, *63*) have studied the corrosion behavior of carbon and low alloy steel pipes below the mud line. Typical results by Larrabee (*16*) for exposures at Wrightsville Beach, North Carolina are given in Figure 5 (*16*) and show that rates of general and pitting attack in the mud are lower than those observed in the submerged zone and are comparable for the carbon steel and the corrosion resistant low alloy steel B01458 (0.54 Ni, 0.52 Cu, 0.12 P). Similar results were reported by Escalante (*63*).

TABLE IV

CORROSION RATES OF CARBON STEEL PILING IN SEAWATER (*62*)

Location	Exposure Period, years	Maximum Corrosion Rates per Exposed Face, μm/yr			
		Buried	Submerged	Intertidal	Splash
Port Adelaide, S. Australia	52	30	50	40	70
Queenscliff, Vic., Australia	17	—	80	109	132
Wrightsville Beach, NC	8	—	110	20	300
Mayport, FL	5.5	—	160	30	580
Mobile Bay, AL	7	—	200	160	330
Dam Neck, VA (bare steel)	6	100	440	240	440
Dam Neck, VA (coated)	6	76	6	44	110
La Costa Island, Caribbean	5	46	420	320	425
Boston Harbor, MA	10	—	320	—	40
Lowestoft, U.K. (coated waterside)	20	15	40	13	15

Kinson et al. (*62*) reported corrosion data for carbon steel pilings at various geographical locations throughout the world. These data, given in Table IV (*62*) shows that corrosion rates for mud exposure varied considerably from one location to the next, as did corrosion rates under submerged, intertidal, and splash exposure, but were consistently lower than rates of attack in the other regions of the piling.

Reinhart (*19*) studied the effect of exposure time and alloy composition on the corrosion rate of carbon and low alloy steels in bottom sediment under deep exposure (2350 and 5500 feet). Results of this study showed that rates of attack decreased with increasing exposure time and were comparable for all of the steels examined.

Although the above results indicate that mud exposure generally is less severe than submerged, tidal, or splash exposure, it has been suggested (*2*) that the presence of sulfate-reducing bacteria may exacerbate attack in the mud zone. Results of limited studies by Mor et al. (*64*) do suggest that the presence of anaerobic bacterial activity can accelerate corrosion of carbon steels. Specimens of carbon steel were exposed under natural anaerobic conditions in the sediment of Genoa Harbor, Italy, where significant biological activity was present. Results of the study showed that rates of pitting and general corrosion accelerated after about 90 weeks of exposure and were somewhat higher than most of those data reported above. Mor et al. (*64*) attributed the increase in rates of attack with exposure time to breakdown of protective films on the metal surfaces as a result of biological activity.

B. Stainless Steels

Stainless steels (Fe base alloys with $\geq$ 12 w/o Cr) have excellent resistance to general corrosion in all marine exposure conditions. The major corrosion problems are pitting and crevice corrosion and, to a lesser extent, stress corrosion cracking (SCC) and intergranular attack (IGA). Pitting occurs in the ferritic alloys ($\geq$ 12 w/o Cr, e.g., 410, 431, 446), the austenitic alloys ($\geq$ 18 w/o Cr plus $\geq$ 8 w/o Ni, e.g., 302, 304, 310, 316), and in precipitation hardened alloys (e.g., 17-4 PH, 17-7 PH, PH 13-8 Mo, etc.). SCC and IGA occur primarily in the sensitized austenitic alloys. The high molybdenum ferritics, duplex, and austenitic alloys (e.g., 29 Cr-4 Mo, 25 Cr-5 Ni-5 Mo, and 20 Cr-25 Ni-6 Mo) are generally more resistant to pitting, crevice corrosion, and SCC.

1. Atmosphere

General Corrosion, Pitting, and Crevice Corrosion

As pointed out in the introduction, corrosion in marine atmospheres is not so severe as in immersed or splash zone areas. Since resistance to pitting is not a major requirement in most atmospheric exposure applications, the high molybdenum ferritic and duplex stainless steels are not required, and the conventional ferritic and austenitic stainless steels are used extensively in marine atmospheres. Long-term atmospheric corrosion data for stainless steels at Kure Beach, North Carolina (15 years) are summarized in Table V

TABLE V

Atmospheric Corrosion of Some Stainless Steels Exposed 15 Years to a Marine Environment (800-foot Lot at Kure Beach, NC) *(65)*

Type	Average corrosion rate, μm/yr	Average depth of pits, cm
301	< 0.03	0.004
302	< 0.03	0.003
304	< 0.03	0.003
321	< 0.03	0.007
347	0.03	0.009
316	< 0.03	0.003
317	< 0.03	0.003
308	< 0.03	0.004
309	< 0.03	0.003
310	< 0.03	0.001

(*65*). Note the very low corrosion and pitting rates of $\leq 0.3\ \mu m/yr$ and $\leq 6 \mu m/yr$, respectively, after 15 years of exposure. Similar results were obtained at Cristobal Panama where, after 16 years of exposure, Types 410 (12 Cr) and 430 (16 Cr) exhibited general corrosion rates of 0.3 and $\leq$ 0.3 $\mu m/yr$ and average pitting rates of $< 8\ \mu m/yr$ (*66*). The high molybdenum 29-4 and AL-6X alloys have been exposed two and five years, respectively, at 800 feet, but at Kure Beach there was no evidence of general attack or pitting (*67*).

Galvanic Corrosion

Galvanic corrosion is not as prevalent in the marine atmosphere as it is in seawater. Nevertheless, galvanic corrosion does occur, particularly over short distances where moist salt deposits serve as the conducting electrolyte.

Because stainless steels are noble to most alloys, they often promote attack of dissimilar metals such as steel, aluminum, zinc, or magnesium in the galvanic couple. However, those materials that are more noble than stainless steel can promote corrosion of stainless steel in galvanic couples, such as the corrosion of metal fasteners inserted in graphite-epoxy composites (*68*, *69*).

Stress Corrosion Cracking

Stress corrosion cracking (or hydrogen stress cracking) has often been a failure mode of stainless steels in marine atmospheres, particularly with those alloys that are heat treated to high strength levels. Stanley (*70*) has summarized the performance of stainless steels in marine atmospheres according to whether they have high resistance, intermediate resistance, or low resistance to SCC (see Tables VI, VII, and VIII) (*70*).

Austenitic Stainless Steels. In the mill annealed (or solution treated conditions), the austenitic stainless steels are very resistant to stress corrosion cracking in marine atmospheres. Money and Kirk (*71*) conducted a five-year exposure at the Kure Beach 80 foot lot with U-bends of a number of austenitic alloys and found that none of the annealed, cold worked, or welded specimens failed by SCC. However, a high proportion of the sensitized specimens failed by SCC, some in as short a time as 35 days.

Several of the austenitic stainless steels can be cold worked to high strength levels. Phelps (*72*) exposed stressed specimens for more than six years at the 80 foot lot at Kure Beach and found that some cold worked alloys with strength levels in excess of 225 ksi are resistant to SCC in marine atmospheres. In these same tests, several of the alloys that were heat treated at 1100°F failed, possibly because they were sensitized by the heat treatment.

TABLE VI

MATERIALS WITH A HIGH RESISTANCE TO STRESS CORROSION CRACKING (*70*)

Materials	Type	Heat treatment*
300 series stainless types 303, 304, 316, 321, 347	Austenitic	Annealing
17-4PH	Martensitic	H1000 and above
17-7PH	Semi-austenitic	CH900
PH13-8Mo	Martensitic	H1000 and above
15-5PH	Martensitic	H1000 and above
PH15-7Mo	Semi-austenitic	CH900
PH14-8Mo	Semi-austenitic	CH900
AM-350	Semi-austenitic	SCT1000 and above
AM-355	Semi-austenitic	SCT1000 and above
Custom-455	Semi-austenitic	H1000 and above
A-286	Austenitic	Solution treated and aged
A-286 (CW and aged)	Austenitic	
Inconel 718	Face centred cubic	Solution treated and aged
Inconel X-750	Face centred cubic	Solution treated and aged
Rene-41	Face centred cubic	Solution treated and aged
MP 35N	Face centred cubic	Solution treated and aged
Waspaloy	Face centered	Solution treated and aged
Low-alloy steels 4130, 4140, 4340, 8740	Martensitic	Quenched and tempered $\leq$ 160 ksi
Maraging Steel	Martensitic	Solution treated and aged to $\leq$ 200 ksi

* For heat treatments refer to *AerospaceStructural Metals Handbook*, *ASM Metals Handbook*, or steel producer's literature.

Martensitic and Precipitation Hardened Stainless Steels. As pointed out earlier, those stainless steels that can be heat treated to high strength levels often are susceptible to SCC (or hydrogen stress cracking) (See Tables VI, VII, and VIII). Long-term exposure results for stressed specimens of these types of alloys may be found in references *73*, *74*, and *75*. These studies found that susceptibility to SCC was highest for the heat treatments that produced maximum strength levels in the alloys. In general, susceptibility to SCC decreased with higher aging temperatures (lower strengths), but some failures have been observed in overaged specimens.

TABLE VII

MATERIALS WITH A HIGH RESISTANCE TO STRESS CORROSION CRACKING IF USED WITH CAUTION (*70*)

Materials	Type	Heat treatment
Low-alloy steels 4130, 4140, 4340, 8740, D6AC, HY-TUF	Martensitic	Quenched and tempered 160 to 180 ksi
Maraging steel	Martensitic	Solution treated and aged
400 series stainless 410, 416, 422, 431	Martensitic	Quenched and tempered > 1100°F
15-5PH	Martensitic	H950 to H1000
PH13-8Mo	Martensitic	H950 to H1000
17-4PH	Martensitic	H950 to H1000
AM-355	Semi-austenitic	SCT950 to H1000

2. SPLASH AND TIDE ZONES

The stainless steels, as a group, usually do well in the splash zone. The availability of well-aerated seawater tends to maintain the passivity of stainless steels without difficulty, provided crevices and shielded areas are avoided in the design. Since the splash zone is above the point at which marine life develops on the metal surface, biofouling is not a factor. As with

TABLE VIII

MATERIALS WITH A LOW RESISTANCE TO STRESS CORROSION CRACKING (*70*)

Materials	Type	Heat treatment
Low-alloy steels 4130, 4140, 4340, 8740, D6AC,HY-TUF	Martensitic	Quenched and tempered > 180 ksi
H-11	Martensitic	Quenched and tempered
17-7PH	Semi-austenitic	All heat treatments except CH900
PH15-7Mo	Semi-austenitic	All heat treatments except CH900
AM-355	Semi-austenitic	Heat treatments below SCT900
400 series stainless 410, 416, 422, 431	Martensitic	Quenched and tempered 500° to 1100°F

atmospheric exposure, the austenitic grades are superior to the martensitic and ferritic grades.

The 300 series, particularly Types 304 and 316, have been used with success for small deck-mounted fittings for seagoing pleasure boats and other craft. Removal of sea-salt deposits by hosing with fresh water plus occasional polishing will keep such fittings in good conditions. When salt deposits are allowed to accumulate, particularly at crevices, there is a tendency for local attack. This has been observed between the strands of stainless steel rope, in the threads of tensioning screws, and at the point where wire ropes enter swaged ferrules.

Even though the tide zone is well aerated a large part of the time, there are factors present in this environmental zone that interfere with the maintenance of passivity. Marine animals, such as barnacles and mollusks, as well as other types of fouling, settle in the tide zone (and below) and shield the surface from the oxygen needed for passivity. Crevice attack at areas of restricted access, such as structural joints, are more apt to take place in the tide zone, where, during the immersed portion of the day, relatively large areas outside the crevice are available to serve as cathodes.

Thus, corrosion performance in the tide zone approaches that of full immersion. The conventional stainless steels exhibit pitting in the tide zone, and only the highly alloyed Cr-Mo and Cr-Ni-Mo alloys that are resistant to pitting under submerged conditions, are resistant to pitting in the tide zone.

Southwell et al. (*76*) have found that, after 16-year exposures in the tide zone in the tropics, general corrosion rates of austenitic stainless steels were low at < 1.0 mpy, and the maximum depth pitting rates were moderately low at 2 to 7 mpy.

3. Immersion

General Corrosion

The general corrosion rates of stainless steels in seawater are so low that metal loss rates are of no consequence to structural integrity. As noted earlier, pitting and crevice corrosion are the most usual forms of attack. These failure modes will be described in the next section.

Pitting and Crevice Corrosion

The principal failure modes of most stainless steels immersed in seawater are pitting or crevice corrosion. As will be discussed later, certain alloys that contain molybdenum are resistant to these forms of failure.

Effect of Alloy Composition. The pitting susceptibility of the common stainless steels was demonstrated as early as 1945 in long-term tests at Kure Beach, North Carolina reported by Larrabee (*77*). Other long-term results have been reported by Alexander et al. (*78*) for exposure in the Panama Canal Zone; Reinhart for deep ocean exposure west of Port Hueneme, California (*79*); Lawson et al. (*80*) in the 110°F to 200°F heat recovery portion at a desalination plant in Freeport, Texas; and Kowaka et al. (*81*) for crevice tests conducted in Japan. It should be noted that even Type 316 stainless steel (2% to 3% Mo) was susceptible to pitting and crevice corrosion in these studies.

Pitting failures have been observed in condenser tubes exposed under service conditions. Baker et al. (*82*) report that tubes at one particular site in England started to leak at 8 months, and all tubes had leaked by 11 months, with an overall ranking of pitting resistance of 316L > 316 > 304L > 304. At another site, these same authors reported perforation of 304 and 304L tubes, but no perforations of 316 and 316L tubes after two years of condenser operations (25.4 mm OD tubing by 1.2 mm wall thickness). Hoskinson and Kuester (*83*) also report that the service performance of Type 316 condenser tubes has been variable, with eight reported failures within two years at one power plant and only slight pitting with no service failures at another.

The critical alloying elements that prevent pitting in the stainless steels are chromium and molybdenum. Hack (*84*) in 30-day tests with some 45 alloys at Wrightsville Beach, North Carolina, concluded that at least 8% Mo is needed in austenitic steels and approximately 25% Cr and 3.5% Mo are needed in ferritic steels to prevent pitting and crevice corrosion.

Effect of Temperature. In general, an increase in temperature in the range of 10° to 60°C would be expected to increase the severity of pitting and crevice corrosion in seawater; but this has not been true in most exposure tests. Data obtained by Lee (*85*) indicate that a maximum in pit depth is observed for Types 304 and 316 at 25°C as compared to 10° and 50°C. Similar, but not so striking, trends also have been observed by Asphahani et al. (*86*) in almost identical tests. The effect of temperature also is inferred from the depth of exposure in tests by Ulanovskii et al. (*87*), where they found an increased rate of pitting attack and general corrosion as the temperature increased from 9° to a maximum of 25°C. Long-term data obtained by Garner (*88*) indicate little effect of temperature on crevice corrosion at four exposure sites ranging in temperature from 0° to a maximum of 32.5°C.

Service experience with the high molybdenum alloys reveal that they have not pitted at temperatures to 40°C (Kovach et al. (*89*) with a 25 Cr-2 Ni-3 Mo alloy) and to 65°C (Eden (*90*) with 20 Cr-25 Ni-4.5 Mo).

It should be noted that the effects of temperature are complex. An increase in temperature would be expected to increase the rate of corrosion and pitting. However, this may be counterbalanced at the higher temperatures by a decrease in oxygen solubility and the killing of microorganisms, both of which would be expected to reduce the rate of corrosion attack.

Biofouling. Biofouling produces slimes and deposits that can lead to pitting of the conventional stainless steels. This can occur by oxygen depletion under the deposit as in conventional crevice attack, or, as pointed out by Liebert et al., (*91*) filamentous growths can trap water and result in high surface temperatures, which would further aggravate corrosion in heat transfer systems.

The buildup of a deposit appears to be critical. For example, Willingham (*92*) found no pitting of fully exposed Type 316 when exposed to 90 days at 20°C in a seawater medium that contained three types of sulfate-reducing bacteria. Lendvai-Lintnor et al. (*93*) state that in the absence of antifoulants, corrosion inhibitors are not effective in preventing pitting in high chloride recirculating cooling water systems. Walston (*94*) observed that Type 316 performed well in cooling tower systems located worldwide if the waters were aerated, but experienced localized corrosion in crevices with poor microbiological control. LaQue (*95*) attributed differences in the corrosion behavior of Type 304 exposed to synthetic seawater and natural seawater to microorganisms in the latter.

The high Mo stainless steels appear to be resistant to pitting by biofouling presumably because they are resistant to crevice corrosion in seawater, as has been discussed in preceding sections. The results of Brennert and Lindh's (*96*) questionnaire revealed that at least 3.5% Mo was required in stainless steel to ensure resistance to corrosion by microorganisms. In biofouling tests of OTEC heat exchanger alloys in the Gulf of Mexico, Little et al. (*97*) found that AL-6X exhibited patchy colonization by microorganisms, but minimal corrosion after 90 days of exposure at a velocity of 1.8 m/sec.

Geometric Effects. The geometry of crevices, exposure conditions, microstructure, and surface condition affect pitting and crevice corrosion. Mankowski et al. (*98*) found that pits in 18 Cr-12 Ni-2 Mo-Ti stainless steel grew more slowly if the specimen was facing up than if it was facing down. They attributed this effect to a higher chloride ion concentration in the upward facing pits which slowed diffusion processes.

Scotto et al. (*99*) have investigated the effects of fabrication orientation on the pitting of 18 Cr-9 Ni and 17 Cr-11 Ni-2 Mo in 3% NaCl. They found that pits occurred more readily on transverse than on longitudinal sections because of the more unfavorable sulfide inclusion shapes in that direction. Eklund (*100*) and Brennert and Eklund (*101*) believe that the prime cause of crevice corrosion in stainless steel is the electrochemical dissolution of sulfide inclusions, the formation of acid conditions, and the resulting exposure of bare active metal.

Several investigators have attempted to quantify the effects of crevice dimensions on the crevice corrosion of stainless steels. Most have used a multiple crevice assembly (MCA) which consists of 20 grooves and plateaus machined in a plastic washer type specimen (*102*). Susceptibility to attack has been related to both crevice depth and crevice gap. In general, crevice corrosion has been found to increase as depth increases and gap decreases.

Wheatfall (*103*) has correlated the amount of area outside the crevice with amount and depth of crevice attack for several depths of exposure in the ocean. The results showed an increase in attack as the area outside the crevice was increased by a factor of five. The amount of attack was greatest and the percent increase in attack was least in surface waters. This would be expected because the higher oxygen content would provide a larger driving force for the concentration cell corrosion and would tend to be as effective on a small external surface area as on a large one.

Velocity. As pointed out earlier, the conventional stainless steels tend to pit in low velocity seawater. However, as a general rule of thumb, velocities greater than 1.5 to 2 m/s are sufficient to prevent pitting on boldly exposed surfaces. This critical velocity is believed to be sufficiently fast to prevent the deposition of suspended material from natural seawater. Thus, deposits do not build up, and crevice-type attack does not take place beyond the critical velocity.

As would be expected, the Mo-containing stainless steels are resistant to pitting at velocities < 1.5 m/s. This has been demonstrated by Kovach et al. (*89*) for 25 Cr-2.5 Ni-3 Mo; by Liebert et al. (*90*) for AL-6X; by Fukuzuka et al. (*104*) for alloys with 4.5 to 6.5 Mo.

In general, the stainless steels are very resistant at very high seawater velocities. Weber (*105*) states that attack rates are very low for velocities through 55 m/s. As shown in Figure 8 (*105*), the rates for stainless steel at 40 m/s are very low, and are about three orders of magnitude lower than those for carbon steel at the same velocity.

Pollutants. Despite the fact that stainless steels are widely recommended where corrosion problems are encountered in polluted seawater, only limited data are available from laboratory or field studies. Larrabee (*77*) mentioned that stainless steel pump impellers have been used successfully in polluted seawater, and Todd (*106*) quotes LaQue that Type 316 has performed well in polluted seawater. Eden (*90*) indicates that 20 Cr-25 Ni-4.5 Mo has been used successfully for at least ten years in heat exchangers that handled polluted

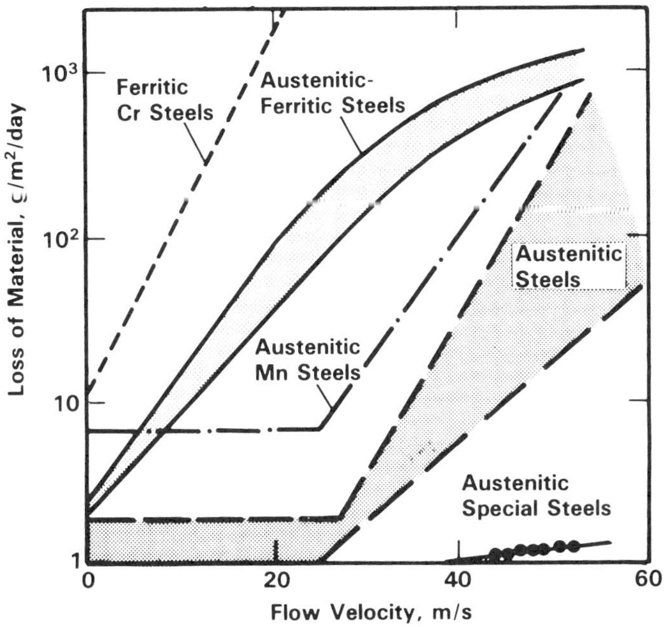

Fig. 8. Influence of flow velocity on weight loss of different materials in media with high H_2S content (500 ppm) and low pH values (pH 4.5) (*105*).

seawater. Hack (*107*) has conducted 120-day tests in flowing (2.4 m/s) seawater contaminated with 0 to 0.2 ppm sulfide and observed only minor attack of approximately 0.1 mm depth, and only at the 0.01 ppm sulfide level. Finally, Deverell and Davis (*108*) have conducted 100-hour tests in Na_2S contaminated synthetic seawater and observed only minor attack (< 1 μm/yr) on Type 304, 26-1, and 29-4 stainless steels (*108*). Types 316, 29-4-2, and AL-6X exhibited no attack in these short-term tests.

Effect of Aeration. Most of the studies on the effect of aeration (dissolved oxygen) have been done in connection with desalination or exposure in the deep ocean. In many experiments, both temperature and oxygen content varied. Hodgkiess and Rigas (*109*) compared oxygen levels at 25°C and 60°C in seawater tests conducted in Scotland. In these tests, crevice corrosion was less severe under deaerated than under aerated conditions.

Kain (*110*) in 30-day tests at Wrightsville Beach, North Carolina found a maximum in crevice attack for several stainless steels at 28°C (approximately 5.6 ppm O_2) compared to 12°C (approximately 7.8 ppm O_2) and 50°C (approximately 4.3 ppm O_2). He was not sure whether this effect could be attributed to oxygen alone.

Depth of Exposure. As pointed out in the section on aeration, the principal effect of depth of exposure is to affect the oxygen content of the seawater. As shown in Figure 2 the dissolved oxygen content in the Pacific Ocean is at a maximum at the surface (approximately 5.6 ml/L), drops off dramatically in the first 1000 feet, reaches a minimum of approximately 0.3 ml/l at approximately 7000 feet. Reinhart and Jenkins (*111*) found that the corrosion rates of stainless steels tend to increase as the oxygen content of the seawater increases, but, as pointed out by Reinhart (*79*), the corrosion rates do not necessarily correlate with pitting depths.

Wheatfall (*103*) did show some correlation between pit depth and oxygen content as a function of depth of exposure. The maximum pit depth in Type 304 stainless steel increased from approximately 10 mils in 0.6 ppm oxygen water at 2340 feet to $>$ 125 mils in 6.77 ppm oxygen at the surface in his tests of approximately 200 days' duration.

Seawater Composition. The constituents of seawater that usually vary, or are varied, are salinity (Cl^- ion), oxygen content, pH, pollutants, and biological activity (biofouling). The effect of the last four of these on pitting of stainless steel have already been discussed in prior sections. The general trend has been that pitting/crevice corrosion susceptibility increases, and pitting potentials decrease (become more negative) with increasing oxygen, decreasing pH, increasing H_2S (pollutant), increased biofouling, and increasing chloride concentration. The effect of salinity has been alluded to in several of these sections and additional information is provided in this section.

Based on laboratory tests and field experience, Kovach et al. (*89*) have determined the critical chloride levels for crevice corrosion of Types 304, 316, and Crucible SC-1 (25 Cr-2.25 Ni-3 Mo). Their results indicate that Types 304 and 316 exhibit crevice corrosion at approximately 1000 ppm chloride (seawater contains approximately 19,000 ppm Cl^-). Kain (*102*) exposed Types 304 and 316 stainless steel to diluted seawater and observed 0.05 to 0.1 mm (2 to 4 mils) crevice corrosion in 90 days at the 1000 ppm Cl^- level.

Stress Corrosion Cracking (SCC)

Effect of Alloy Composition. Stress corrosion cracking has not been a problem with the unsensitized austenitic and ferritic stainless steels in ambient temperature seawater. On the other hand, SCC in seawater could be a problem with the martensitic and precipitation-hardened stainless steels, based on their performance in marine atmospheres. Reinhart (*79*) exposed 15-7, 17-4, 17-7, and 14-8PH-type alloys in the deep ocean and observed SCC in only four conditions from a variety of heat treatments. The SCC failures were:

(a) 15-7 AMV RH950, 4-point loaded beams, 123 days at 5640 foot depth and 197 days at 2340 depth

(b) 15-7 AMV RH1150, 4-point loaded beams, 197 days at 2340 foot depth

(c) 15-7 AMV RH1150, at a drilled hole, 1064 days at 5300 foot depth

(d) 17-7PH RH1050, heat-affected zone on a welded specimen, 197 days at 2340 foot depth.

Veingarten et al. (*112*) have exposed welded specimens of 17-7 PH (103 ksi YS), 15 Cr-3Ni (91 ksi YS), and 17-6 Ti (131 ksi YS) to Baltic seawater for 1000 hours and observed no SCC.

Stress corrosion cracking of the austenitic grades becomes more of a problem as the temperature of the seawater is increased to greater than 60°C. Sedriks (*3*) reports no SCC of U-bends of Type 316 or 2RK65 after three months' exposure in boiling seawater, but he noted that 316 has cracked in the vapors above boiling seawater. The effect of molybdenum on SCC behavior in strong NaCl solutions at approximately 105°C has been summarized by Sedriks (*3*) and by Dundas and Bond (*113*). Although one-to-one comparisons may not be possible between NaCl and seawater, their results indicated that the resistance to cracking increased as the Mo level increased, particularly at $> 2\%$ Mo. Redmond and Miska (*114*) also have observed a similar beneficial effect of Mo. The effect of nickel on SCC susceptibility to aerated 22% NaCl at 105° C has been summarized by Speidel (*115*), who found that susceptibility to SCC increased in the range 0% to 20% nickel, but that resistance was restored at nickel levels $> 35\%$.

The ferritic molybdenum steels are resistant to SCC in most high-temperature chloride solutions, as determined in studies by Cramer and Carter (*116*), where E-Brite 26-1 was resistant to Salton Sea geothermal brine (155,000 ppm Cl^-), whereas several austenitics and Type 430 exhibited SCC even under deaerated conditions.

Heat Treatment and Welding. Heat treatment and/or welding can affect the SCC susceptibility of stainless steels by varying the strength levels in martensitic and precipitation hardening grades or causing grain boundary sensitization (usually in the austenitic grades). The cracking of 15-7 and 17-7 in the deep ocean has already been alluded to in the section on effects of alloy composition.

There is an apparent gap in the data for the SCC behavior of welded or sensitized austenitic stainless steels in seawater. There are a number of reported incidences of intergranular SCC in sensitized stainless steels that have been exposed to marine atmospheres or to even dilute (> 20 ppm) chloride solutions at pH 3.5 to 5.0. However, only limited experience has been reported for immersion in seawater. Sedriks (*3*) reports that some unstressed sensitized specimens of Type 316, 316L, and 317 exhibited intergranular attack in a one-year exposure at Wrightsville Beach. Speidel (*115*) has reported that furnace sensitization produced a strong decrease in SCC resistance in Type 304 in 22 % NaCl solution, but no definitive data have been found for seawater immersion of stressed sensitized austenitic stainless steels. Until definitive data are developed, it would appear to be prudent to avoid sensitization of any form in austenitic alloys that are to be immersed in seawater or chloride solutions.

Effect of Temperature. Surprisingly, no systematic studies have been performed on the effect of the temperature on the SCC of stainless steels in seawater. Some studies have been conducted in NaCl solutions which at equivalent chloride levels would be expected to produce results similar to those for seawater. Truman (*117*), in studying the effect of temperature on the SCC of welded Type 304 tubing, found that cracking occurred at ≥ 80°C in neutral chloride solutions containing 1000 to 100,000 ppm Cl^-. Herbsleb and Popperling (*118*) have conducted limited temperature studies on 22.5 Cr-5.5 Ni-3 Mo alloy specimens uniaxially loaded to 90 % of yield strength in 4.3 M NaCl (approximately 150,000 ppm Cl^-). Their results revealed no SCC in this alloy after exposure of approximately 1000 hours at temperatures to 90°C.

Lin et al. (*119*) have investigated the SCC susceptibility of sensitized Type 304 in 0.01 M NaCl (approximately 350 ppm Cl^-) at temperatures ≥ 100°C. Their results indicated that SCC occurs at more negative potentials as the temperature is increased.

Mechanical Effects. Somewhat surprisingly, few data have been found for the effects of mechanical treatments on the SCC of stainless steels in seawater. This lack of data is probably related to the good resistance of the austenitic alloys to SCC at temperatures < 80°C. Since many stainless steels with some degree of cold work are being used in seawater, it must be concluded that cold work does not have a strong adverse effect, at least at temperatures < 80°C. This conclusion tends to be borne out by the results of Radhakrishnan and Iyer (*120*). Working in a chloride medium, they found no effect of plastic strain (in the range of 0 % to 24 %) on the mechanical properties of stainless steel (hence no SCC) after 60 days exposure in 3.5 % NaCl. They did not give the composition of the stainless steel.

Geometry Effects. The very limited data on the effects of geometry on SCC of stainless steels have been obtained in NaCl solutions instead of seawater. As mentioned in several previous sections, a crevice in conjunction with stress tends to increase the tendency for SCC. Gurevich (*121*) has examined the effects of preexisting fatigue cracks on the SCC of 15 Cr-8 Ni-3 Mo in 3% NaCl and concluded that the plastic zone at the crack tip reduced the crack propagation rate by a factor of two. However, the crack propagation rates in his three-point loaded bent beams were inordinately high at 10^{-2} mm/sec.

Effect of Pollution. H_2S is the principal agent of pollution that affects corrosion behavior. Several studies have been conducted in NaCl solutions that indicate trends. Herbsleb and Popperling (*118*) exposed a 22 Cr-5.5 Ni-3 Mo duplex alloy to H_2S-contaminated and noncontaminated 4.3 M NaCl solution at 50°, 70°, and 90°C. No SCC was observed in specimens exposed at 0.9 yield strength for 1000 hours in noncontaminated solution. However, SCC was observed at high H_2S concentrations at both ambient temperature and 60°C. SCC was observed at the 100 ppm H_2S level at ambient temperature, which is probably an upper limit for H_2S associated with pollution. The absence of SCC at the 100 ppm H_2S level at 60°C suggests a hydrogen stress-cracking mode of failure, since it is known that increasing temperature decreases the propensity for this type of failure.

McGuire et al. (*122*) observed a similar effect of temperature on the SCC of Type 410 U-bends in 3% NaCl that contained 0.25 g/l H_2S (approximately 250 ppm). At 25°C, specimens with yield strength $\geq$ 120,000 psi exhibited cracking, whereas at 100°C, only those specimens with yield strength $\geq$ 175,000 psi failed.

Erosion/Cavitation

There is a moderate amount of information on the erosion corrosion and cavitation of stainless steels in seawater. The two deterioration processes are quite different. The erosion data are for clean seawater, where the shear stresses from the high velocity fluid remove a portion of the protective film and thereby increase the corrosion rate of the stainless steel. Erosion effects may be general over a broad area or may be local, such as at sharp changes in direction or at reduced cross sections. On the other hand, cavitation is the result of bubble collapse in the fluid as the result of rapid movement or vibration of the component or rapid pulsations in the fluid. The usual effect of the high implosive force of the collapsing bubbles is to rupture the protective oxide film on the stainless steel and, in severe cases, to damage the metal itself.

Cavitation is almost universally restricted to small areas and is often found in pumps and high velocity systems.

Erosion. Erosion effects have been observed in high velocity pumping systems and in high velocity surface ships (hydrofoils). As noted in the section on pitting, the conventional stainless steels pit at seawater velocities < 5 fps but are resistant to corrosion at velocities > 5 fps through upwards of 100 fps. At very high velocities, the corrosion rate increases as the protective film is removed by the high velocity fluid. There are conflicting data on the upper limit of velocity. A number of studies (*105*, *123*, *124*) indicate that the stainless steels are relatively resistant to flowing seawater at velocities up to 150 fps. On the other hand, Pini and Weber (*125*) indicate that Type 316 and the conventional stainless steels exhibit increased attack at seawater velocities > 70 fps.

Cavitation. Lee (*126*) has rated the austenitic and precipitation hardened stainless steels among the most resistant to cavitation attack in seawater (see Table IX) (*126*). This ranking appears to be confirmed by studies conducted by Weber (*105*), Hohman and Kennedy (*127*), and Werchniak and Gudas (*124*). The data reported from these sources indicate a moderate amount of attack, but it must be remembered that laboratory cavitation tests are designed to accelerate attack to permit relative rankings of materials.

Corrosivity of the environment is a factor in cavitation. Simoneau et al. (*128*) have shown that the cavitation loss for Type 304 is about twice as great in 3% NaCl solution as in distilled water in tests conducted at 20 KHz and 17 μm amplitude. Simoneau et al. (*128*) also report that the application of cathodic protection decreases the damage due to cavitation, which further indicates a contribution of corrosion to cavitation damage.

Intergranular Attack (IGA)

Welding and Heat Treatment. Welding can sensitize stainless steels to grain boundary attack in many corrosive environments. In the common austenitic stainless steels, the normal carbon grades (0.04 to 0.08C) will sensitize adjacent to welds in the so-called heat-affected zone (HAZ). In an exhaustive study, Samans (*129*) showed, as early as 1964, the time-temperature relationships that produce sensitization in these alloys. When heated in the temperature range of about 500° to 800°C, chromium carbides precipitate at the grain boundaries. This precipitation depletes the adjacent area of chromium to the extent that the chromium level is below that required for stainless steel (12%) and, hence, these narrow zones become sensitized to

TABLE IX

MATERIALS OF VARIOUS RESISTANCE TO CAVITATION DAMAGE IN SEAWATER (*126*)

Resistance to Cavitation Damage Rating	Metals
Group I — Most resistant. Little or no damage. Useful under highly cavitating conditions.	Cobalt-base hardfacing alloys Titanium alloys Austenitic (series 300) and precipitation hardening stainless steels Nickel-chromium alloys such as Alloy 718 Nickel-chromium-molybdenum alloys such as Alloys 625 and C-276
Group II — These metals are commonly utilized where a high degree of resistance to cavitation damage is required but where some metal loss under the most severe conditions of cavitation is acceptable	Nickel-copper-aluminum Alloy K-500 Nickel-copper Alloy 400 Nickel-aluminum bronze Nickel-aluminum-manganese bronze
Group III — These metals have some degree of cavitation resistance but are generally limited to low speed applications.	70/30 copper-nickel alloy Manganese bronze G Bronze and M Bronze Austenitic nickel cast irons
Group IV — These metals are normally not used in applications where cavitation damage may occur.	Carbon and low-alloy steels Cast irons Aluminum and aluminum alloys

intergranular corrosion. During welding, an area on either side of the molten weld bead is heated into this critical temperature range, which then results in sensitization in the heat-affected zones.

The "L" grades of stainless steel contain $\leq 0.03\%$ carbon and do not sensitize during normal welding operations. However, long-term ($>$ 12-hour) exposure in the 500° to 800°C temperature range can sensitize the "L" grades.

The austenitic alloys that contain Mo or very high Cr contents also can be sensitized by the precipitation of an Fe-Cr phase (sigma phase) at grain boundaries. As noted by Sedriks (*3*), sigma phase formation is possible at temperatures between 565° and 925°C in austenitic stainless steels containing more than 16% Cr and $< 32\%$ nickel. In one year tests in ambient seawater at Wrightsville Beach, North Carolina, he observed intergranular attack (IGA) when specimens were heat treated to produce the sigma phase, but no attack in as-welded 316 and 317 stainless steel components.

As pointed out by Steigerwald (*130*), the ferritic stainless steels also exhibit intergranular corrosion because of chromium carbide and chromium nitride precipitation at grain boundaries. However, nitrides are more of a problem than carbides in the ferritics, and the heat treatments that lead to sensitization are different for the ferritics than for the austenitics.

Despite sensitization, several investigators (*79*, *81*, *131*, *132*) have shown that the performance of welded stainless steel components in condensers and desalination equipment has generally been as good as that of nonwelded components. Evidence of selective attack (IGA) has been reported by Reinhart (*79*) in an annealed and pickled Type 405 (12 Cr) alloy after 123-day exposures at the 5640 feet depth in the Pacific and by Henrikson et al. (*133*), who reported attack at welds in loop tests in 3200 to 3700 ppm Cl^- in Sweden, but who attributed this attack to oxides and defects in the welds that caused crevice attack.

Thus, the available data tend to indicate that welding *per se* does not adversely affect the performance of stainless steels in seawater. However, elevated temperatures or higher chloride solutions might be more aggressive. Heat treatments that produce sigma phase are harmful, and, as pointed out by Sedriks (*3*), it seems prudent to avoid sensitization in the welding of Fe-Cr-Ni and Fe-Cr-Ni-Mo alloys.

Galvanic Corrosion

Considerable information is available relative to the galvanic corrosion of stainless steel in seawater and sodium chloride environments. The information falls into four broad categories: (1) cathodic protection to prevent crevice corrosion of stainless steels; (2) galvanic series in seawater; (3) galvanic corrosion of the dissimilar metal of the couple; and (4) galvanic corrosion of the stainless steel member of the couple. Cathodic protection will not be discussed here except in passing to note that Lee and Tuthill (*134*) and others have found that galvanic coupling to carbon steel prevents seawater crevice corrosion in the conventional 300 series stainless steel.

Galvanic Series. Several galvanic series have been developed based on the corrosion potentials of the various metals and alloys in seawater. One such series is shown in Figure 1 (*3*). Note that the stainless steels are very noble (positive) and thus would be expected to cause galvanic corrosion on the large group of alloys that exhibit more negative potentials. In turn, stainless steels would be expected to experience galvanic corrosion when coupled to the few metals that exhibit more positive corrosion potentials. Also, note that the corrosion potentials in stainless steel crevices may be several hundred millivolts more negative than those for fully exposed surfaces, which accounts for the driving force for crevice corrosion.

Velocity in the range of 3 to 15 m/s has only a minor effect on corrosion potential (*3*) and thus does not create a major change in ranking in the galvanic series. A galvanic series also has been developed by Henrikson and Asberg (*135*) for Baltic Sea water, which is lower in salinity than most ocean waters. Their results revealed that Type 317 stainless steel was more noble than titanium in the Baltic Sea water.

Smith and Compton (*136*) have determined galvanic series as a function of temperature (30°, 100°, 160°, and 200°C). The effect of increased temperature as determined by Smith and Compton (*136*) was to make the potential more negative and to reduce the spread of the values for most metals.

The corrosion effects of galvanically coupling Types 304 and 316 to a number of dissimilar alloys in seawater have been studied by Ebihara (*137*) and Waldron et al. (*138*). Their results indicated that coupling to stainless steel increased the amount of attack on steel and 6061 and 7075 aluminum. Negligible, if any, increase in attack was observed for 360 brass, Monel 400, or titanium when coupled to Type 316.

Several galvanic corrosion studies have been conducted with the high Mo ferritic and duplex stainless steels. Gehring and Maurer (*139*) report considerable galvanic attack of copper tube sheet materials when coupled to AL-6X tube material. The galvanic attack was more severe on Muntz metal than it was on aluminum bronze. Gehring and Kyle (*140*) reported somewhat similar galvanic effects but noted a marked decrease in galvanic attack when the seawater temperature was reduced from 22° to 6°C.

Couples between stainless steels and dissimilar metals also have been studied for:

(a) Aluminum: References *141* to *146*
(b) Steel: References *141* to *143*
(c) Copper alloys: References *141–143*, *147*
(d) Nickel alloys: References *141*, *142*, *147*
(e) Titanium: References *141*, *142*

Galvanic Corrosion of Stainless Steels. In most galvanic couples, stainless steel is noble to the dissimilar metal and thus is protected from corrosion. However, Types 304 and 316 are attacked in some galvanic couples. Ebihara (*137*) found a slight increase in attack in some of his 316 specimens that were coupled to Monel 400 and Ti-6A1-4V in seawater in the Panama Canal Zone. Pelensky et al. (*141*) did not observe incrased attack of Type 316 when coupled to Monel. Pelensky et al. (*141*) and Charlot et al. (*142*) also found no increased attack on Type 316 when coupled titanium in seawater.

Miller and Lee (*147*) coupled several stainless steels to graphite-epoxy composites and exposed them to 3.5% NaCl solution. Their results indicated that only Type 440 C experienced increased corrosion as the result of the galvanic couple, whereas 301, 304, PH 17-7, and AFC-77 were compatible with the graphite-epoxy.

4. Mud

General Corrosion and Pitting

Corrosion data are limited for stainless steels exposed to the mud zone in the ocean. Bottom sediments vary somewhat in composition and in marine activity. Sulfate-reducing bacteria have been identified in the mud zone. Since oxidizing agents are absent, one would expect stainless steels to lose passivity at local sites and, as a result, to undergo pitting and crevice attack. Such is actually the case. However, in most cases, the corrosion behavior in the mud zone (Pacific Ocean off California) is similar to that observed for the same alloy in the water just above the ocean bottom.

Thus, pitting and crevice corrosion occur in the mud zone in the conventional ferritic and austenitic alloys, and only the high molybdenum ferritics and duplex alloys are resistant to this type of attack. General corrosion rates are low for all the stainless steels, except that pitting may be the primary failure mode.

The results of mud zone exposure of stainless steels at depths in the Pacific Ocean have been summarized by Reinhart (*79*). Reinhart concluded that the corrosivity of the mud zone was about the same as that of the seawater just above the ocean floor. However, in his study, there was less pitting and crevice corrosion in the mud zone at the 2300 foot depth than at other depths. This behavior can be attributed to the lower oxygen level in the water at the 2000 foot depth than elsewhere, and, in this respect, the behavior parallels that observed for stainless steels in the seawater itself.

C. Nickel-base Alloys

The corrosion behavior of unalloyed nickel, high-nickel alloys, and Ni-Cr-Fe alloys with more than 30% nickel (such as Alloy 800), is discussed in this section. The reader is also referred to the section on stainless steels where data for several nickel alloys may be included with those presented for stainless steels. In particular, data for Alloys 625, 825, and C-276 are often included for comparison.

The corrosion behavior of nickel-base alloys is somewhat analogous to that of the stainless steels. The principal modes of failure of Ni-Cr alloys are pitting and crevice corrosion, and the addition of molybdenum to these alloys improves their resistance to these forms of attack. However, the Ni-Cr alloys are more resistant to SCC and hydrogen stress cracking than are their stainless steel counterparts.

1. Atmosphere

General Corrosion and Pitting

Unalloyed nickel and the nickel-base alloys have excellent resistance to corrosion in the marine atmosphere. After 16 years of seacoast exposure in the Panama Canal Zone (*148*), unalloyed nickel and Alloy 400 exhibited no visible attack, no pitting, and weight loss penetration of < 0.02 mpy. The weight losses increased linearly with time over the 16-year exposure.

Miska (*149*) reports the following corrosion rates and pit depths for nickel-base alloys exposed seven years at the Kure Beach, North Carolina, 25-m lot:

Alloy	Corrosion Rate, mpy	Maximum Pit Depth, mils
Unalloyed	0.01	—
600	0.0016	—
800	0.06	0.9
825	0.06	0.7

He also reports that Alloys F, G, 700, 718, Elgiloy, and Illium R have excellent resistance to marine atmospheres, but the most resistant alloys are Alloy C, C-276, X-750, 625, MP35N, Chlorimet 3, and René 41.

Galvanic Corrosion

Southwell et al. (*148*) report on galvanic couples between nickel/steel, Alloy 400/steel, and Alloy 400/phosphor bronze. In all cases, the nickel alloys were noble to the dissimilar metal, and their corrosion rates after 16 years of exposure at the seacoast in the Panama Canal Zone were < 0.03 mpy.

Johnson and Abbott (*150*) exposed disc-type galvanic couples of steel to a number of nickel-base alloys in a marine atmosphere. Their results showed that Alloys 200, 400, 600, 800, 825, and B increased the rate of attack of the carbon steel as a result of the galvanic coupling.

Bomberger et al. (*151*) coupled alloys 400 and 600 to titanium in studies conducted at Kure Beach. Their results indicated that, after more than four years' exposure, the attack on both alloys was negligible (penetration < 0.03 mpy), but there was some slight increase in attack of the Alloy 400 when the Ti/400 area ratio was 7/1.

Stress Corrosion Cracking and Intergranular Attack

Money and Kirk (*71*) conducted five-year stress corrosion tests at Kure Beach on U-bends of annealed, welded, cold worked, or sensitized stainless steels and nickel-base alloys. No SCC or IGA was observed for any of the nickel-base alloys in any condition studied, which included Alloys 600, 800, 825, and 20Cb3.

2. Splash and Tide

General Corrosion and Pitting

Miska (*149*) states that the corrosion resistance of nickel-base alloys in the splash zone is virtually as good as in the marine atmosphere, but may be reduced in tide zones. For example, Southwell and Alexander (*148*) found that, after 16 years at mean tide in the Panama Canal Zone, the average penetration and maximum penetration for unalloyed nickel and Alloy 400 were 4.4 and 2.6 (average) and 121 and 36 mils, respectively, whereas corrosion was negligible for these same alloys after 16 years' exposure in the marine atmosphere.

Galvanic Corrosion

Ballantine (*152*) has prepared a guide for galvanic coupling of fasteners in exposure above the water line, and only those materials close to the nickel alloys in the immersed and atmospheric galvanic series can be safely used without undue galvanic corrosion. Anderson and Ross (*153*) report that Alloys 301, 400, 500, and X-750 were in excellent condition after 15 to 25 years' service as fasteners in Alloy 400 sheathing at Wrightsville Beach, North Carolina.

Stress Corrosion Cracking and Intergranular Attack

Alternate immersion tests conducted by Davis (*154*) in 3.5% NaCl solution produced no SCC in Alloys C, 718, and TDNiCr. Montano (*155*) reported no

SCC of strengthened and aged MP35N exposed 180 days in alternate immersion tests at loads of 50%, 75%, and 100% of yield strength. Deel and Mindlin (*156*) found no SCC of alloys 702, 706, 710 that were exposed 1000 hours to the alternate immersion test at loads of 80% of yield strength.

3. Immersion

As with the conventional stainless steels, the most frequent failure mode of the conventional nickel-base alloys in seawater is pitting or crevice corrosion rather than general attack. As a rule, the weight change from pitting attack is so low that corrosion rates obtained from calculations based on the entire surface area are very low and can be misleading as to the service life of the material. Since pitting and general corrosion are so intertwined in the reported literature, the general corrosion rates will be discussed in conjunction with pitting.

General, Pitting, and Crevice Corrosion

Alloy Composition. Unalloyed nickel exhibits a relatively low overall rate of attack in seawater with calculated penetration rates ranging from 0.05 to 1 mpy. It does exhibit pitting in quiescent seawater, and this pitting rate can range up to 50 mpy. However, Southwell and Alexander (*148*) have shown that most of the deep pitting attack occurs within the first year or so, and the depth of penetration of the deepest pits increases very little during subsequent exposure.

Alloying improves the resistance of nickel to pitting. Alloy 400 (66 Ni-32 Cu-2 Fe) is more resistant to pitting in seawater than is unalloyed nickel, although Alloy 400 (Monel B and C) does pit. The major application of Alloy 400 is in the splash zone where it performs very well.

As with the stainless steels, the major alloying elements in nickel that affect resistance to pitting in seawater are chromium and molybdenum. Additionally, as with the stainless steels, the chromium alloys are subject to pitting, while the most resistance alloys are those that contain molybdenum in addition to chromium. Boyd and Fink (*2*) have ranked the relative seawater corrosion resistance of the nickel alloys in Table X (*2*). This ranking is based on data such as those presented in references *148*, *157*, and *158* and reflects the importance of increasing molybdenum content in the alloys on their resistance to pitting and crevice attack.

Temperature. As pointed out elsewhere, increasing temperature can have different effects on corrosion, depending upon whether it reduces the amount

TABLE X

CLASSIFICATION OF NICKEL ALLOYS THAT MAY FIND SPECIAL APPLICATION IN MARINE ENVIRONMENTS (*2*)

Alloy	Composition, Weight Percent	Remarks
	Class I: Most Resistant	
Hastelloy C	57Ni-16Cr-17Mo plus Fe, W, Co	Completely resistant except at welds.
Hastelloy C-276	57Ni-16Cr-17Mo plus Fe, W, Co	Low carbon (< 0.02 C) can be welded.
Inconel 625	61Ni-22Cr-9Mo plus Fe, Cb	Approaches Hastelloy C in total resistance.
MP35N	35Ni-35Co-20Cr-10Mo	Excellent, so far, in preliminary experiments.
Chlorimet-3	60 Ni-18Cr-18Mo plus Fe, Si	Cast alloy; excellent for pumps, etc.
René 41	56Ni-11Co-19Cr-10Mo-3.1Ti	Good-to-excellent resistance to pitting.
Hastelloy X	52Ni-22Cr-9Mo-19Fe plus W, Co	Good-to-excellent resistance to pitting.
	Class II: Very Resistant	
Hastelloy F	46Ni-22Cr-7Mo-21Fe	
Hastelloy G	45Ni-21Cr-7Mo-20Fe-2Cu-25Co	Usually satisfactory; molybdenum
Illium R	68Ni-21Cr-5Mo-3Cu	content provides resistance
Inconel 700	46Ni-28Co-15Cr-4Mo plus Ti, Al	to pitting.
Elgiloy	15Ni-40Co-20Cr-7Mo-15Fe-2Mn	
Inconel 718	53Ni-19Cr-3Mo-18Fe plus Co, Ti, Al	
	Class III: Resistant, Some Pitting	
Inconel 600	76Ni-16Cr-7Fe	
Inconel X750	73Ni-15Cr-7Fe plus Cb, Ti, Al	
Incoloy 800	32Ni-21Cr-46Fe	May be some pitting at sites
Incoloy 825	42Ni-22Cr-30Fe-3Mo-2Cu	where seawater is stagnant.
Monel-400	66Ni-32Cu-2Fe	
Monel-K500	65Ni-30Cu-1Fe-3Al	

of dissolved oxygen in the seawater, decomposes bicarbonates to form insoluble carbonate scales, or kills biological species. All of these latter effects would tend to reduce the corrosion attack.

Tipton and Kain (*159*) have shown that weight losses for Alloy 400 in seawater reach a maximum at 30°C and drop to very low values at 50°C, whereas, in this same exposure, pit depths were greatest at 18°C and decreased to negligible values at 50°C. Asphahani et al. (*160*) report that Alloy G exhibits a maximum in depth of crevice attack at 28°C. In 30-day tests the average crevice attack for Alloy G was 0.03 mm at 12°C, 0.05 mm at 28°C, and 0 mm at 50°C.

In studying Alloy 825, Hodgkiess and Rigas (*109*) found no crevice corrosion after 31 weeks' exposure at 25° and 60°C, but did observe crevice corrosion after 20 weeks' exposure at 100°C. Temperature had no effect on crevice corrosion or general corrosion of C-276 in tests reported by Asphahani et al. (*160*).

Aeration. The principal studies on the effect of aeration have been conducted as part of the deep sea exposures by Reinhart and coworkers (*23*, *158*). For the conventional alloys, the general corrosion rate increased with increasing oxygen content in the seawater over the range of 0.4 to 5.8 ml O_2/l seawater. For most of the alloys, the general corrosion rates were < 1 mpy, even at the highest oxygen level of 5.8 ml/l. However, several of these alloys exhibited crevice attack even at the lowest oxygen level studied (0.4 ml/l), namely, Alloy 600 to 28 mils, X-750 to 9 mils, and 825 to 15 mils after 402 days of exposure. Alloys 625, 700, and a modified Alloy C did not exhibit crevice attack at any of the depths (O_2 levels) studied. Asphahani et al. (*160*) also have observed no crevice attack in C-276 after 112 days' exposure to seawater containing 0.03 or 5.2 ppm dissolved oxygen at 35° to 37°C and pH 3.5.

Velocity. For the nickel alloys of interest, seawater velocities to 40 m/s (131 fps) do not have an adverse effect on corrosion behavior. In fact, those alloys that exhibit crevice corrosion in stagnant seawater, such as 400, 600, and X-750, perform very well in high velocity seawater and exhibit corrosion rates < 0.05 mpy at 120 fps velocity (*123*).

Boyd and Fink (*2*) have indicated that, over the velocity range of 0.5 to 128 fps, Alloy C exhibited corrosion rates of < 0.6 mpy. Asphahani et al. (*160*) also reported a low corrosion rate for Alloy C-276 of 0.2 mpy in 10°C seawater at 38 m/s (125 fps).

Geometrical Effects. Some data are available on the effect of crevice geometry on crevice attack. Wheatfall (*103*) observed that the depth of crevice attack in Alloy 400 increased monotonically with increasing area outside the crevice at three exposure depths in the ocean that represented a range of dissolved oxygen concentration. Anderson (*161*) also observed a more or less linear increase in depth of attack in Alloy 400 at bold/sheltered area ratios of 0.1/1 to 100/1, but at ratios above 100/1, the depth of attack increased sharply.

Kain (*162*) predicted hours to breakdown for two nickel alloys as a function of crevice gap and crevice depth, as he did for stainless steels. His results indicated the high resistance of 625 and an 18 Cr-18 Mo alloy to crevice attack and showed that the tendency for crevice attack decreases with increasing crevice gap and crevice depth.

Pollution. Limited data obtained by Hack (*107*) show that Alloy 400 exhibits increased corrosion when the sulfide content of seawater increases from 0 to 5 mg/l (0 to 5 ppm). The conventional nickel-chromium alloys are not subject to increased attack by sulfides in ambient seawater.

Erosion and Cavitation

As described in the section on velocity, the nickel-chromium alloys are resistant to flowing seawater at velocities in excess of 140 fps (40 m/s). The results of jet erosion tests conducted by Hohman and Kennedy (*127*) for 30 days at 90 knots (152 fps, 46 m/s) revealed rates of attack for Alloys C and 718 that were less than 4 mpy, and welding appeared to have no adverse effect on the rate. Alloy 500 exhibited a 9 mpy rate of attack in these tests. Hohman and Kennedy (*127*) also conducted cavitation tests in seawater. Their results showed that the nickel alloys are more resistant to cavitation than are most other alloy systems, even though, in these severe tests, Alloys C and 718 exhibited cavitation rates that approached 0.5 in/yr. The cavitation rate for Alloy 500 in these tests was about 1 in/yr, with the annealed material having a slightly higher rate than the material heat treated to the strength needed for hydrofoil craft.

Galvanic Corrosion

Several nickel alloys are included in the galvanic series presented in Figure 1 (*3*). Note that the more resistant nickel alloys are very noble. Thus, they would be expected to promote galvanic corrosion of those alloys that are more active and experience little or no corrosion when coupled to the few alloys that are more noble; such is the case. Johnson and Abbott (*150*) found that all the nickel alloys increase the corrosion of steel when galvanically coupled in seawater. Ebihara (*137*) found that Alloy 400 promoted galvanic attack of A231 magnesium, 6061 and 7075 aluminum, 4340 steel, and 260 brass, but had little or no effect on 316 stainless steel. Southwell et al. (*148*) also report Alloy 400 produced galvanic attack on carbon steel as well as on phosphor bronze. Hack (*109*) has shown that Alloy 400 produces galvanic corrosion of 90-10 Cu-Ni and Ni-Al-Bronze, but that Alloy 625 produces galvanic attack of Alloy 400.

Alloy 400 has been coupled to a number of alloys close to it in the galvanic series with little or no adverse effect. Vreeland (*164*) reports that Alloy 400 has little galvanic effect on a number of cobalt-base (Stellite) alloys, while Anderson and Ross (*152*) report that the following fasteners were in excellent condition after being in Alloy 400 sheathing for 15 to 25 years: nickel 301, Alloy 400, Alloy K500, Alloy X-750, and Types 410, 431, 305, and 316 stainless steels.

Bomberger et al. (*151*) have coupled Alloy 400 and 600 to titanium. Their results indicate no galvanic attack of Alloy 600 after one year of exposure at Ti/Alloy 600 ratios of 7:1. On the other hand, the Alloy 400 did exhibit some slight galvanic attack when the Ti/Alloy 400 ratio was 7:1, but not when the ratio was 1:7.

Stress Corrosion Cracking

The nickel-base alloys are known for their resistance to SCC in chloride-containing waters and are often used as alternative materials when stainless steels fail by chloride SCC. However, there are few documented cases of the performance of stressed nickel-base alloys. Reinhart (*158*) included stressed Alloy C and 825 in his deep ocean tests. The results of his tests revealed no SCC in exposures to 731 days at stresses to 75% of yield strength. Sedriks (*3*) has reported no SCC of U-bends of 904L, G3, 625, and C276 after 300 to 550 hours' exposure in synthetic seawater at 35° and 80°C.

SCC may occur with nickel-base alloys in very high temperature chloride environments, Cramer and Carter (*116*) exposed Alloys 625 and C-276 to geothermal type brines at 232°C and detected SCC in both alloys in 7- to 30-day tests under deaerated conditions and with 100 ppm dissolved oxygen.

Intergranular Attack

Intergranular attack (IGA) has been reported to occur in heat-sensitized areas of welded Alloy C in heated seawater, and thus Alloy C-276 is preferred if welded equipment is to be used under these conditions (*3*). An extensive discussion of IGA of Fe-Cr-Ni alloys is given in the section on stainless steels.

4. MUD

Mud zone exposure tests in the deep ocean have been reported by Reinhart (*158*). General corrosion rates for all nickel base alloys were low, at < 0.1 to 0.3 mpy, whether exposed in the mud or in the seawater itself. However, for those alloys that exhibited crevice attack (Alloys 600, 610, X-750, 88, 825, and 902), the attack appeared to be deeper in the water than in the mud zone. In these tests, Alloys 625, 700, 804, 825SS, 825Cb3, 901, B, G, F, X, and a modified G did not exhibit crevice attack in the mud zone.

TABLE XI

ALUMINUM ALLOY DESIGNATION SYSTEM

Major alloying element	Alloy series
None 99% +	1000[a]
Copper	2000[b]
Manganese	3000[a]
Silicon	4000[a]
Magnesium	5000[a]
Magnesium/Silicon	6000[b]
Zinc/Magnesium	7000[b]
Others	8000

[a] Non-heat treatable
[b] Heat treatable

D. *Aluminum Alloys*

Since the corrosion resistance of aluminum alloys depends on the presence of a passive oxide film, they are generally most resistant in well-aerated marine environments. Failure by localized attack such as pitting, crevice corrosion, exfoliation corrosion, or stress corrosion cracking is characteristic of many alloys, in particular the high strength alloys.

Table XI briefly summarizes the aluminum designation system. The seven principal series contain the major alloying elements shown and are either heat treatable or non-heat treatable, as indicated. The 5000 series aluminum alloys are widely used for marine applications. Examples are alloys 5006, 5083, and 5456. If higher strength is required, alloys in the 6000 series, such as 6061-T6, are often selected. High strength alloys in the 2000 and 7000 series are commonly used aerospace alloys, but have relatively poor performance in marine environments.

1. ATMOSPHERE

Pitting

Since the corrosion resistance of aluminum alloys depends on the presence of a thin protective oxide film, many of these alloys tend to pit in marine atmospheric environments. Long term exposure studies of 1000, 2000, 3000, 5000, 6000, and 7000 series aluminum alloys (*165*, *166*) have indicated that

Alloy 6061 had the highest resistance to pitting. These studies also show that the pitting rate decreases with time and in some cases pitting actually stops. Furthermore, it was indicated that position, upward versus downward facing, can have a significant effect on the pitting rate of some aluminum alloys. In general, the surface facing the ground contained deeper pits than the surface facing the sky. This was attributed to rainwater washing over the surface facing the sky and to the fact that this surface dried more readily.

Galvanic Corrosion

Galvanic corrosion of aluminum alloys may occur when they are joined in metal-to-metal contact to other more noble materials. Figure 1 (*3*), which shows the galvanic series of a range of materials commonly used in marine environments, indicates that aluminum is active to most other materials.

Fifteen year exposure tests were performed in a marine atmosphere at Bohus Malmon on different aluminum alloys (A199.5, AlMn1, AlSiMg, AlCu4Mg1, AlMg2.5, AlSi12, and AlSi5Cu3) coupled with copper, steel, stainless steel, and zinc (*168*). The results of the marine exposure tests after 1 and 15 years are shown in Table XII and indicate a significant increase in corrosion rate of the aluminum alloys when coupled to more noble metals. In particular, coupling to Cu resulted initially in an order of magnitude increase in the corrosion rate of aluminum. In all cases, the corrosion rate decreased with time, likely as a result of the formation of corrosion products on the aluminum surface. Moreover, zinc, which is more active than aluminum, provided effective cathodic protection.

TABLE XII

ATMOSPHERIC CORROSION RATE OF ALUMINUM COUPLED TO DIFFERENT METALS IN A MARINE ATMOSPHERE (*168*)

	Mean corrosion rate, mg/yr Marine (Smögen, Sweden)	
Al coupled with metals	1 Yr	16 Yr
Al (normal corrosion)	28	7.6
Cu	313	79
Fe	187	41
Stainless steel	120	21
Zn	3	0.3

Stress Corrosion Cracking

In marine environments, many aluminum alloys, in particular the high strength aluminum alloys, can fail by SCC at stresses far below the yield stress (*1*). Alloys 2024-T351, 7074-T6, and 7075-T651 are highly susceptible to SCC, while alloys 2024-T851 and 2124-T851 exhibit moderate resistance to SCC, with failures occurring when stressed to 75% of their yield strengths (*1*). However, alloys 2048-T851, 2214-T87, 7049-T7351, 7050-T73651, 7075-T7351, and 7475-T7351 are very resistant to SCC.

Sprowls et al. (*169*) studied the SCC susceptibility of aluminum alloys using fracture mechanics techniques. Table XIII shows the crack velocities in a sea coast environment (Point Judith, Rhode Island) of double cantilever

TABLE XIII

STRESS CORROSION CRACK VELOCITY IN ALUMINUM ALLOYS EXPOSED TO VARIOUS ENVIRONMENTS (*169*). SHORT-TRANSVERSE (S-T) DOUBLE CANTILEVER BEAM SPECIMENS WERE BOLT LOADED TO POP-IN

	Average Initial Crack Velocity in/hr Outdoor Atmosphere[a]	
Alloy	Seacoast (Point Judith, RI)	Industrial
	Alloys with High Resistance to Stress-Corrosion Cracking	
5456-H117	0	0
6061-T651	0	0
7075-T7351	0	0
2219-T87	4×10^{-5}[b]	1×10^{-5}[b]
2021-T81	7×10^{-5}[b]	2×10^{-5}[b]
2024-T851	1×10^{-5}[b]	4×10^{-6}[b]
	Alloys with Low Resistance to Stress-Corrosion Cracking	
2014-T651	2×10^{-4}	2×10^{-4}
2024-T351	2×10^{-4}	2×10^{-4}
2219-T37	3×10^{-4}	3×10^{-4}
7075-T651	2×10^{-4}	1×10^{-4}
7039-T6351	2×10^{-4}	3×10^{-4}
7079-T651	4×10^{-4}	4×10^{-4}
5456-Sens.	5×10^{-4}	5×10^{-5}

[a] Based on the first six months of exposure.

[b] Transgranular mechanical fracture rather than intergranular SCC.

beam (DCB) bolt specimens in the short transverse direction of aluminum alloys with low and high resistance to SCC. The table indicates that the performance of the resistant alloys was markedly superior to that of the low resistance alloys. No crack growth was observed on specimens of 5456-H117 and 6061-T651 after 24 months' exposure. The other four alloys showed small amounts of crack growth that initiated in three to eight months. However, metallographic examination revealed, only in the case of 7075-T7351, cracking was due to intergranular SCC.

It was further noticed that corrosion products formed in the cracks, which resulted in so-called self loading. This can significantly increase the stress-intensity factor at the crack tip and be a potential driving force for SCC of moderately sussceptible alloys. Studies by Dorward et al. (*170*) of 2000 and 7000 series aluminum alloys at a marine-atmosphere exposure site near Daytona Beach, Florida, indicated that self loading is enhanced by copper but is not limited to copper-bearing alloys. The copper-free alloy 7039-T64, for example, showed self loading effects after three to four years' exposure. Naturally aged 2000 series alloys and peak aged, copper containing 7000 series alloys were most sensitive to self loading, usually showing the effects within six months.

Intergranular Corrosion

Intergranular corrosion of aluminum alloys can seriously affect their strength. Susceptibility to intergranular corrosion, which is parallel to the susceptibility to stress corrosion cracking, strongly depends on alloy composition, heat treatment, and grain orientation. A common form of intergranular corrosion is exfoliation, which occurs in plates or forgings where intergranular attack occurs along elongated grains, and subsequent formation of corrosion products lifts the metal surface. Some alloys are extremely prone to attack where machining occurs across the grain of the material.

The exfoliation performance of 7000 series alloys was evaluated at two seacoast locations, one at Point Judith, Rhode Island and another at Daytona Beach, Florida (*171*, *172*). After 11 years' of exposure the 7075-T73 did not show, and 7075-T76 and 7178-T76 showed very little exfoliation corrosion. Relatively short-term (three-year) exposure tests indicated similar excellent performance of the T76, T736 tempers of the newer alloys 7049 and 7050.

Shipboard exposure tests by Jankowski et al. (*172*, *173*), which were performed to study the combined effect of stack gas exhaust and sea spray on the corrosion of aluminum alloys, indicated that the naturally aged type T3 temper of the 2124 plate was susceptible to exfoliation, whereas the artificially aged T8 temper was resistant. Further, it could be concluded from this work

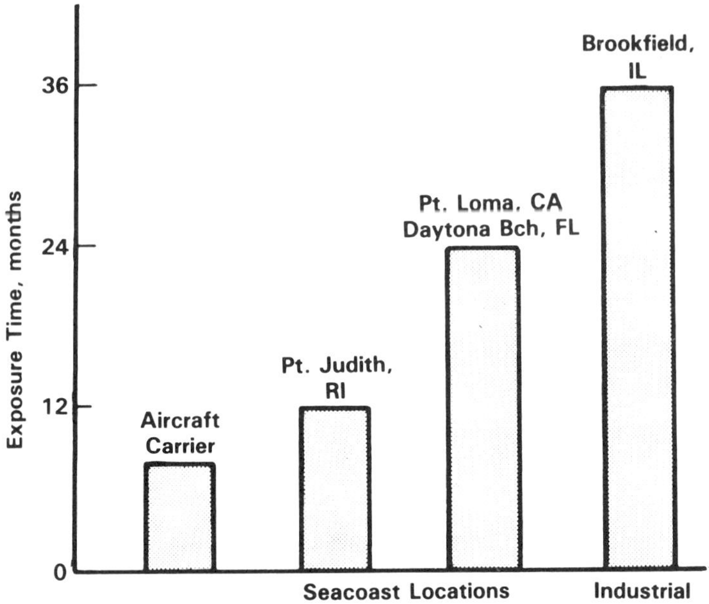

Fig. 9. Time required at various marine and industrial sites to produce severe exfoliation in 0.5 inch thick aluminum 2124 plate heat treated to be susceptible to exfoliation (*173*).

that an increase in ambient temperature resulted in more severe corrosion and that the environment of an aircraft carrier, however powered, was considerably more corrosive than sea cost or industrial environments (see Figure 9).

The aluminum-magnesium alloys 5086, 5083, and 5456 are widely used for naval construction because of their resistance to marine corrosion, high strength-to-weight ratio, and weldability. The alloys have been used for light-weight, super structures and as primary structural metal for high-speed, high-performance ships and craft. The most desirable combination of properties is achieved in a mildly cold-worked temper designated as 5086-H32, 5083-H321, and 5456-H321. However, under certain conditions, exfoliation corrosion problems have been experienced with the standard temper. To solve the exfoliation problem, the aluminum industry developed rolling procedures for tempers meeting the mechanical property requirements specified for the H32 and H321 tempers and having a structure which is resistant to exfoliation corrosion (*174*). These tempers for sheet and plate 5000 series alloys are designated H116 (Reynold Metals Company) and H117 (Alcoa). Czyryca and Hack (*175*) performed studies exposing the exfoliation resistant tempers to three marine environments at Wrightsville Beach, North Carolina, for periods of six months, one year, and two years. The results of two years' of exposure to the marine environments (marine atmosphere, splash and spray, and fully submerged in seawater) indicated that alloys 5086, 5083, and 5456 in the H116 and H117 tempers have good corrosion resistance. No exfoliation attack was evident on any of the test panels.

2. Splash and Tide

The weight loss for aluminum alloys under tide exposure levels tends to be intermediate between the low atmospheric weight loss and the somewhat higher values observed under fully immersed conditions. However, a study by Southwell et al. (*176*) indicated that pit depths tend to be greater under tide exposure than those observed under either atmospheric or immersed exposure. Thus, one can interpret the results of this study to suggest that initiation of pits in the tide zone is somewhat less likely than under submerged conditions, but once pits initiate they propagate at higher rates in the tide zone.

3. Immersion

General Corrosion

Effect of Alloy Composition. Aluminum alloys usually have good resistance to general corrosion in seawater. In particular, Al-Mg (5000 series) and Al-Mg-Si (6000 series) are widely used for seawater applications. Exposure studies by Reinhart and Jenkins (*177*) in the Pacific Ocean have shown that the corrosion rates of these alloys, based on weight loss, were less than 3 mpy.

A study by Beberdick (*178*) comparing Al-Mg alloys to 7075-T6 in a 3.5% NaCl environment indicated that the binary 8% Mg and 10% Mg alloys are highly resistant to general corrosion. Alloying with Cu (0.4 weight percent), and Cu and Mn (0.4 and 0.5 weight percent), to reduce the tendency to intergranular corrosion, was found to result in an increase in the weight loss over the binary alloys, but the weight loss of the Cr and Mn containing alloys was less than that of 7075-T6 (see Figure 10) (*178*).

Zinc additions to aluminum alloys decrease the tendency for localized attack and promote general corrosion (*179*). Zinc is used as an alloy addition in aluminum alloy 7072, which finds applications as a sacrificial anode. A successful application of this alloy is cladding over Al 3003 or 3004.

Zamin (*180*) showed that Fe had a detrimental and Mn had a beneficial effect on general corrosion as well as pitting of Al containing 0.4% to 1.2% Mn, 0.4% to 0.8% Fe, 0.15% Si, and a small amount of Mg. The beneficial effect of Mn was explained in terms of it reducing the difference in electrochemical potential between the matrix and the intermetallics.

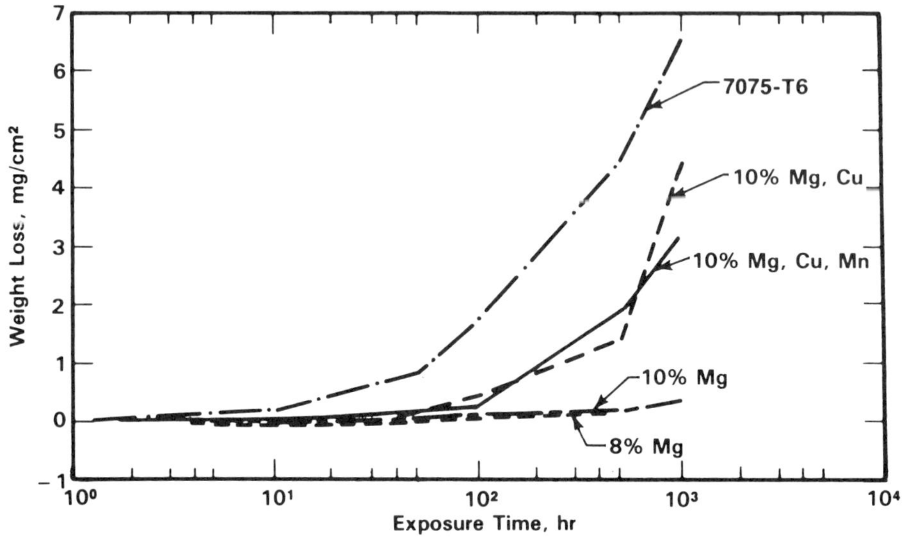

Fig. 10. Effect of Cu, Mn, and Mg alloy additions on the general corrosion rate of aluminum in 3.5% NaCl spray (*178*).

Micro additions of Ti, Zr, and Cr (0.3 weight percent) were found to increase the resistance of Al 6 Mg alloys (*181*) to general corrosion. The effect of Zr was found to be the most significant, which was attributed to a more even distribution of Mg_2Al_3 through the bulk of the alloy. Addition of 0.3 weight percent Ti or Cr increased the corrosion resistance of the alloy after quenching. However, addition of these elements promoted decomposition of the solid solution, which caused a more heterogeneous structure after aging at 150° to 200°C. This could have a detrimental effect on the corrosion resistance of the alloys.

Effect of Heat Treatment. Results of exposure studies by Reinhart of various Al alloys in the Pacific Ocean (*177*) indicated that the corrosion rates of 5000 series alloys, based on weight loss, were low (less than 1 mpy) and appeared to be independent of heat treatment. The highest rate was found for Al5052 in the annealed condition. Furthermore, no effect of standard heat treatments on general corrosion of 6000 and 2000 series alloys was detected (*177*).

Effect of Temperature, pH, and Aeration. Studies by Reinhart et al. (*79*, *176*) have shown that aluminum alloys tend to corrode at higher rates in the deep ocean than at the surface. The rates that were reported for 1000, 3000, and 5000 series alloys at 6000 feet reached a maximum of about three times that of the alloys exposed to surface water. It has generally been assumed that the low dissolved oxygen concentration was responsible for the high corrosion rates. However, Dexter (*182*) has argued that the decrease in corrosion rate could be explained in terms of a decrease in pH rather than a decrease in dissolved oxygen concentration. It can be concluded from these studies that, for reasons not fully understood, corrosion rates of aluminum alloys are higher at depth than near the surface.

Studies on the effect of temperature have mainly been concerned with temperature ranges associated with seawater depth effects, 5° to 25°C, and 25° to 125°C, which is associated with desalination. In these studies it is often difficult to separate effects of temperature and the effects of seawater composition associated with temperature effects such as oxygen solubility and fouling.

Larsen–Basse (*183*) studied alloys 3004, 5052, Alclad 3003, and Alclad 3004 and found a strong temperature effect on uniform and pitting corrosion. It was found that in warm water (24.5° to 28°C) under free fouling conditions, weight loss was much higher than in cold water (9.5° to 10.5°C). Microscopic examination of the test specimens showed that, in the warm water, the metal loss was uniform and no pitting was observed. In contrast, examination of the cold water specimens showed that all the specimens corroded by pitting.

Pitting and Crevice Corrosion

Effect of Alloy Composition. Alloy composition has a significant effect on the susceptibility to pitting and crevice corrosion of aluminum alloys. In early studies by Godard and Booth (*184*), it was observed that alloy 7075-T6 was very susceptible to pitting, whereas alloys 1100-H14 and 3003-H14 were somewhat more resistant after ten years' exposure at three marine sites. Additional studies (*79*, *185*) in various seawater environments confirmed this general behavior and further indicated that Alclad and the 5000 (Al-Mg) series alloys were the least susceptible to pitting and crevice corrosion.

Several investigators (*185–191*) had found significant effects of Cu, Mg, and Zn alloy addition on the pitting behavior of Al. Addition of Cu had a beneficial effect, whereas addition of Zn promoted pitting and crevice corrosion. Addition of Zn did not appear to significantly affect the pitting behavior. Further, Holtan and Sigurdsson (*192*) found that small amounts of Cr, Mn, and Sb added to an Al-Mg (3.5 weight percent) alloy did not greatly affect pitting performance.

Electrochemical tests on four aluminum-lithium alloys in 3.5% NaCl under several aging conditions performed by Niskanen et al. (*193*) indicated that the anodic Al-Li (delta) phase formed in the alloy deleteriously affected pitting behavior. Alloying with either Cu or Mg retarded delta formation by tying up Li in transition phases and improved pitting performance.

Effect of Heat Treatment. Although no systematic studies on the effect of heat treatment on pitting are reported, some studies were performed that indicated an effect of heat treatment. Groover et al. (*191*), who compared the pitting and crevice attack behavior of aluminum alloy 1100 in the T and H14 temper in surface water at Key West, Florida, found that pitting was relatively minor for both heat treatments but the H14 heat was found to provide better resistance to crevice corrosion.

Early studies by Galvele and coworkers (*190*, *194*) on the pitting behavior of high purity Al-Cu alloys indicated that heat treatment of these alloys affected the pitting potential because of the formation of copper-depleted zones in the aged alloys. Specifically, when an Al-4Cu alloy was aged to peak hardness, the pitting potential of the alloy in a 1 M NaCl solution decreased, indicating a higher susceptibility to pitting corrosion. Sugimoto et al. (*195*) confirmed this by showing different pitting potentials for the matrix and grain boundary region in Al-4% Cu alloys.

Effect of Depth and Aeration. There is considerable controversy in the literature concerning the relationship between pitting of aluminum and depth and aeration. Work on the subject by Rynewicz (*196*), Wheatfall (*103*), and early data by Reinhart (*177*) indicated that the severity of pitting increased with increasing depth. This effect was attributed to a decrease in oxygen with increasing depth. Typical data for 5000 series aluminum alloys are given in Figure 11 (*177*). On the other hand, results of electrochemical studies by Dexter (*182*, *198*) indicated that, when all other factors are held constant, a decrease in dissolved oxygen reduced the severity of pitting and crevice corrosion. Thus, Dexter (*198*) attributed the variation in corrosion behavior of aluminum alloys with depth to the variation in pH, temperature, and oxygen concentration. He observed that combinations of low temperature, low pH, and high dissolved oxygen were most corrosive toward aluminum alloys. Based on the relationship between oxygen, temperature, pH, and depth for the Pacific Ocean test site (see Fig. 2) (*4*)) Dexter predicted that the most corrosive condition will occur at intermediate depths, between 150 and 400 meters; while in the Atlantic Ocean, the most corrosive conditions may be found at greater depths due to the overall higher oxygen levels.

Results of more recent data by Reinhart and Jenkins (*23*) for a series of aluminum alloys at the Pacific Ocean test site near Port Hueneme, California are consistent with this hypothesis, at least for some of the alloys, see Figure 12 (*23*). Data for 2024, 6061-T6, and 2219-T81 exhibited maximum pitting at an intermediate depth of 250 meters, while 5083-H113 and 5086-H34

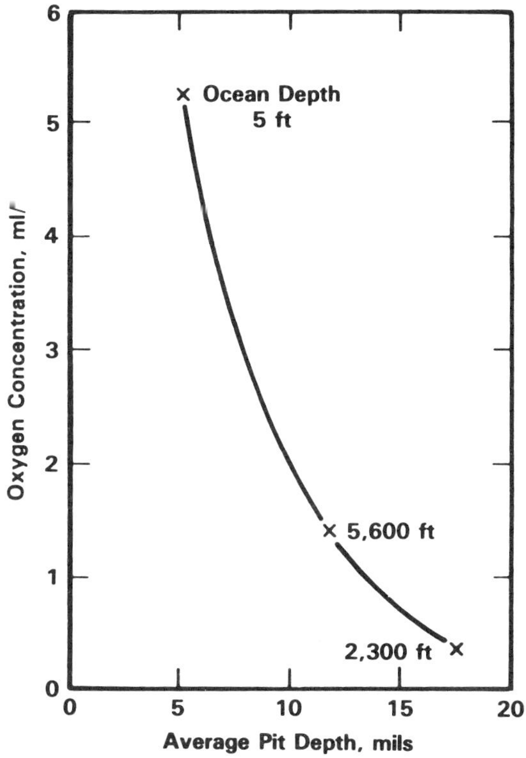

Fig. 11. Relation between average pit depths of aluminum-mangesium alloys (5000 series) and oxygen concentration of seawater (*177*).

exhibited a minimum in attack at this depth, and attack of 3003-H14 increased monotonically with increasing depth.

It can be concluded from all of the above studies taken together, that for reasons which are not fully understood, pitting of aluminum alloys is likely to be more severe at depth than at the surface and that the specific relationship between rates of pitting and depth is probably dependent upon the specific location and aluminum alloy being considered.

Effect of Temperature. Studies of pitting of aluminum alloys have generally been concerned with two temperature ranges: 5° to 25°C, which is associated with seawater depth effects, and 25° to 125°C, which is associated with desalination. Dexter (*182*, *198*) found that a decrease in seawater temperature promoted a noble shift in the corrosion potential, while not greatly influencing the pitting potential; thus, pitting is more likely at lower temperatures. Results of the deep ocean studies by Reinhart (*23*, *177*) are

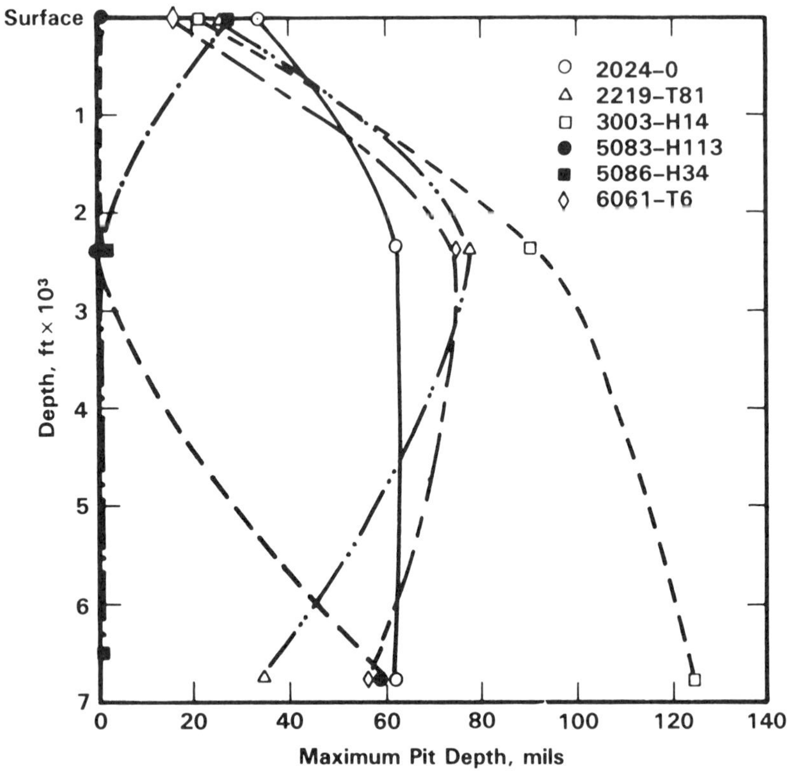

Fig. 12. Maximum depths of pits of aluminum alloys versus depth after 1 year of exposure in the Pacific Ocean near Port Hueneme (*23*).

essentially in agreement with Dexter's results, although the authors differ in the interpretation of the data (see section on effect of depth). Reinhart et al. (*23*, *177*) observed more severe pitting of most aluminum alloys at depths where temperatures were colder rather than near the surface waters.

Ahmed (*199*) and Alier (*200*) studied the effect of temperature on pitting of aluminum alloys for desalination applications, and both found that the propensity for pitting decreased with increasing temperatures. Ahmed found that the pit density of 5052 aluminum decreased dramatically with increasing temperature from 82° to 108°C in flowing seawater, while the general corrosion rate of this alloy increased with increasing temperature.

Effect of Time. Fundamental considerations of pit propagation predict that pitting will follow a power law,

$$d \propto t^b$$

where d is depth of pitting, t is time, and b is an exponent, that is usually less than one. For example, $b = \frac{1}{3}$ for a hemispherical pit (*201*). Aluminum alloys frequently follow power law behavior with exponents less than one, which is desirable since this indicates that rates of pitting decrease with time. Unfortunately, some long-term exposure data for aluminum alloys in seawater suggest that the exponent is greater than one. Data by Reinhart and Jenkins (*23*) for several aluminum alloys indicates that only Alclad 3003-H12 exhibited a decrease in the rate of pitting with time. The authors attributed the good performance of the latter alloy to the sacrificial nature of the Alclad, which promoted lateral attack, rather than deep pitting.

Effect of Pollution. Little work has been done on the effect of seawater pollution on pitting and crevice corrosion of aluminum alloys. However, it is well know that various metallic ions in seawater such as copper, nickel, or lead can have a detrimental effect on the pitting resistance. In particular, Herrigel (*201*) described the severe pitting of aluminum in a desalination plant as a result of the presence of copper in the feedwater. Raising the pH of the feedwater to about 7.5 and increasing the oxygen concentration to about 50 μg/l eliminated the pitting attack.

Erosion-Corrosion

Aluminum alloys are generally resistant to erosion-corrosion in low velocity seawater, below velocities of 20 fps (*202–204*). However, at higher seawater velocities, considerable weight loss is experienced.

Studies by Gehring et al. (*205*) on 5000 series alloys indicated that the rate of corrosion of specific aluminum alloys was strongly dependent on seawater velocity. It was concluded from the studies that the basic mechanism for high velocity corrosion of a 5000 series alloy (5456-H117) involves film breakdown, leading to enhanced micropitting near intermetallics.

Cavitation damage to aluminum can occur in the presence of turbulent liquid. During turbulence, voids form and collapse, releasing bursts of energy. If this occurs at the metal surface, the protective oxide may break which results in a pit. Pitting can be sufficiently dense to cause roughening of the surface. Although cavitation damage of aluminum has been reported, little quantitative work has been performed on cavitation of aluminum in actual or simulated seawater.

Galvanic Corrosion

Marine structures are usually designed with the strength requirement as the primary parameter for selection of materials of construction. The use of more than one type of metal in a structure may be necessary to obtain the desired mechanical properties. Accelerated corrosion of aluminum alloys commonly occurs when these alloys are electrically coupled to other alloys such as stainless steel, titanium, and Monel. Usually, uncoupled materials are ranked in a galvanic series according to their corrosion potential in a specific environment, e.g., seawater, see Figure 1 (*3*). Although this type of classification could be useful as an indication of general trends in galvanic corrosion, it cannot provide information on the extent of galvanic corrosion when dissimilar materials are electrically coupled. As pointed out by Mansfeld et al. (*206*) the extent of galvanic corrosion does not only depend on potential difference between the dissimilar materials but also on kinetic parameters such as corrosion rates of the uncoupled materials and Tafel slopes as well as on area ratios.

Mansfeld (*206*) measured galvanic current densities of aluminum-dissimilar metal couples in an aerated 3.5% NaCl solution and found that they were in good agreement with the increase in dissolution rates due to galvanic coupling. Mansfeld ranked various alloys in three classes based on their compatibility with aluminum, see Table XIV.

Seawater exposure tests of galvanic couples of aluminum alloys 6061-T6 and 7075-T6 with various alloys were performed in a seawater environment at the U.S. Army Tropical Testing Station, Panama Canal Zone (*176*, *207*). These studies indicated that coupling of the aluminum alloys with either active or noble metals tended to result in galvanic attack of the aluminum.

The accelerating effect of the cathode-anode area ratio is illustrated in Table XV (*208*). If the anodic area (aluminum) is small compared to the cathodic area, severe corrosion of the aluminum may result. The area ratio effects should be considered when galvanic corrosion is controlled by applying protective coatings. A coating should never be applied to the anode only. In that case, if the coating is damaged, the galvanic current will concentrate, resulting in localized high corrosion rates. The cathodic surfaces should be coated and the anodic surface should be left bare.

Stress Corrosion Cracking

Many of the high strength aluminum alloys can, under certain conditions, fail in marine environments at stresses far below the yield strength due to intergranular SCC. The susceptibility of aluminum alloys to SCC greatly depends on alloy composition, microstructure, orientation, and heat treatment as well as on environmental variables such as pH and temperature. The effect of each of these variables on the SCC behavior of aluminum alloys is discussed in the following sections. The information on SCC is based on

TABLE XIV

COMPATIBILITY OF AL ALLOYS AND DISSIMILAR METALS (*206*)

Class	Al 1100	Al 2024	Al 2219	Al 6061	Al 7075
Compatible	Cd	Cd, 7075, 1100, Sn, Haynes, 2219, Ti-6-4, 6061, Zn, Inco 718, PH13-8Mo, SS347, A286, SS304, Ni, SS301	Cd, 2024, 6061, 1100, Haynes, 7075, Zn, PH13-8Mo, SS301, SS304, A286, Inco 718, Ti-6-4, SS347	7075, 2219 2024	1100, 6061, 2024, Zn 2219
Borderline	2024, 6061, 2219	4130, Cu, Ag	Ni, Zn, 4130, Cu, Ag	1100, Cd, Sn, Inco 718, Zn, Haynes, Ti-6-4	Cd, Haynes, Sn, Ti-6-4, Inco 718
Incompatible	Sn, 7075, Haynes, Ti-6-4, Zn, Inco 718, A286, SS347, Ni, SS304, PH13-8Mo, SS301, 4130, Cu, Ag			SS304, SS301, A286, PH13-8Mo, SS347, 4130, Ni, Cu, Ag	SS304, SS347, PH13-8Mo, SS301, A286, Ni, 4130, Cu, Ag

TABLE XV

CALCULATED CORROSION RATE OF ALUMINUM IN AERATED STATIC SEAWATER WHEN GALVANICALLY COUPLED AT THREE AREA RATIOS TO OTHER METALS (*208*)

	Corrosion Rate, mm/yr M:Al[b]		
M[a]	10:1	1:1	1:10
Carbon	39	2 to 10	0.4
Titanium	—	—	—
Monel 400	2.2	0.12	—
EN58J	—	—	—
Copper	4.5	0.66	0.05
Silicon bronze	2.0	0.6	—
NiAl bronze	2.4	0.15	0.09
10 percent Al bronze	2.4	0.6	—
CN30	1.7	0.1	—
LG4 gunmetal	—	—	—
Lead	0.5	0.15	—
Tin	2.2	0.14	—
HY80	1.6	0.3	0.2
Ni resist	0.7	0.06	—
Mild steel	1.7	0.07	0.06
H30 aluminum	0.09	0.01	0.001

[a] Cathodic member of couple.
[b] Anodic member of couple.

service experience, marine exposure of experimental panels, and laboratory testing in actual or simulated seawater.

Effect of Alloy Composition. Metallurgical variables such as alloy composition and heat treatment have a strong effect on the SCC resistance of aluminum alloys. The major alloy systems for commercial application that have suffered SCC include Al-Cu, Al-Cu-Mg, Al-Mg, Al-Mg-Zn, Al-Zn, Al-Zn-Mg-Cu, and Al-Mg-Si.

Speidel (*209*) lists some general rules for SCC of these alloys in aqueous environments. Pure aluminum is generally considered not susceptible to SCC and, for any given alloy system, the susceptibility to SCC tends to increase with increasing concentration of alloying elements that can be put in supersaturated solid solution. SCC resistance of ternary or higher order systems is not only influenced by the sum but also by the ratios of the alloying elements. Minor alloy additions of Cr, Mn, Zr, Ti, V, Ni, and Li to the alloys mentioned above can increase the resistance to SCC.

5000 Series. Aluminum-Mg (5000 series) alloys containing more than 5 weight percent Mg are susceptible to SCC in seawater environments. At these concentrations, Mg preferentially precipitates along the grain boundaries in the form of a continuous film of Mg_2Al_3 phase. This phase is anodic with respect to the matrix and preferential dissolution of this phase is generally thought to be the mechanism of SCC in this alloy (*210*). However, if the Mg content is kept at values less than 4 weight percent, no continuous precipitate can form along the grain boundaries and the alloy becomes resistant to SCC (*211*). Various Mg containing commercial Al alloys, 5000 series alloys, have less than 4% Mg and are resistant to SCC.

In order to improve the resistance of Al alloys containing more than 5% Mg to SCC, the effects of additional alloying elements have been investigated. These elements are Bi, (*212*) B, and Be (*213*), and Cu and Zr (*214*). Baba et al. (*212*) found that additions of 0.2 to 0.4 weight percent Bi greatly improved the resistance to SCC of Al-8% Mg in a 3.5% NaCl solution at room temperature. Addition of Bi did not change tensile properties, grain size, or aging characteristics. The effect of Bi was attributed to the formation of the insoluble compound Bi_2Mg_3, which may act as a sacrificial anode with respect to the base metal. Chromium and Zr were also shown to have a beneficial effect.

The trace elements B and Be were found to have a beneficial effect on the SCC resistance of Al-Mg alloys containing 8% Mg (*213*). Addition of 0.02% B and 0.02% Be were found to improve SCC resistance by promoting general precipitation inside the grains rather than at the grain boundaries. A 5000 series alloy was developed containing 6% to 8% Mg, 0.001% to 0.02% Be, and 0.001% to 0.05% B; this alloy was designated 5090 and showed excellent resistance to SCC and exfoliation corrosion in marine environments (*215*).

6000 Series. Of the Al-Mg-Si (6000 series) alloys, 6061 is the most common alloy and, in a fully aged condition (T6 temper), the precipitate is present as small discrete particles and the alloy is resistant to SCC.

2000 Series. Aluminum-Cu and Al-Cu-Mg (2000 series) alloys are heat-treatable, high strength, aluminum alloys. Aluminum-Cu alloys derive their strength from precipitation hardening. Addition of Cu also increases the quench sensitivity of the alloy (*216*).

Addition of trace amounts of Ti (0.2%) to a Al-5% Cu alloy was found to improve the resistance to SCC in a 1 M NaCl solution (*217*, *218*). The improved SCC behavior was attributed to a reduction in grain size which promoted general precipitation of an intermetallic phase.

The addition of Mg to Al-Cu alloys serves to improve the mechanical properties of the alloy. Commercial Al-Cu-Mg alloys are designated 2000 series alloys. The SCC resistance of common 2000 series alloys is summarized in Figure 13, which shows the stress corrosion crack velocity as a function of stress intensity factor K_I during alternate immersion in 3.5% NaCl (*219*). Three groups of alloys can be distinguished in the diagram. The alloys most

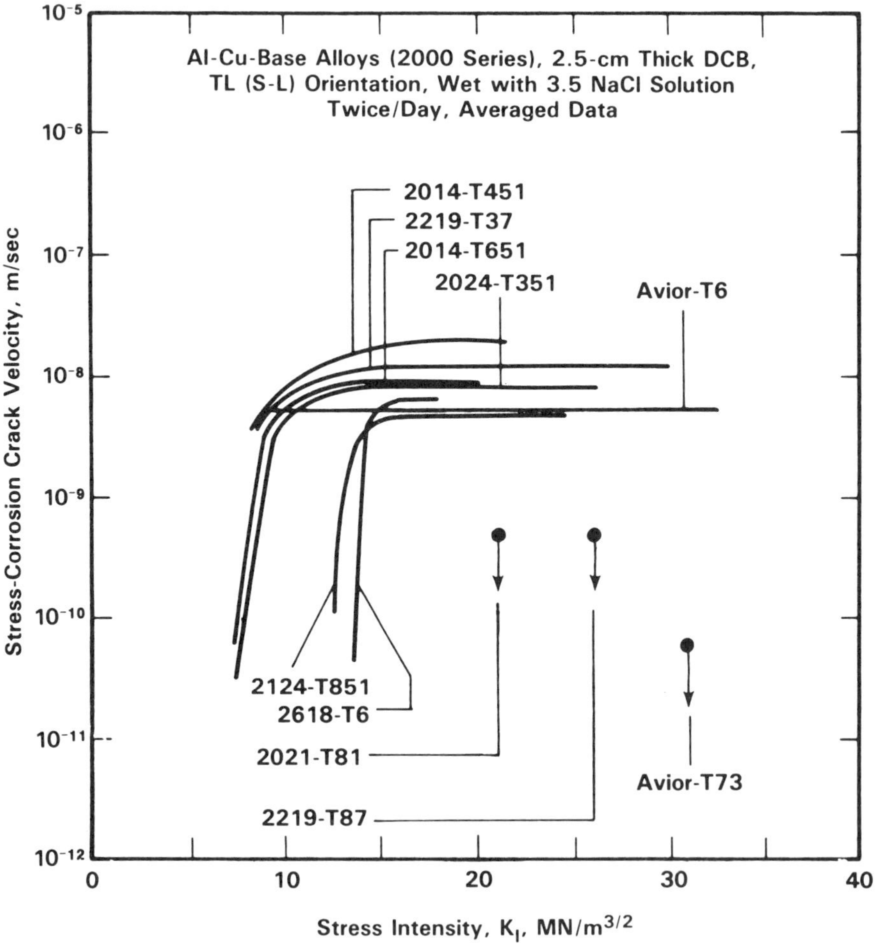

Fig. 13. Summary of stress corrosion crack growth rates in various aluminum alloys based on the aluminum copper system (*219*).

susceptible to SCC are 2014-T451, 2219-T37, 2014-T651, and 2024-T351. These alloys formed cracks at very low levels of K_I, below 10 MN m$^{-3/2}$.

The second group of alloys consists of those alloys which are susceptible to SCC but cracked at much higher stress intensity levels (between 13 and 15 MN m$^{-3/2}$). These alloys include alloys 2124-T851 and 2618-T6 (RR58). The third group of alloys in the 2000 series are those which are highly resistant to stress corrosion cracking. These alloys include 2021-T81 and 2219-T87.

The addition of trace amounts of Ag appears to have a beneficial effect on the SCC resistance of 2000 series alloys. As indicated in Figure 13, the high strength cast Ag-containing aluminum alloy developed in France principally for aerospace application, Avior-T73, is very resistant to SCC, as measured by passing an alternate immersion test for 30 days at 80% of the yield strength. It should be noted that, in the T6 condition, Avior belongs to the most susceptible alloys, whereas, in the T73 condition, the alloy has excellent resistance.

7000 Series. The highest strength aluminum alloys (7000 series) are based on Zn and Mg alloying additions. Exposure studies by Reinhart (*177*) in the Pacific Ocean near Port Hueneme showed several 7000 series alloys to be susceptible to SCC. Experimental studies by Miller and Scott performed in synthetic seawater (*220*) showed that the susceptibility to SCC increased with increasing Mg content (2.5 to 3.0 weight percent) but was little affected by changing the Zn level (4.0 to 5.0 weight percent). Hence, the best combination of strength and SCC resistance may be achieved by maintaining a high Zn content and keeping the Mg level to a minimum. The SCC resistance of several 7000 series alloys is summarized in Figure 14 (*209*). Addition of minor alloying elements can also have a significant effect on the SCC resistance of Al-Mg-Zn alloys. The resistance to SCC of Al-Zn-Mg alloys in a 3.5% NaCl aqueous solution was improved significantly with the addition of small amounts of Ti. Chen and Judd (*221*, *222*) found that a significant improvement in the SCC resistance was obtained for the maximum strength condition in a Al-Zn-Mg alloy with a 0.041 weight percent Ti addition when compared to alloys without Ti. Increasing the Ti concentration to higher than 0.2 weight percent considerably lowered the SCC resistance. Chen and Judd (*221*) hypothesized that Ti acts as a grain refiner and retards the precipitation kinetics at the grain boundaries.

Some controversy exists on the effect of Ag on the SCC resistance of 7000 series alloys. On the one hand, Ag was shown to have a beneficial effect on the SCC resistance (*223–225*). As a result of these findings one 7075 type alloy containing Ag is available commercially, i.e., the German Alloy AZ74.

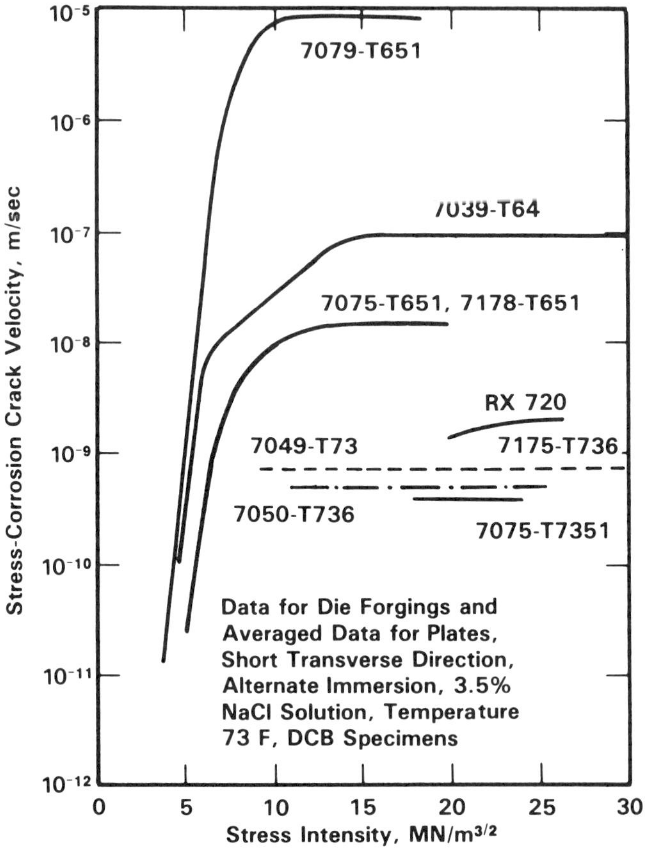

Fig. 14. Comparison of stress corrosion crack growth rates of Al-Zn-Mg-Cu (7000 series) alloys in 3.5% NaCl solution (*209*).

However, other workers investigating Ag additions have found no significant increase in SCC resistance due to the presence of this element (*226*, *227*). The confusion regarding the effect of Ag was eliminated when it was pointed out that much of the earlier work was carried out at high temperature aging conditions using thin products that could be heated up rapidly. It was shown (*227*) that, under these conditions, Ag containing alloys attained higher strength and resistance to SCC. Virtually the same effect could be achieved in Ag free alloys by using appropriate heat treatments (*227*).

Chromium is added to Al-Zn-Mg alloys to suppress recrystallization. Commercial alloys containing Cr are 7049, 7075, 7175, 7178, and RX720. Chromium does not appear to have an effect on K_{ISCC} of 7000 series alloys but some effect is noted on the stress corrosion crack velocities as indicated in

Figure 14 (*209*). Addition of Cr to 7000 series alloys increases the quench sensitivity of the alloys. To reduce the quench sensitivity Cr has been replaced by Mn or Zn in alloys which contain a Cu concentration greater than 2 weight percent. The effects of Mn and Zr on recrystallization and grain growth resemble those of Cr, but substantially more Mn than Cr is required to achieve the same results.

Aluminum alloys containing Li offer attractive characteristics. Lithium is the only element which, when added to Al, simultaneously increases strength and elastic modulus, and decreases the density. However, a major drawback of Al-Li alloys is low fracture toughness. Various trace elements such as Mn, Mg, Cr, Zr, and Cu have been added in an attempt to improve the toughness (*193*). A study by Pizzo (*228*) on Al-Cr-Cu, Al-Li-Cu, and Al-Li-Cu-Mg indicated that the Al-Li alloys were susceptible to SCC as well.

Effect of Heat Treatment. Heat treatment has a significant effect on the SCC resistance of aluminum alloys in seawater environments. In the following sections the effect of heat treatment on different aluminum alloy groups are discussed: Al-Mg (5000 series alloys), Al-Mg-Si (6000 series alloys), Al-Cu (2000 series alloys), and Al-Zn and Al-Zn-Mg (7000 series alloys).

5000 Series. The SCC behavior of Al-Mg (5000 series) alloys depends on the precipitation state of the alloy. In particular, the precipitation of Mg_2Al_3 along the grain boundaries strongly affects the susceptibility. The grain boundary precipitation reaction is sluggish and is accelerated by deformation, by exposure to elevated temperatures, and by higher Mg content. Following solution treatment, various aging treatments can be performed. The SCC susceptibility is little affected by different solution treatment temperatures (*229*). However, aging time and temperature have a strong effect on the susceptibility to SCC.

A study of the effect of nine different tempers of Alloy 5456 with and without sensitization (to simulate welding) on the SCC resistance in seawater near Wrightsville Beach, North Carolina was performed by Neiderberger et al. (*230*). Failures occurred with the H12, H16, H32, and H36 tempers, whereas no failures were detected in sensitized or unsensitized samples in the O, H22, H24, and H321 tempers. However, in studies by Reinhart (*177*) alloy 5456-H32, as well as a series of other magnesium-bearing alloys, showed no signs of stress corrosion failure after exposure for about 400 days in the deep ocean. Although no explanation can be offered for this discrepancy it may be argued that, in the studies by Reinhart, the stress was below the threshold for SCC.

6000 Series. The commercial Al-Mg-Si alloys (6000 series) which are strengthened by precipitation hardening, are generally highly resistant to SCC (*226*). Alloy 6061 is one of the commercial 6000 series alloys that can be made susceptible to SCC in the T4 heat treatment condition if a high heat treating temperature is followed by a slow quench (*226*). In the fully aged condition, T6 temper, the precipitates are present as small discrete particles and the alloy is highly resistant to SCC.

2000 Series. Al-Cu and Al-Cu-Mg alloys derive their strength from precipitation hardening, which is achieved by solution heat treatment, followed by rapid cooling and either natural aging at room temperature (T4 temper) or artificial aging at an intermediate temperature (T6 temper). Cold working after quenching further increases the strength, resulting in the T3 temper and, when artificially aged, the T8 temper.

Al-Cu-Mg alloys such as 2024, 2014, and 2219 in the naturally aged T3 and T4 tempers can be highly susceptible to SCC in seawater (*231*). The susceptibility of Al-Cu and Al-Cu-Mg alloys to SCC is greatly affected by heat treatment. In particular, the susceptibility to SCC decreases when the rate of quenching following solution treatment and aging increases. However, very fast quench rates may introduce stresses in relatively complex forgings. Therefore, controlled quenching with polyalkylene glycol has been used when quench distortion needs to be avoided or low residual stress combined with resistance to SCC is required (*232*). During aging following queching, the susceptibility to SCC increases with aging time initially (the rate of increase depending on aging temperature), reaches a maximum, and then decreases (*195*).

7000 Series. The resistance of Al-Zn-Mg and Al-Zn-Mg-Cu (7000 series) alloys to SCC in seawater strongly depends on heat treat parameters such as solution treatment temperature, quench rate, and aging temperature and time. In particular, the 7000 series alloys are highly susceptible to SCC when they are aged to peak strength (T6 temper).

As in the case of the 2000 series alloys, the resistance to SCC can be improved by increasing the quench rate. Rapid quenching (greater than 200 F/sec) was found to improve the SCC resistance of the Cu containing 7000 series alloy 7075-T6 in the long transverse direction (*226*). However, the beneficial effects of these high quenching rates on the resistance to SCC of 7075-T6 plate is nonexistant for plates and forgings where stressing in the short transverse direction is possible.

The resistance of Al-Zn-Mg alloys to SCC can be improved by aging at relatively high temperatures (325° to 350°F). A drawback of the aging treatment is that the increase in resistance to SCC is achieved at the expense of the strength of the alloys.

A number of commercial overaging heat treatments have been developed for several high strength 7000 series aluminum alloys (*211, 227, 233–235*). The commercially available tempers for these alloys are T76 and T73. Overaging has no effect on the cracking velocities of the low Cu alloys 7019, 7039, and 7007, but has a strong effect on K_{ISCC} of these alloys after long aging times (*209*). Since these long aging times also result in a considerable decrease in tensile yield strength, overaging of these alloys is consistently impractical. On the other hand, in the Cu-rich alloys 7075 and 7178, a considerable reduction in crack velocity to be achieved after 15 hours at a temperature of 160°C, which results in a T76 temper. Moreover, the decrease in strength as a result of overaging to the T76 temper is minimal. The overaging from T6 to T73 results in a moderate reduction in tensile yield strength of 10% to 15% (*227*) and a considerable improvement in fracture toughness (*235*). Recently, alloys have been developed that combine the strength of the 7000 series T6 temper and the SCC resistance of the T73 temper (*209, 236–239*). These alloys are 7049-T73, 7175-T736, 7075-T736, and 7010-T736.

An alternative method to achieve a combination of high strength and SCC resistance includes thermomechanical treatments. Reimann and Brisbane (*240*) showed that a combination of heat treatment and forming could improve the SCC resistance and increase the strength compared to the T6 temper.

Geometric Effects. Since SCC in aluminum alloys is almost always intergranular, the resistance of these alloys to SCC is strongly related to the local grain shape and orientation with respect to the applied stress. The grain orientations in standard wrought forms are indicated schematically in Figure 15 (*241*). Because of the importance of relating the grain orientation to the direction of the applied load, two index systems have been devised. In one system, used in the absence of cracking, the three stressing directions are denoted by longitudinal (L), long transverse (LT), transverse (T), or short transverse (ST) as shown in Figure 15. The second system, which considers the presence of cracks, specifies both the cracking plane and the direction of crack propagation. It uses (L, T, W) to indicate the three perpendicular directions, L for the longitudinal direction, T for the thickness direction, and W for the width direction. The crack propagation is identified by the direction normal to the crack. The crack propagation direction is identified by one of the directions, L, T, or W, see Figure 16 (*241*).

Aluminum alloys are most susceptible when the stress is applied in the ST direction, i.e., perpendicular to the grain flow. During forging, the grains are flattened in the parting plane, and cracking may occur along that plane.

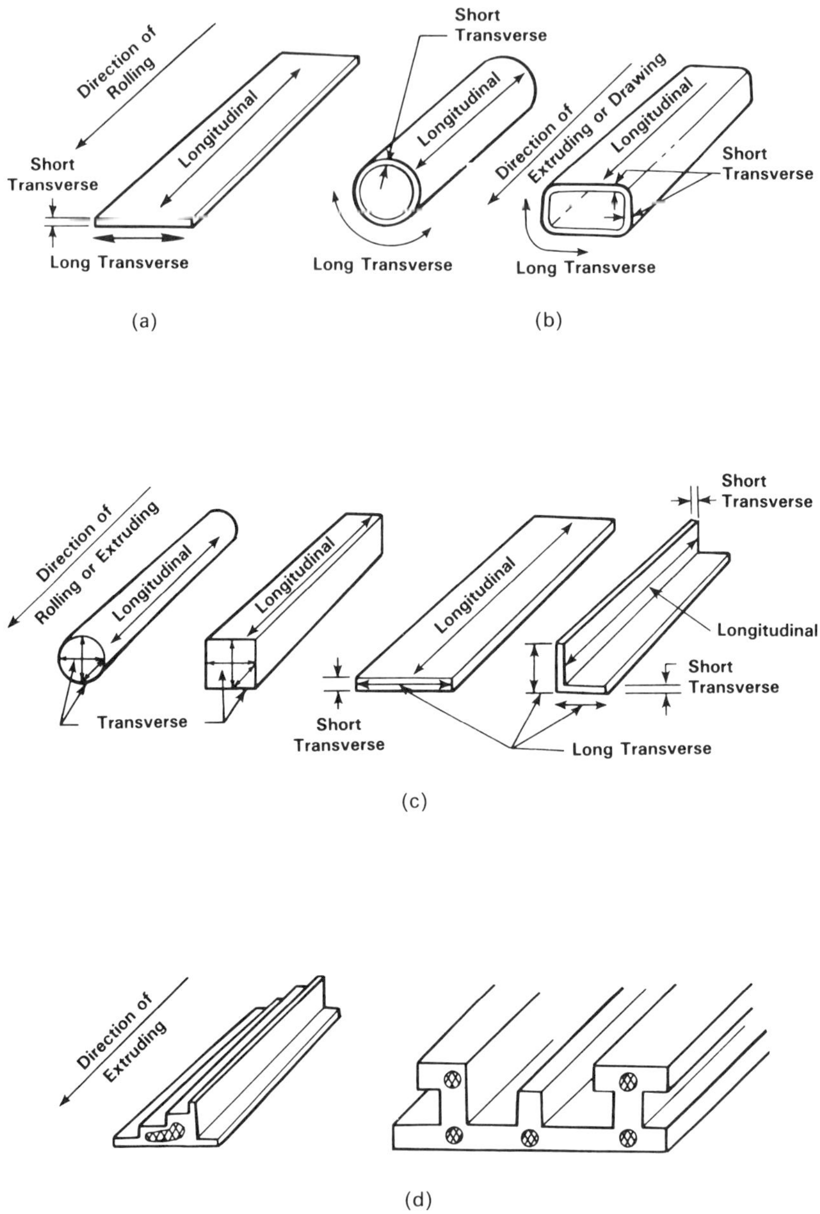

Fig. 15. Grain orientations in standard wrought forms (*241*): (a) Sheet and plate, (b) extruded and drawn tube, (c) rolled and extruded rod bar and thin shapes (hand forgings generally the same), and (d) extruded thick and complex shapes (crosshatched areas are transverse; other areas same as (c)).

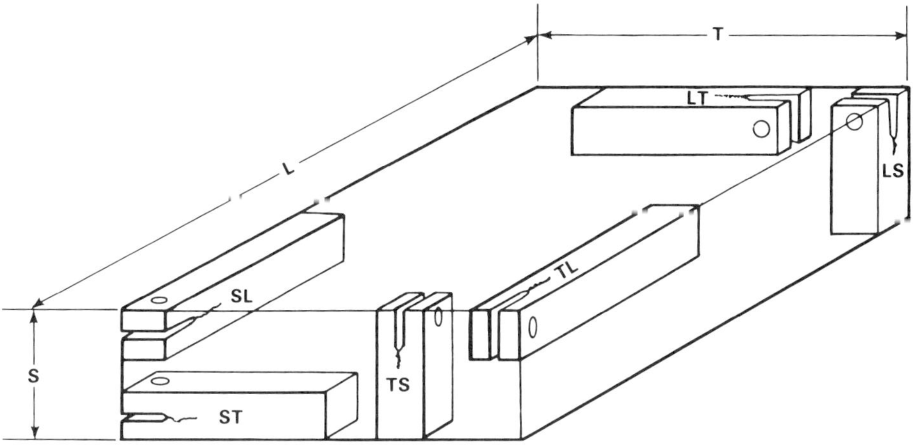

Fig. 16. Some possible double-cantilever-beam (DCB) specimen orientations in plate material (*24*): T = long transverse, L = longitudinal (rolling or extrusion) direction, S = thickness, first letter = normal to the fracture plane, second letter = direction of propagation in fracture plane.

Hyatt (*241*, *242*) showed that residual stresses in susceptible materials can be sufficient to initiate cracks and can also play an important role in accelerating crack growth rates in artificial seawater. It was further shown that residual stresses in precracked specimens could easily be eliminated by plastically deforming the specimens after quenching.

Surface deformation of aluminum structural components also has a significant effect on stress corrosion cracking initiation. Ratke (*243*) indicated that superficial plastic deformation such as that produced by milling or drilling could increase the susceptibility to SCC and should be avoided. However, shot peening can greatly increase the SCC resistance (*243*, *244*). In addition to introducing compressive stress on the surface, shot peening results in smearing of the surface, which greatly reduces the susceptibility to SCC. Further, the velocity of existing stress-corrosion cracks can be reduced significantly by applying stress overloads, which introduce crack tip residual compressive stresses (*245*).

Effect of Temperature. Although no field data are available on the effect of temperature on SCC, data obtained in simulated seawater have indicated that temperature has a significant effect on SCC resistance of aluminum alloys. Several investigators (*209*, *246*, *247*) showed that an increase of test temperature of a 3.5% NaCl solution decreased the time to failure of the aluminum alloy 7039-T64.

Intergranular Corrosion

Intergranular corrosion is selective attack of the grain boundary zone along the grain boundary with no appreciable attack of the matrix. The mechanism of attack is electrochemical and is the result of local cell action in the grain boundaries. Localized electrochemical cells are formed between precipitates along the grain boundary and the depleted aluminum solid solution from which the precipitates formed. In some alloys, such as the Al-Mg and Al-Zn-Mg-Cu alloys, the precipitates Mg_2Al_3, $MgZn_2$, and Al_x-Mg are more anodic than the adjacent matrix, whereas in other alloys such as Al-Cu, the precipitates $CuAl_2$ and Al_xCu_xMg are more cathodic than the adjacent matrix.

Exfoliation corrosion or layer corrosion is a form of intergranular attack that proceeds along multiple narrow paths parallel to the surface of the metal. Exfoliation, which occurs predominantly in relatively thin products with highly worked elongated grain structures, is characterized by leafing of relatively uncorroded metal and thicker layers of corrosion products. The layers of corrosion products cause the metal to appear to swell.

As in the case of intergranular stress corrosion cracking, intergranular corrosion is strongly dependent on alloy composition and heat treatment. Alloys that do not form second-phase precipitates or those in which the precipitates have corrosion potentials similar to the matrix, such as $MnAl_6$, are not susceptible to intergranular corrosion. Examples of alloys of this type are 1100, 3003, and 3004. A discussion of the intergranular corrosion of susceptible aluminum alloys is given below.

5000 Series Alloys. Aluminum-magnesium alloys containing less than 3 weight percent magnesium are resistant to intergranular corrosion in sea-water environments. At higher Mg concentration, the 5000 series alloys may become susceptible to intergranular attack depending upon the precipitation of Mg_2Al_3 along the grain boundary. The commercial 5000 series alloys (e.g., 5083, 5086, 5454, and 5456) generally show acceptable resistance to inter-granular corrosion (*23*, *177*). A study by Reinhart (*177*) where various aluminum alloys were exposed in the Pacific Ocean near Port Hueneme, California, showed only intergranular attack in the heat-affected zone adjacent to welds in 5454-H32 and 5083-H113.

A method to achieve resistance of Al-Mg alloys containing 4 to 9 weight percent Mg is a combination of heat treatment and cold working or thermomechanical treatment (*178*).

Exfoliation has been observed in some highly worked 5000 series materials such as 5456-H321 boat hull plates (*171*). These plates developed a highly elongated grain structure and selective grain boundary precipitation. Thus, special boat hull plate tempers H116 and H117 were developed for these alloys which have high resistance to exfoliation corrosion.

6000 Series Alloys. 6000 series alloys (Al-Mg-Si) are usually resistant to intergranular corrosion (*248*). However, when the alloy contains an excessive amount of Si, more than needed to form Mg_2Si, susceptibility to intergranular corrosion increases because of the strong cathodic nature of the insoluble Si constituent.

2000 Series Alloys. In 2000 series (Al-Cu) alloys, thermal treatments that cause selective grain boundary precipitation can lead to susceptibility to intergranular corrosion. Factors that affect the susceptibility are quenching rate, aging, and plastic deformation. It has been shown that rapid quenching during heat treatment, and subsequent aging to peak or slightly averaged strength results in high resistance to intergranular corrosion (*190*). On the other hand, slow cooling from solution treatment temperatures results in susceptibility to intergranular corrosion. As-quenched or naturally-aged tempers (T3, T4) may be susceptible to SCC, but only slightly susceptible to intergranular corrosion (*249*). However, when artificially aged (tempers T6, T8), the same alloy may show a substantial increase in the resistance to both SCC and intergranular corrosion. Examples of 2000 series alloys in T6 or T8 tempers that have been successfully used without exfoliation problems are 2024, 2048, 2124, and 2219 (*171*). A relatively new alloy in this series is 2048-T851, which was primarily developed to provide high toughness (*171*).

7000 Series Alloys. Intergranular corrosion and exfoliation corrosion in Al-Zn-Mg alloys can also be affected by thermal treatments. As in the case of 2000 series alloys, rapid quenching from the solution, heat treatment temperature, and overaging will produce alloys that are resistant to intergranular corrosion. Addition of Cu generally increases the susceptibility to intergranular corrosion, which does not arise from preferential dissolution of grain boundary precipitates but from attack of Cu-depleted zones (*250*). Of the commercial 7000 series alloys (Al-Zn-Mg-Cu), alloys 7075-T73, 7075-T63, and 7178-T76 were found to provide very high resistance to intergranular corrosion (*171*, *251*). The T76 temper designation is that which denotes the highest strength obtainable with high resistance to exfoliation corrosion. Also, recently developed alloys 7050 and 7010 in the T73 temper (*236*, *251*) are resistant to exfoliation.

4. Mud

Little work has been performed on the corrosion performance of aluminum alloys in mud. Results of experiments conducted by Reinhart et al.

(*177*) in the Pacific Ocean off Port Hueneme have indicated that pit depths of aluminum alloys exposed in mud were generally greater than those of the same alloys exposed under immersed condition at a similar depth. The poorer performance of the aluminum alloys under mud exposure is not surprising; the presence of deposits and deaeration tends to promote local breakdown of passive films on the aluminum alloys. Since the composition of the ocean sediments varies, caution should be exercised in applying the data to other geological locations.

E. Titanium and Titanium Alloys

Titanium and its alloys are among the most corrosion resistant metallic materials to seawater. Although there are limited data available on atmospheric, splash and tide, and mud exposure, the data indicate that resistance to corrosion extends to these environments as well.

However, under specific conditions, titanium and its alloys may be susceptible to crevice corrosion and stress-corrosion cracking. Alloy modifications and heat treatment can effectively increase the resistance of titanium to these types of failures. Discussions of these corrosion modes are given below.

1. Pitting and Crevice Corrosion

Effect of Alloy Composition

Titanium is highly resistant to pitting and crevice corrosion in seawater. Immersion tests with titanium and its alloys have never shown pitting or crevice corrosion in ambient seawater (*103*, *252–254*). However, in seawater and brines at temperatures above 70°C (158°F), crevice corrosion of unalloyed titanium has been observed (*255*, *256*).

Griess (*256*) studied the crevice corrosion of titanium and its alloys in chloride solutions at elevated temperatures (above 70°C) and found that the localized attack was associated with the development of low pH conditions. Palladium (Pd), Mo, and Ni alloy additions were found to be beneficial in inhibiting this crevice attack. Commercial alloys Ti grade 7 (Ti-Pd) and Ti grade 12 (Ti-Ni-Mo) have been developed to make use of the beneficial effects of palladium, molybdenum, and nickel alloying elements in improving pitting and crevice corrosion. The Ti-Pd alloy contains up to 0.25% palladium, whereas titanium grade 12, its low-cost alternative, contains 0.8% nickel and 0.3% molybdenum.

Effect of Depth and Aeration

Depth or aeration have not been found to influence the pitting or crevice corrosion of titanium at ambient temperature. Reinhart and Jenkins (*22*) exposed several titanium alloys to different depths, which corresponded to oxygen concentrations of 0.4, 1.35, and 5.75 ml/l. The titanium alloys were not attacked by seawater under any of these conditions after one year of exposure. The alloys in this investigations were TiCP, Ti75A, Ti-0.15Pd, Ti-5Al-2.5Sn, Ti-6Al-4V, Ti-7Al-2Cb-1Ta, and Ti-13V-11Cr-3Al.

Effect of Temperature and pH

As discussed above, pitting or crevice corrosion of titanium have not been observed at temperatures below 70°C (158°F) (*256*). The boundary conditions for crevice corrosion for different titanium alloys as a function of temperature and pH were obtained by immersion testing in saturated sodium chloride brines and are shown in Figure 17 (*257, 258*). The diagram indicates that, at neutral pH, no crevice corrosion occurs in commercial pure titanium below approximately 120°C (250°F). Addition of nickel, molybdenum (grade 12) or palladium (grade 7) further extends the resistance to titanium to crevice corrosion to higher temperatures and lower pH.

Effect of Biofouling

Titanium is not toxic to marine organisms and, therefore, biofouling can easily occur on titanium surfaces immersed in seawater. Cotton and Downing reported extensive biofouling on titanium after 800 hours' immersion in shallow seawater (*254*).

However, the integrity of the corrosion resistant oxide film is fully maintained under marine deposits, and no pitting or crevice corrosion has been observed. It has been shown that biofouling on titanium heat exchanger tubes can be reduced by maintaining water velocities greater than 1.2 m/s (*259*). Franson and Covington (*260*) indicated that titanium is also highly resistant to corrosion by chlorine and hypochlorite solutions, which can be used to reduce biofouling. Finally, because of the high resistance of titanium to erosion corrosion, abrasive particles can be injected into seawater to control biofouling of titanium heat exchanger surfaces.

2. Erosion-Corrosion

Titanium alloys have the ability to withstand high seawater velocities with negligible attack. Several investigators have established that titanium

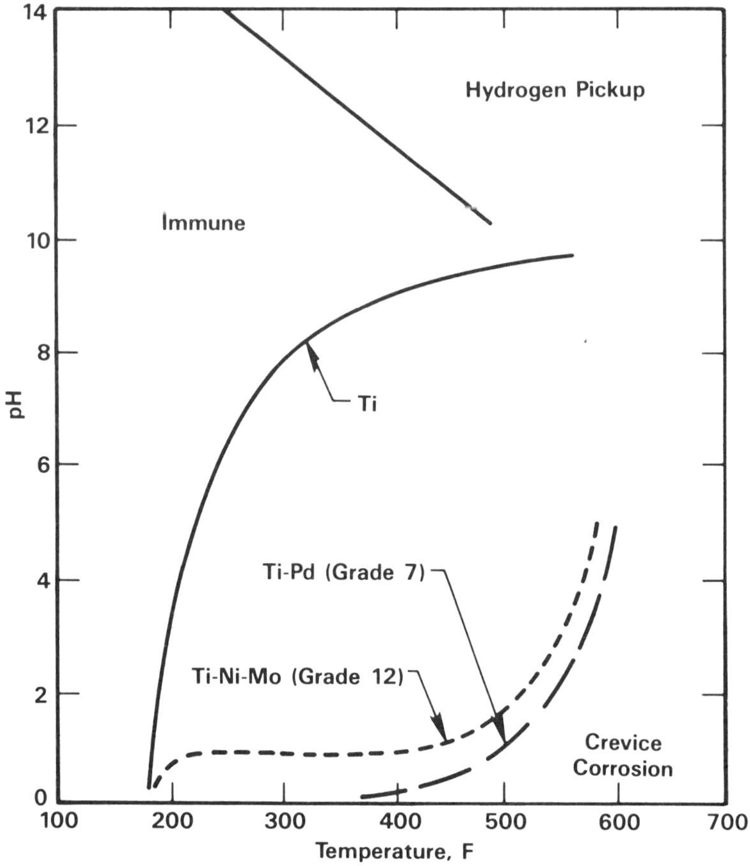

Fig. 17. Corrosion resistance of titanium alloys in saturated NaCl brines (*257*, *258*).

alloys will withstand seawater at velocities that are of technical interest (*123*, *127*, *253*, *254*, *260–264*). Velocities as high as 36.5 m/s (120 ft/s) have been found to cause only minimal erosion rates in commercially pure titanium and Ti-6Al-4V (*123*). These high velocities are of importance for the selection of materials of construction of hydrofoils or craft operating at speeds of 60 knots (101 fps) or more (*127*, *263*). In addition, high velocity information is important for other applications involving rapidly flowing seawater, e.g., high speed pumps and high velocity piping systems where titanium is used to reduce the size and weight of these components in deep diving submarines (*265*).

Titanium also has been found to have excellent resistance to flowing seawater containing sand or silt (*254*, *266*). Titanium is at least 20 times more resistant to erosion by water containing sand than copper base materials, and metal loss under these conditions was equivalent to only 0.1 mpy penetration.

Finally, titanium has excellent resistance to impingement attack in seawater (*267*) and an unusually high resistance to cavitation corrosion, which makes titanium an ideal material of construction for tube inlets (*268*).

3. Galvanic Corrosion

The coupling of titanium with dissimilar metals usually does not accelerate the corrosion of titanium alloys. In aerated environments, however, titanium will accelerate the corrosion of less noble materials. Figure 1 (*3*) lists the galvanic series of various metals in flowing seawater.

Because of its compatibility with graphite, titanium is frequently used for fasteners in graphite-epoxy composite materials (*269*). An increasing number of carbon fibre reinforced epoxy matrix components are now being used in the construction of high performance aircraft to reduce weight.

The accelerating effect on the corrosion rate of various metals when they are galvanically connected to titanium in seawater is illustrated in Figure 18 (*254*). The figure also indicates the significance of the anodic to cathodic area ratio effect on galvanic corrosion. If the cathodic area (titanium) exposed is small compared to the anodic area, galvanic corrosion may be relatively low. However, if the cathodic area greatly exceeds the anodic area, severe corrosion of the anodic metal may result. These area-effect principles apply to efforts to control galvanic corrosion by applying protective coatings. Coatings should never be applied to the anode only. The cathodic surfaces should be coated as well or the anodic surfaces should be left bare. If only the anodic surfaces are coated, galvanic current will concentrate at damaged areas, resulting in localized high corrosion rates.

4. Stress Corrosion Cracking

Early exposure test results which were based on smooth specimens indicated that titanium alloys are very resistant to stress corrosion cracking in both seawater and marine atmosphere environments (*270*). In one field test, titanium alloys were stressed to values equivalent to 35%, 50%, and 75% of their respective yield strength and exposed for different times at nominal depths of 2500 and 6000 feet in the Pacific Ocean near Port Hueneme, California (*271*). No stress corrosion cracks were detected in most of the alloys, both in unwelded and butt welded specimens. However, the butt welded Ti-13V-11Cr-3Al alloys failed at stresses equivalent to 75% of the

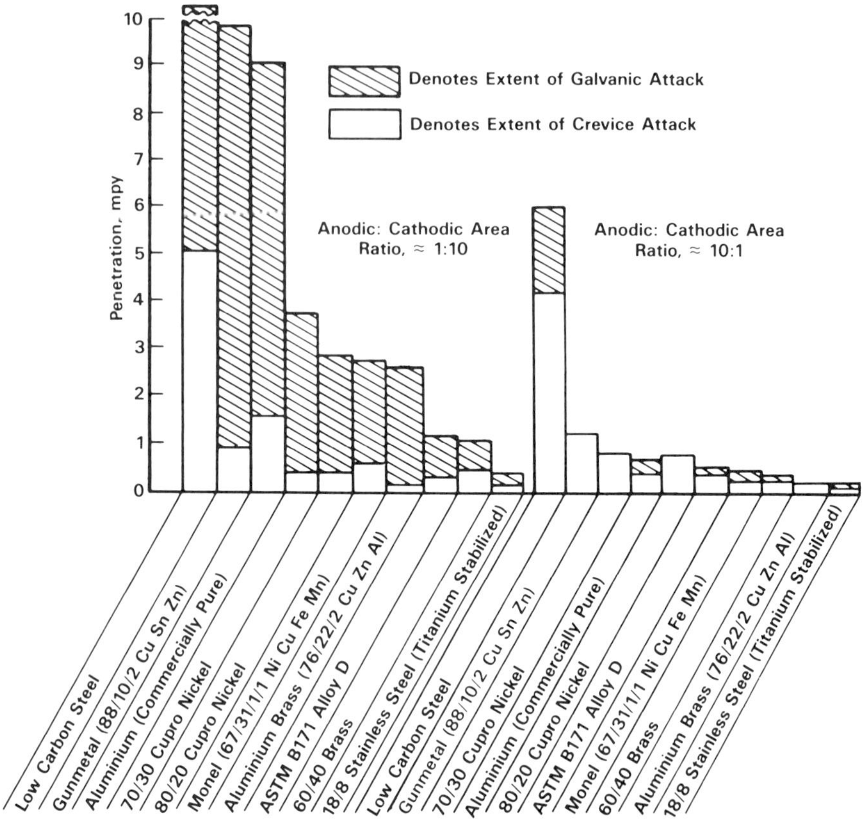

Fig. 18. The corrosion behavior of various metals coupled to titanium and immersed in aerated seawater for 2,500 hours (*254*).

yield stress after as little as 35 days of exposure at the surface of the Pacific Ocean.

The resistance of titanium to stress corrosion cracking was considered to be related to its resistance to pitting and crevice corrosion; these defects can create stress raisers and stimulate the formation of cracks in susceptible materials. Surface defects are often present in actual structures. Brown (*270*) considered this situation and first showed, in 1964, the susceptibility of titanium alloys to cracking in aqueous sodium chloride solutions in the presence of a notch. Subsequently, a great amount of research has been devoted to stress corrosion cracking of titanium alloys in seawater and in aqueous sodium chloride solutions. Extensive research on these alloys has indicated that, in seawater and marine atmospheres, environmental factors do not play a dominant role in the susceptibility to stress corrosion cracking.

Thus, no difference has been found in cracking behavior of titanium alloys in seawater and in marine atmospheres. However, alloy composition, heat treatment, and mechanical variables such as stress concentrators, stress state (plane strain or plane stress), and strain rate have a significant effect on stress corrosion susceptibility. The effects of these variables on the stress corrosion cracking behavior of titanium alloys is discussed below.

Effect of Alloy Composition

The composition of titanium alloys has a significant effect on their stress corrosion cracking susceptibility. Titanium alloys can be grouped into three classes based on alloy composition and microstructure, α, $\alpha + \beta$, and β phase alloys. The α phase alloys have an hexagonal close packed (HCP) microstructure, whereas the β phase alloys have a body centered cubic (BCC) microstructure. The α alloys contain one or more of the α phase stabilizing elements, Al, O, Sn, or Zr. So called near α alloys such as Ti-6Al-5Zr-0.5Mo-0.2Si (IMI 685) or Ti-8Al-1V-1Mo contain a high proportion of α and little β phase stabilizer.

The stress corrosion susceptibility of commercial α and $\alpha + \beta$ alloys in aqueous sodium chloride solutions and in seawater is summarized in Tables XVI and XVII (*2*). These tables show that, in general, all α and near α alloys such as Ti-8Al-1Mo-1V, Ti-7Al-3Cb-2Sn, and Ti-8Al-2Cl-1Ta are susceptible to SCC in seawater environments. The $\alpha + \beta$ alloy Ti-6Al-4V with low oxygen concentration is considered the most resistant commercial titanium alloy for seawater service. The alloy elements which affect the SCC resistance of aluminum alloys are discussed below.

Aluminum, which is added to increase the strength of titanium by affecting slip (*272*, *273*), has a significant influence on the stress corrosion cracking behavior of titanium alloys (*274–278*). It was found for binary Ti-Al alloys that, a critical Al concentration of about 5 weight percent must be exceeded for SCC to occur in aqueous solutions. At higher Al concentrations, the intermediate α_2 phase (Ti_3Al) can form, which both lowers the K_{ISCC} and increases the crack velocity (*279–281*). Ti-Al base alloys have been found to have K_{ISCC} values that vary from a high of more than 90% of K_{IC} to a low of 19% of K_{IC} (*229*).

Trace amounts of O are deleterious to the stress corrosion susceptibility of α alloys. In studies of binary Ti-O systems, it was shown that the critical O concentration was in the range of 0.2 to 0.4 weight percent (*282*). Similar results have been obtained on various grades of commercially pure titanium. For example, Ti-50A (0.12 weight percent O) shows a K_{IC} and K_{ISCC} of approximately 60 ksi $\sqrt{\text{inch}}$, whereas Ti-70A (0.38 weight percent O) shows a K_{IC} of 100 ksi $\sqrt{\text{inch}}$ and a K_{ISCC} as low as 30 ksi $\sqrt{\text{inch}}$ (*283*). In the case of

TABLE XVI

TITANIUM ALLOYS WITH SCC SUSCEPTIBILITY IN SEAWATER AND AQUEOUS SODIUM CHLORIDE SOLUTIONS (*2*)

Ti with O > 0.32 weight percent
Ti-8Mn
Ti-2.5Al-1Mo-11Sn-5Zr-0.2Si
Ti-3Al-11Cr-13V
Ti-4Al-4Mn
Ti-5Al-2.5Sn
Ti-6Al-2.5Sn
Ti-6Al-4V
Ti-6Al-3Cb-2Sn
Ti-6Al-4V-1Sn
Ti-6Al-4V-2Co
Ti-6Al-6V-2Sn
Ti-6Al-6V-2.5Sn
Ti-7Al-2Cb-1Ta
Ti-7Al-3Cb (as received and beta annealed)
Ti-7Al-3Mo
Ti-7Al-2Cr-1Ta
Ti-7Al-3Cb-2Sn
Ti-7Al-6V-2Sn
Ti-8Al-1Mo-1V
Ti-8Al-3Cb-2Sn
Ti-13V-11Cr-3Al

TABLE XVII

TITANIUM ALLOYS WITH SCC RESISTANCE IN SEAWATER AND AQUEOUS SODIUM CHLORIDE SOLUTIONS (*2*)

Ti with O < 0.3 weight percent
Ti-2Al-4Mo-4Zr
Ti-4Al-3Mo-1V
Ti-5Al-2Sn-2Mo-2V
Ti-6Al-2Mo
Ti-6Al-2Nb-Ta-0.8Mo
Ti-6Al-2Sn-1Mo-1V
Ti-6Al-2Sn-1Mo-3V
Ti-6.5Al-5Zr-1V
Ti-7Al-2.5Mo (as received and beta annealed + WQ + 1100°F age for 2h)
Ti-11.5Mo-6Zr-3.4Sn

commercial α phase alloys, the O level below which SCC does not occur decreases with increasing Al concentration. Systematic investigations performed by Wood et al. (*281*) have shown for Ti-1.5Mo-0.5V base alloys that material containing 4 weight percent aluminum was immune to SCC with oxygen concentrations less than 0.2 weight percent.

One might anticipate that oxygen would have a detrimental effect on $(\alpha + \beta)$ alloys similar to that on α alloys. However, Sorkin et al. (*284*) found for Ti-6Al-4V that O concentrations ranging from 0.06 to 0.12 weight percent had no significant effect on the susceptibility to SCC in seawater, while a decrease in fracture toughness with increasing O content was observed. At higher O concentration, 0.18, (*282*), 0.15, and 0.3 weight percent (*283*) Ti-6Al-4V was found to be susceptible to SCC in aqueous salt solutions. Gastelow and Weaver (*285*) concluded that O levels approaching 0.1 weight percent are damaging for many $\alpha + \beta$ alloys. This number is significant since an oxygen concentration of 0.15 weight percent is typical for U.S. and U.K. commercial alloys. A maximum O content of 0.1 weight percent is specified for U.S. materials used for deep submerged vessels. The effect of β phase stabilizers on the susceptibility to seawater SCC is not clear. Several investigators have shown that isomorphous stabilizers such as Mo, V, and Nb have a beneficial effect on the SCC resistance of titanium alloys (*283*, *286*). However, studies by Boyd et al. (*274*) have indicated that, in alloys with 7 and 8 weight percent Al, addition of a β stabilizer could be detrimental; whereas, in alloys with 6 weight percent aluminum, additions of the β stabilizer V at concentrations greater than 2 weight percent could be beneficial. In titanium alloys containing large amounts of the β stabilizer Mn, cleavage of the β phase has been observed (*277*, *278*). Further, $\alpha + \beta$ alloys containing Si were also found to be susceptible to SCC. (*284*, *287*, *288*). These authors have suggested that addition of Si, added to improve the creep strength, promotes the formation of Ti_3Si, which effectively embrittles the β-phase.

Hydrogen has been found to have a considerable effect on the SCC behavior of a number of titanium alloys (*289–293*). Most work has been concentrated on Ti-8Al-1V-1Mo and Ti-6Al-4V. In Ti-8Al-1V-1Mo, lowering the hydrogen content to below $5\,g/m^3$ (ppm) rendered the $(\alpha + \beta)$ equiaxed microstructure immune to SCC (*289*). Howe and Goode (*290*) showed that maintaining hydrogen levels blow 30 ppm raised the resistance to SCC in both β annealed and $\alpha + \beta$ alloys. Similar results were reported for Ti-6Al-4V (*290*, *292*).

Effect of Heat Treatment

Heat treatment of α and $(\alpha + \beta)$ alloys has a marked effect on their SCC behavior in seawater. In α-phase alloys, the effect of heat treatment is to alter

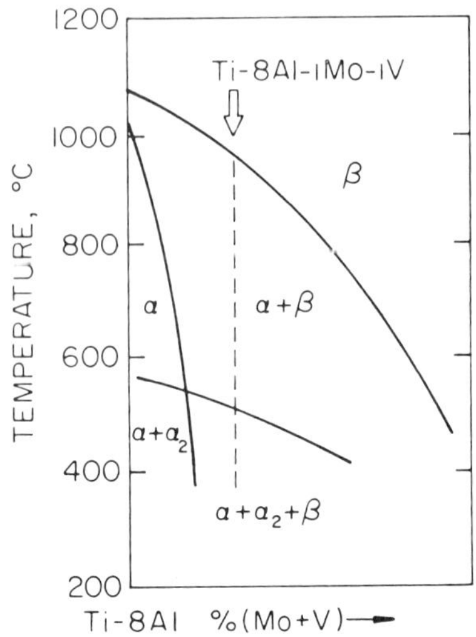

Fig. 19. Graph showing the phases present in Ti-8Al-Mo-V as a function of percent (Mo + V) (*293*). The composition vertical for Ti-8Al-1V-1Mo is indicated.

the morphology of the α-phase, see phase diagram in Figure 19 (*293*). Alpha annealing in the α-phase field produces an equiaxed grain structure irrespective of cooling rate. Low temperature aging in the α phase field (< 500°C for Ti-1Al-1V-Mo) of alloys with aluminum contents of 5% and above will lead to the formation of α_2 phase ($TiAl_3$), which results in an increase in stress corrosion cracking susceptibility. Extended aging in the $\alpha + \alpha_2$ phase field has been shown to produce a partial recovery of the resistance (*275*), which indicates an apparent dependency of SCC on the particle size of the α_2 precipitates. Heating α alloys in the β-region (> 1000°C for Ti-8Al-N-1Mo) reduces SCC resistance by allowing grain growth to occur (*283*).

In susceptible (α + β) alloys such as Ti-7Al-2Mo-1V it has been observed that equiaxed microstructures are very sensitive to stress corrosion cracking (*275*, *284*, *295*, *296*), whereas materials heated in the high temperature β phase field and rapidly quenched are immune. After aging in the $\alpha + \alpha_2$ phase field, many of the advantages of the β solution heat treated and quenched material are lost. On the other hand, stress corrosion resistant alloys such as Ti-4Al-3Mo-1V are not greatly affected by heat treatment. Processing at or just below the β transus to maximize strength with minimal grain growth produces the most desirable microstructure (*274*).

Beta and (α + β) alloys may either be solution treated (β-ST) or solution treated and aged (β-STA). At the low aging temperatures, the microstructure consists of fine needle-like or Widmanstätten plates of α phase in a β phase matrix. The effect of β-phase on stress corrosion susceptibility of titanium

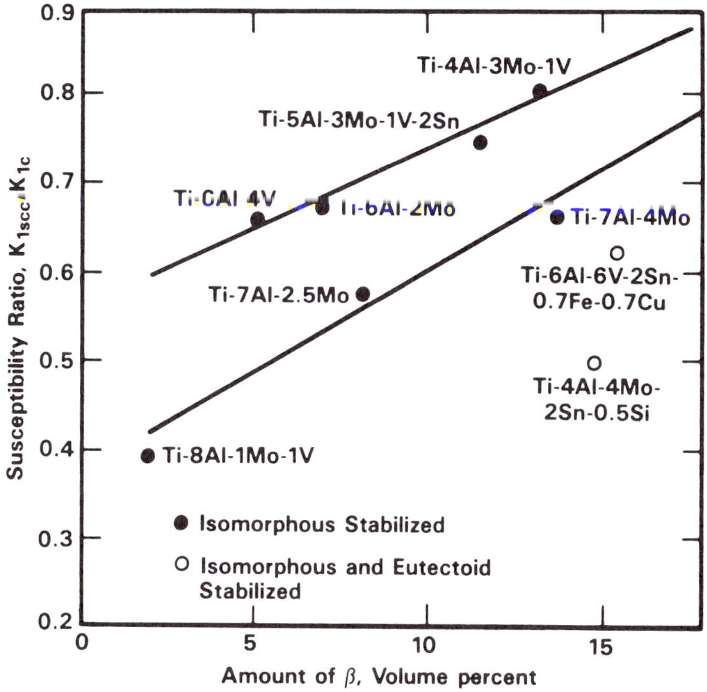

Fig. 20. Effect of β phase on stress corrosion susceptibility of titanium alloys in a β-STA heat treatment condition (283).

alloys in β-STA heat treatment condition in a 3.5% NaCl aqueous solution is shown in Figure 20 (*283*). The diagram indicates that an increase in β phase increases the resistance to SCC. Metastable β-phase alloys are solution treated above the β transus and then either water cooled or water quenched. The ω phase, a finely dispersed intermetallic phase (*297*), might form on cooling from the aging temperature and cause embrittlement of the alloy.

Mechanical Variables. Titanium alloys that are susceptible to stress corrosion cracking can be divided into two general groups.

(a) Those which experience stress corrosion cracking without a stress concentration. These are mostly alloys such as Ti-13V-11Cr-3Al and Ti-8Mn and very susceptible α and $(\alpha + \beta)$ alloys such as Ti-8Al-1Mo-1V.

(b) Those which will only experience stress corrosion cracking in the presence of a stress concentration such as a notch or a precrack.

For both groups of alloys, stress-corrosion-cracking susceptibility was shown to be dependent on component or specimen thickness (*276*, *287*). The increase in susceptibility to SCC with increasing thickness can be attributed to plane strain conditions that develop as the result of increasing thickness. The transition from plane stress to plane strain may be affected by preferred orientation, straining rate, and heat treatment.

As discussed above, titanium alloys fail in seawater transgranularly by a cleavage-like process, which suggests that stress-corrosion cracking susceptibility will depend on the orientation of the fracture plane with respect to the tensile stress. Such a preferred orientation effect is most pronounced in α and $(\alpha + \beta)$ alloys for which the cleavage plane is $(10\bar{1}7)$ which is about 15 degrees from the basal plane (*298–300*). In β alloys, the cleavage plane is (100), and thus the anisotropy of properties is less pronounced.

F. Copper and Copper-Base Alloys

1. Atmospheric Corrosion

Copper and copper alloys exhibit low rates of attack in marine atmospheres. On the basis of weight-loss measurements, corrosion rates over periods of up to 20 years are 0.01 to 0.17 mpy. The low corrosion rates of copper and its alloys are the result, in part, of the formation of protective corrosion product films. Soon after installation, the bright copper surface takes on a dull tan color (tarnish) which, after a few years, gradually changes to dark brown or black. At a later stage, depending on the level of sulfur contaminants in the air, the copper corrosion products turn green, which is referred to as a patina. The corrosion reactions involved in the formation of the green patina usually are more rapid on boldly exposed surfaces facing south (northern hemisphere) or in the direction of the prevailing wind (where the air is carrying pollutants) than they are for other orientations.

In a marine atmosphere containing low sulfur dioxide concentrations, the green patina may be largely basic copper chloride $CuCl_2{\cdot}3Cu(OH)_2$ (atacamite) (*301*). Basic sulfates are also present. If the atmosphere is at all contaminated, sulfate will be the major constituent. Osborn (*302*), for example, has established that the patina on the bronze Statue of Liberty is predominantly basic copper sulfate, even though it is surrounded by the seawater in New York harbor.

From the standpoint of alloy additions, slightly beneficial trends are indicated for aluminum, nickel, and zinc, whereas those alloys containing silicon and tin tend to be somewhat less resistant than copper alone. These trends are, in general, not so significant as to dictate the use of one alloy over another except for an extremely critical application.

The high-zinc brasses (naval brass, Muntz metal, and manganese brass), although exhibiting low rates of penetration on the basis of weight loss, were found in studies by Hummer (*303*) to have lost considerably more strength than other alloys in the series. Although not indicated by Hummer et al. (*303*) these materials may have exhibited dezincification as a result of exposure to the marine environment. Indeed Landegren and Mattsson (*304*) reported significant dezincification of Cu-37Zn deep drawn cups following exposure for 121 months at a marine test site in Sweden. For single-phase brasses, the attack can be controlled by small additions of arsenic, antimony, or phosphorus. About 0.02% arsenic is very effective. Note that in the study by Hummer (*303*), Muntz metal, which contained 0.19% arsenic, showed significant loss of strength; in this two-phase alloy, the inhibitor apparently is not effective in preventing dezincification.

Copper-zinc alloys also are susceptible to stress corrosion cracking in marine atmospheric environments. Landegren and Mattsson (*304*) exposed highly stressed specimens (deep drawn cups) of copper 37 zinc to indoor, urban, rural, and marine environments for up to 121 months. Results of the exposures indicated that 95% of the unannealed specimens failed within about 120 months when exposed without shelter in a marine atmosphere. The sheltered marine atmosphere was more aggressive with 100% failures within 87 months. In general, the rural and urban environments were much more aggressive than the marine environments with 100% failures within a few months. For all environments, annealing was beneficial, presumably because the residual stresses within the cups were reduced, although the change in microstructure as a result of the heat treatment may also have affected the cracking susceptibility.

2. Splash and Tide

The corrosion behavior of copper alloys in the splash zone more nearly follows their performance in the atmosphere than in immersed conditions. As a general rule, any copper alloy that has good resistance to a severe marine atmosphere will also do well in the splash zone.

Alloys exposed at mean tide show corrosion rates ranging from about 20% to 60% of those observed for alloys under fully immersed conditions. Thus, the behavior under fully immersed conditions can be used as a guide for the selection of a resistant alloy in a half-tide application. The resistance of an isolated[2] panel of copper in the tide zone is superior to that in the fully immersed condition and is in marked contrast to the relatively poor behavior

[2] Isolation in this context refers to a specimen exposed in one particular zone as opposed to a piling which passes through several zones.

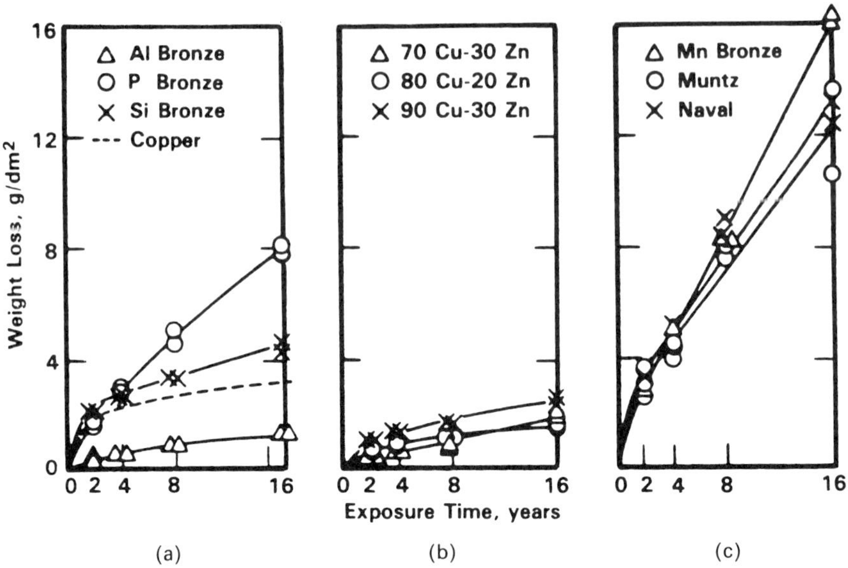

Fig. 21. Corrosion weight loss as a function of time for copper alloys exposed at mean-tide elevation in tropical seawater (303).

of an isolated steel panel at the half-tide level. Weight-loss curves for the half-tide performance of copper alloys after 16 years are given in Figure 21 (*303*).

3. Immersion

General and Pitting Corrosion

Effect of Alloy Composition. Studies by Hummer et al. (*303*) and Reinhart (*305*) show that the corrosion rates calculated from weight loss for a wide range of copper base alloy decrease with time and are generally less than 1 mpy. For the brasses, depth of attack associated with dezincification are generally much greater (see section on dealloying).

The low and decreasing corrosion rates are attributed to the formation of protective films on the metal surfaces. These films are composed of copper oxides as well as oxychlorides of copper, copper carbonate, and calcareous material (*303*).

The copper base alloys also are susceptible to pitting under immersed conditions. However, few pitting failures of copper base alloys have been reported in heat exchanger application in nonpolluted seawater (*306*) suggesting that pitting of copper base alloys is not a serious problem for most fresh seawater applications.

Effect of Temperature. The corrosion rates for copper alloys usually increase, other factors being the same, as the temperature is raised. Pump impellers, condenser tubes, valves, etc. tend to corrode at higher rates in seawater in the summer than in the winter (*307*). Copper alloys also show higher rates of attack in the warm waters of the Pacific Ocean near the Panama Canal Zone as compared with those obtained in waters off the coast of California. However, when the seawater is heated above ambient, corrosion rates of copper base alloys may decrease as a result of the decrease in oxygen concentration and the retrograde solubility of mineral scales.

Effect of Depth and Aeration. A comparison of corrosion rates for copper and copper-based alloys in the Pacific Ocean at three depths (surface, 2340 feet, and 5640 feet) indicates that the corrosion rates were generally a minimum at the 2340 foot depth, e.g., for Muntz metal, corrosion rates were 2.2 mpy at the surface, 0.8 mpy at 2340 feet, and 1.5 mpy at 5640 feet (*4*, *305*). These minimums in corrosion rate correspond to a minimum in the oxygen concentration-depth curve, see Figure 2 (*4*); the oxygen concentration was less than 0.5 ppm at 2500 feet, about 1 ppm at 5500 feet, and about 6 ppm at the surface. A similar correlation between oxygen content of seawater and corrosion of copper alloys was reported by Schreiber et al. (*308*) for studies of heat exchanger tube materials in flowing seawater at elevated temperatures.

Effect of Biological Fouling. It is well established that the by-products generated by fouling organisms or their decay products will promote the localized attack of copper-base alloys. Rogers (*309*) evaluated the effect of a variety of microorganisms and of their by-products, such as cystine on the corrosion of copper base alloys, and found, in all cases, that localized attack was promoted. However, the attack only occurs if the organisms are able to attach and grow on the metal surface. LaQue and Clapp (*310*) found that pitting of copper-nickel alloys occurred in quiescent seawater under macro-organisms, following their death. The severity of fouling and of pitting increased with increasing Ni content of the alloys. Pitting only occurred in alloys containing 40% or more Ni; whereas, no fouling occurred on copper and on Cu-20 Ni and only incipient fouling occurred on Cu-30 Ni. The alloys that fouled showed increases in corrosion rates with exposure time, whereas the opposite was true of the alloys that did not foul. The relationship between fouling, pitting, and the Ni content of the alloys was associated with the general corrosion rate which provided sufficient Cu in solution to poison the organism and prevent their attachment to the metal surface. On alloys of low nickel content where the general corrosion rate was greater, fouling and resulting pitting did not occur.

In flowing seawater, fouling and pitting that is associated with fouling is less severe. A discussion of the influence of water velocity on fouling is given by Clapp (*311*), where it is stated that the minimum velocity required to prevent macrofouling is between about 1 to 2 m/s.

Effect of Pollution. The corrosion performance of copper based alloys is greatly affected by seawater pollution. The presence of sulfide in the polluted seawater has a pronounced adverse effect. At low concentration, the sulfide promotes pitting; whereas, at higher concentration, high rates of general attack are observed. Data showing the corrosion rates of several copper alloys in polluted and unpolluted seawater are given in Table XVIII (*312*). These data indicate that even the normally corrosion resistant Cu-Ni alloys are highly susceptible to accelerated attack in the presence of H_2S.

Gudas and Hack (*313*) performed a parametric study of the effect of sulfide concentration, seawater velocity, and duration of sulfide exposure on corrosion of Cu-10Ni and Cu-30Ni. All exposures were conducted in fresh seawater, with the sulfide concentration varying from 0.007 to 0.25 mg/l, seawater velocity after sulfide exposure of 0.5 to 5.3 m/s, sulfide exposure times of 1 to 90 days. Results of exposures carried out at 0.01 to 0.35 mg/l sulfide concentration indicated that the sensitivity to attack increased with increasing sulfide concentration over the range studied; subsequent tests at

TABLE XVIII

CORROSION OF CONDENSER TUBES IN POLLUTED AND CLEAN SEAWATER (*312*)
VELOCITY, 7.5 fps; DURATION, 64 days.

	Corrosion Rate, mpy	
Alloy	Clean Seawater	Polluted Seawater (3 pm H_2S)
90Cu-10Ni[a]	3	34
70Cu-30Ni[a]	5	26
2% Al brass	3	22
6% Al brass	5	21
As Admiralty brass	13	35
Phosphor copper	14	105

[a] Of the two alloys, 70Cu-30Ni alloy is usually preferred for more extreme conditions because of its superior corrosion resistance. However, in stagnant seawater, it has a greater tendency to pit than does the 90Cu-10Ni alloy.

lower concentrations, down to 0.007 mg/l, indicated significant localized attack of both alloys, although Cu-30Ni was more resistant in long-term exposures to low concentrations of sulfide. These results also indicated that where sulfide exposure conditions were sufficient to initiate attack, the seawater velocity dependence was minimal over the range of 0.5 to 5.3 m/s.

A number of researchers have investigated the mechanism of accelerated attack of copper base alloys by sulfide (*314–320*). Although details of mechanisms proposed by the different researchers vary, the general picture which emerges is that the sulfide interferes with the formation of the protective films on the surface of the copper base alloys and, in addition, catalyzes the oxygen reduction reaction. The morphology of the attack is dependent upon the concentration and distribution of the sulfide which, in turn, is dependent upon the concentration of the sulfide in the seawater and the prior history of the specimen.

4. Erosion-Corrosion

Effect of Alloy Composition

Extensive research has been done on the effect of alloying elements on the resistance of copper-base alloys (particularly of copper-nickel alloys) to erosion-corrosion. Freeman and Tracey (*321*) investigated the performance of eight copper-base alloys, including bronzes and brasses, for their resistance to erosion-corrosion in a model condenser using water at 3.6 m/s. They ranked the alloys in order of decreasing resistance to impingement attack:

(a) Copper-30 nickel (0.45 % iron)
(b) Copper-30 nickel (0.3 % iron)
(c) Copper-10 nickel (0.08 % iron)
(d) 8 % phosphor bronze
(e) Aluminum brass (Ambraloy 927)
(f) Aluminum bronze (Ambraloy 901)
(g) Arsenical Admiralty
(h) Uninhibited Admiralty

Note that the copper-nickel alloys were found to have greater resistance to impingement attack than the brasses and bronzes.

The relative erosion-corrosion resistance of seven candidate copper alloys —noninhibited Admiralty brass, arsenical Admiralty brass, aluminum brass, Albrac, Cu-10 Ni, Cu-30 Ni, and aluminum bronze—was investigated using a model condenser which operated as an auxiliary unit of the main condenser at Meiko Power Station (*322*). Admiralty brass tubes, both noninhibited and arsenical, suffered the worst erosion-corrosion damage. Extensive impingement attack occurred at the inlet end, outlet end, and in other parts near deposits. Impingement attack at the inlet end occurred in the upper part of the tubes, probably because of release of entrained air bubbles. Impingement attack on the other alloys was not reported.

Contrary to the findings of Otsu and Sato (*322*), Pruvot (*323*) reported extensive erosion-corrosion in aluminum brass tubes in a model condenser at Dunkerque. The difference in performance of aluminum brass at the two locations may be the result of the difference in operating conditions on the quality of seawater at Dunkerque and Japan. In agreement with the Japanese authors, however, Pruvot found no attack on copper-30 nickel.

Results of testing of experimental copper-nickel alloys containing 0% to 30% Ni with varying amounts of Fe and Mn showed that the addition of Fe to the alloys improved resistance against erosion-corrosion (*324*). The amount of Fe needed in Cu-Ni alloys for the optimum resistance increased with decreasing Ni content of the alloy as follows: 0.5% Fe was required for Cu-30 Ni, 0.6% Fe for Cu-20 Ni. Addition of 0.75% Fe for Cu-2 Ni improved the performance of the alloy only marginally and showed no improvement in pure copper. No critical relationship was found between the Fe and Mn contents of the alloys and their resistance to erosion-corrosion.

LaQue et al. (*325*, *326*) used rotating spindle and jet impingement tests to study the erosion corrosion resistance of Cu-30 Ni and Cu-10 Ni, along with some brasses in seawater. It was found that the performance of Cu-30 Ni improved with increasing (from 0.03% to 0.49%) Fe content of the alloy. Similarly, for Cu-10 Ni, weight loss dropped with increasing (up to 2%) Fe (Figure 22). The Cu-10 Ni alloys with 0.7% or 2% Fe performed as well as or better than Cu-30 Ni with 0.47% Fe and Admiralty brass, and much better than aluminum brass and low Fe Cu-3 Ni (*325*).

However, a few specific cases are reported in which the addition of Fe to Cu-Ni alloys reduced the overall performance of Cu-Ni alloys (*327*, *328*). Thus, the role of Fe in Cu-30 Ni alloy appears to be ambiguous and appears somewhat velocity dependent in a few specific cases.

Chromium modified Cu-Ni alloys have been investigated for their resistance to erosion-corrosion at high water velocities (up to 40 m/s) (*329*). Addition of 0.5% to 3% Cr to Cu-Ni alloys reduced remarkably the depth of attack in seawater jet impingement tests. Results for Cu-15 to 18 Ni and Cu-30 Ni alloys as a function of Cr content of the alloys show that, at about 0.5% Cr, the impingement attack becomes almost negligible at 7.5 m/s water velocity (*328*). Results for velocities from 3 to 18 m/s for different Cu-Ni alloys were obtained in a multi-velocity parallel flow apparatus rather than in a jet impingement apparatus (*329*). The data show that Cu-30 Ni-3 Cr has superior erosion-corrosion resistance to that of the Cu-Ni-Fe alloys (CA-706

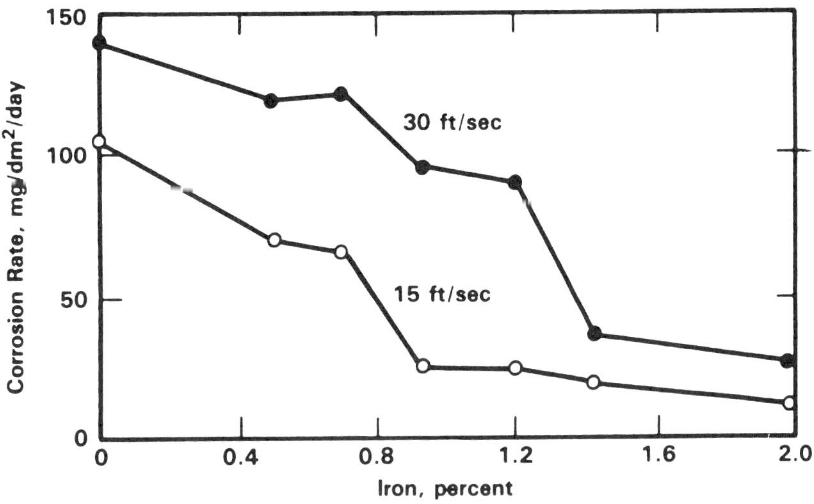

Fig. 22. Effect of iron content and seawater velocity on erosion-corrosion resistance of copper-10 nickel (326).

and CA-715), and Cu-15 Ni-0.5 Cr (In 838) has about the same resistance as CA-715.

Effect of Velocity

The effect of water velocity on the erosion-corrosion of copper-base alloys has been investigated by several authors (*321*, *326*, *329–334*). Davis et al. (*336*) conducted a general review of the literature pertaining to copper-base alloys. Syrett (*327*) and Beavers and Boyd (*337*) directed reviews specifically towards Cu-Ni alloys.

The corrosion rate of copper-base alloys generally increases with increasing velocity until a critical or breakaway velocity is reached, beyond which the rate increases rapidly up to a certain value and then it starts to taper off with further increases in velocity. According to Efird (*334*) the accelerated corrosion, due to high water velocity, occurs when the surface shear stress on the protective passive film of the alloy exceeds the bonding strength of the film to the alloy substrate. Similar explanations for the erosion-corrosion have been given by several authors previously (*326*, *327*).

Sato (*330*) found that the wall shear stress in the entry region of a condenser tube can be twice as high as at other parts of the tube for a given velocity. This implies that in order to avoid inlet impingement attack, the normal flow velocity should be one half or less that of the critical velocity for the alloy under the fully developed flow conditions. Sato (*330*) has also shown that the lodgement of foreign bodies, such as mussels, in tubing or piping can increase the local velocity at restrictions to several times the normal flow velocity. Thus, if the designed normal flow velocity is kept substantially lower than the highest acceptable velocity, or by selecting a material with high surface shear stress tolerance, the erosion-corrosion of tubes can be avoided.

Effect of Temperature

Data compiled from several studies on the erosion-corrosion of copper-base alloys at different temperatures show a mixed trend with regard to the effect of temperature on the erosion-corrosion rate (*327*). The rate increased in some cases, decreased in others, and changed little with temperature in still others. The lack of a trend in the data was attributed to the differences in the oxygen content of the waters used by different authors.

Results of a study of the effect of water temperature on the erosion-corrosion of Cu-30 Ni containing two different Fe contents (*324*) indicated that the weight loss of the high-Fe (0.48% to 0.69%) alloy at 30°C was about one half of that at 15°C, whereas the reverse was the case for the low Fe (0.03% to 0.04%) alloy. According to the authors, the drop in weight loss of the high-Fe alloy was due to an improvement in the quality of the protective film at the high temperatures.

Efird (*334*) calculated the surface shear stress as a function of seawater velocity for different temperatures. Results showed that the shear stress drops for a given velocity with an increase in temperature. The drop is due to the changes in the physical properties of the water, e.g., viscosity and density. Since high shear stress may be responsible for the erosion-corrosion, a drop in its value due to an increase in temperature should lower the erosion-corrosion rate. These findings are in agreement with the observations of Anderson (*338*).

Effect of Aeration

According to Syrett (*327*), "It is generally agreed that when air or other gases are present as bubbles in flowing water, the erosion-corrosion properties of copper alloys are adversely affected." Anderson (*338*) recommended, on the basis of a pilot plant study, that for long life of copper-10 nickel in hot seawater service, the water should be deaerated, because the presence of dissolved air accelerates the corrosion rate.

Evidence to the contrary has been presented by Stewart and LaQue (*326*) for Cu-Ni alloys. These authors have observed a reduction in erosion-corrosion rate in seawater when air was present in the test media. Their observations are based on tests using either a rotating spindle or disc apparatus. Syrett (*327*) is of the view that the discrepancy between these observations and those of others is due to the test techniques employed by the latter authors. With rotating spindle or disc methods any air bubbles present in the water do not strike the surface of the test alloy directly; the bubbles are held back by the boundary layer. As a result, the surface of the sample is not damaged when the bubbles break; on the contrary, the increased oxygen supply promotes film growth and reduces the corrosion rate.

Results from two jet impingement studies demonstrated that when either air bubbles or nitrogen bubbles were present in water, the severity of impingement attack increased over that of plain water (*327*). In cavitation erosion, the rapid breaking of vapor bubbles can create shock waves with pressures high enough to cause damage to many metal surfaces (*339*).

Effect of Chlorination

In model condenser studies, chlorination had a moderate (*342*) or negligible (*325*) effect on corrosion up to velocities of 3 m/s and chloride concentration of 1 ppm. However, under more severe erosion-corrosion conditions, such as with jet impingement tests, chlorination exacerbated the attack. The depth of attack on all the copper alloys tested increased with increasing chlorine concentration in tests (*340*) performed at 5 m/s with 3% NaCl containing 5 volume percent air, e.g., the corrosion rate of Cu-10Ni increased from about 1.5 mg/cm^2-100 hours at 0 ppm Cl_2 to about 28 mg/cm^2-100 hours at 2 ppm Cl_2.

Effect of Particulates

The erosion-corrosion of copper base alloys is generally exacerbated by the presence of particulates in either seawater or NaCl solutions. Thus, in rotating electrode studies by Tanabe (*341*, *342*) and Sato (*330*, *343*) in 3% NaCl, it was found that the corrosion rate of aluminum brass increased linearly with the increasing grain size of the sand from about 55 mg/dm^2-day at 75 μm diameter to about 75 mg/dm^2-day at 200 μm diameter at 2 m/s water velocity and 10 g/l sand. The increase in sand content of the media also resulted in increased corrosion rates. Moreover, when the rotation speed was increased at a fixed sand content, the rates increased almost logarithmically with speed.

Sato et al. (*343*) investigated the sand erosion-corrosion in seawater of aluminum brass, Cu-30 Ni, and Ti tubes using a model condenser. The study included effects of ferrous ion injection, surface pretreatment, and cathodic protection. Cathodic protection with or without Fe^{++} injection was effective in lowering the rate when the grain size was 50 μm but not when the grain size was 250 μm. The latter behavior probably indicates that the damage at the larger grain size is primarily mechanical in nature. The erosion-corrosion rate of Cu-30 Ni was about twice that of aluminum brass under identical conditions of testing. Polluted seawater containing sand was less aggressive than nonpolluted seawater on aluminum brass. This effect was attributed to the formation of an adherent sulfide film in polluted seawater.

5. Galvanic Corrosion

A galvanic series for commercial alloys and metals in seawater is shown in Figure 1 (*3*), where it can be seen that the copper-base alloys are noble to common construction materials such as steel cast iron and aluminum, but active to the noble metals and passive stainless steels and nickel-base alloys. Results of long-term galvanic corrosion tests (*344*), are consistent with the galvanic series. The copper-base alloys were galvanically protected by coupling to active metals such as carbon steel; whereas, they experienced accelerated corrosion as a result of coupling to the stainless steels and the nickel base alloy Monel. In general, the copper-base alloys were found to be galvanically compatible, but galvanic corrosion occurred under some circumstances. The Monel 400-phosphorus bronze couple was surprisingly aggressive, producing more severe galvanic corrosion of phosphorus bronze than any other couple. The corrosion rate of the phosphorus bronze increased from 0.3 to 4.5 mpy as a result of the couple.

Seawater velocity can have a dramatic effect on the galvanic corrosion behavior of copper base alloys. For instance, where a copper alloy is coupled to a more noble metal, the galvanic attack of the copper is greatly accelerated with increase in seawater velocity. Typical data showing this effect for several copper alloys coupled to titanium are given in Figure 23 (*345*). This effect is responsible for the severe attack reported for copper-base alloy tube sheets in surface condensers where stainless steel or titanium tubes or tube inserts have been installed (*346*).

The severe galvanic attack observed in condensers also appears to be highly temperature dependent. In recent condenser studies by Gehring and Maurer (*347*), the rate of galvanic attack of Muntz metal, when coupled to titanium, doubled for every 10°C rise in temperature.

Galvanic corrosion of copper-base alloys also appears to be promoted by the presence of sulfides. Hack (*107*) performed galvanic corrosion tests on two piping alloys, Cu-30Ni and Cu-10Ni, and four fitting materials, bronze (composition M), cast Monel, wrought nickel-aluminum bronze, and cast Cu-30Ni. The experiments were performed at an area ratio of 3:1 (piping

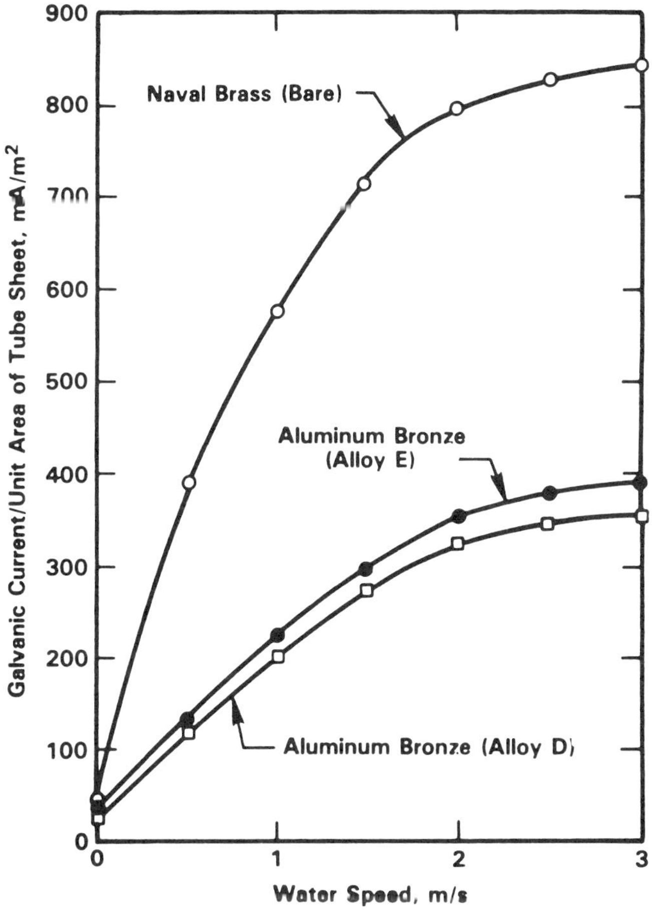

Fig. 23. Effect of water velocity on the galvanic current between titanium tubes and tubes sheet in seawater (345) a. ASTM-B171 No. 628; b. ASTM-B171 No. 614.

material/fitting material) in seawater at 2.4 m/s and three sulfide concentrations: 0.0, 0.01, or 0.05 mg/l. Results of the experiments indicated that significant galvanic effects occurred in the presence of sulfide with alloys that are compatible in unpolluted seawater.

6. Stress Corrosion Cracking

In general, the copper base alloys are highly resistant to stress corrosion cracking under submersed exposure in seawater. For example, Reinhart (*305*) exposed a series of stressed copper-base-alloy specimens, including unalloyed Cu, Muntz metal, Admiralty Brass, phosphor bronze, Al bronze, Si bronze, and Cu-Ni alloys, at a depth of about 2000 feet in the Pacific Ocean and observed no stress corrosion cracking failures in 197 days.

Isolated cases of SCC failures of copper base alloys in seawater have been reported. Bulow (*348*) reported SCC of manganese bronze propellers having high residual stresses, while Beavers et al. (*337*) reported SCC of Admiralty brass condenser tubes which were exposed to brackish river water. In the latter failure, the cracking was attributed to nitrite, which formed as a result of decay of organic matter, in conjunction with high residual stresses from the tube rolling operation.

Most all the SCC failures of copper-base alloys in service have been attributed to ammonia or other nitrogen compounds such as nitrates and amines. These species may form beneath deposits in seawater, as discussed above. In these environments, Cu-Zn alloys are highly susceptible to SCC, whereas the Cu-Ni alloys are resistant. A summary of the relative susceptibilities of the copper base alloys to SCC is given in Table XIX (*349*).

TABLE XIX

RANKING OF COPPER BASE ALLOYS ACCORDING TO THEIR RESISTANCE IN CONCENTRATED AMMONIA ENVIRONMENTS (*349*)

Highly susceptible to ammonia-SCC
brass with 10% zinc and more
aluminum bronze with 1% to 8% aluminum
silicon bronze with 0.8% to 1.2% silicon
all copper alloys with more than 20% zinc
Moderately susceptible to ammonia-SCC
beryllium copper, alloys 170, 172
brass with less than 10% zinc
aluminum bronze with 9% or more aluminum
copper-nickel with less than 10% nickel
phosphorus deoxidized copper, alloy 122
Resistant to ammonia-SCC
pure copper, e.g., alloys 101 through 113
copper-iron, alloy 194
copper-nickel-iron with 10% and more nickel
copper-tin alloys with 2% and more tin

7. MUD

Corrosion studies (*305*) for copper-base alloys in bottom sediments at two depths in the Pacific Ocean indicate that the corrosion rates in the sediments were lower than those in the seawater, at the same depth, for most of the alloys studied. These studies also showed that the corrosion rates in the sediment were lower at a depth of 2340 feet than at a depth of 5630, as were the corrosion rates in the seawater. As discussed in the section on depth and aeration under general and pitting corrosion, this behaviour was the result of the lower oxygen concentration of the seawater at the 2340 foot depth. Similarly, the lower corrosion rate in the sediment than in the seawater, at a given depth, is also probably attributable to the lower concentration of oxygen in the sediment. However, it must be cautioned that the high sensitivity of copper base alloys to sulfide must be considered when they are used in sediments in polluted seawater or where sulfate-reducing bacteria are present.

References

1. Beavers, J. A., Koch, G. H., and Berry, W. E. (1986). *Corrosion of Metals in Marine Environments.* Battelle Columbus Division, Metals and Ceramics Information Center, Report MCIC 86-50.

2. Boyd, W. K., and Fink, F. W. (1978). *Corrosion of Metals in Marine Environments.* Battelle Columbus Laboratories, Metals and Ceramics Information Center Report MCIC-78-37.

3. Sedriks, A. J. (1982). "Corrosion Resistance of Austenitic Fe-Cr-Ni-Mo Alloys in Marine Environments." *International Metals Review* **27** (6), 321-353.

4. Reinhart, F. M. (1966). *Corrosion of Materials in Hydrospace.* U.S. Naval Civil Engineering Lab., Port Hueneme, California, Technical Report R504.

5. H. H. Uhlig, editor (1948). *The Corrosion Handbook.* John Wiley & Sons, New York.

6. Temperley, T. G. (1965). "Corrosion Phenomena in the Coastal Areas of the Persian Gulf." *Corrosion Science* **5**, 581-589.

7. Copson, H. R. (1960). "Long-Time Atmospheric Corrosion Tests on Low Alloy Steels." *ASTM Proceedings* **60**, 650-665.

8. Southwell, C. R., et al. (1958). "Corrosion of Metals in Tropical Environments. Part 2. Atmospheric Corrosion of Ten Structural Steels." *Corrosion* **16** (11), 435t-439t.

9. Larrabee, C. P. (1953). "Corrosion Resistance of High-Strength Low-Alloy Steels as Influenced by Composition and Environment." *Corrosion* **9** (8), 259-271.

10. Fyfe, D., et al. (1970). "Atmospheric Corrosion of Fe-Cu Alloys and Cu-Containing Steels." *Corrosion Science* **10** (11), 821.

11. Buck, D. M. (1919). "The Influence of Very Low Percentages of Copper in Retarding the Corrosion of Steel." *ASTM Proceedings* **19** (2), 224.

12. Copson, H. R. (1952). "Atmospheric Corrosion of Low Alloy Steels." *ASTM Proceedings* **52**, 1005.

13. Larrabee, C. P., and Coburn, S. K., (1967). "The Atmospheric Corrosion of Steels as Influenced by Changes in Chemical Composition." *Proc. First International Congress on Metallic Corrosion, Butterworth, London,* p. 276.

14. Ambler, H. R., and Bain, A. A. (1955). "Corrosion of Metals in the Tropics." *Journal of Applied Chemistry* **5**, 427–467.
15. Humble, A. A. (1949). "The Cathodic Protection of Steel Piling in Seawater." *Corrosion* **5** (9), 292–302.
16. Larrabee, C. P. (1958). "Corrosion Resistant Experimental Steels for Marine Applications." *Corrosion* **14** (11), 501t–504t.
17. Nose, J., et al. (1979). "The Characteristic Properties of NK Marine G: Seawater Corrosion Resistant Steel for Pipe Piles and Sheet Piles." *Nippon Kokan Technical Report* (26), 1–9.
18. Schultze, W. A., and Van der Wekken, C. J. (1976). "Influence of Alloying Elements on the Marine Corrosion of Low Alloy Steels Determined by Statistical Analysis of Published Literature Data." *British Corrosion Journal* **11** (1) 18–24.
19. Reinhart, F. M. (1967). "Corrosion of Materials in Hydrospace. Part I. Irons, Steels, Cast Irons, and Steel Products." U.S. Naval Civil Eng. Lab. Tech. Note N-900, p. 71.
20. Southwell, C. R., and Alexander, A. L. (1969). "Corrosion of Metals in Tropical Environments. Part 9. Structural Ferrous Metals—Sixteen Years' Exposure to Sea and Fresh Water". Naval Research Laboratory Report NRL-6882, p. 28.
21. Bultman, J. D., et al. (1977). "Biocorrosion of Structural Steels in Seawater." *Revue of Coatings and Corrosion* **2** (2–3), 187–214.
22. Reinhart, F. M., and Jenkins, J. F. (1971). "The Relationship Between the Concentration of Oxygen in Seawater and the Corrosion of Metals." *U.S. Naval Civil Eng. Lab. Report*, 562–577.
23. Reinhart, F. M., and Jenkins, J. F. (1972). "Corrosion of Materials in Surface Seawater After 12 and 18 Months of Exposure." *Final Report NCEL-TN*-1213, p. 106.
24. Dosey, P. A., and Palko, A. A. (1979). "Corrosivity of Carbon Steel in Concentrated Chloride Solution." *Corrosion* **35** (1), 38–43.
25. Togano, H., et al. (1972). "Corrosion Behavior of Mild Steel and Low Alloy Steel in Deaerated Hot Sodium Chloride Solution." *5th International Congress on Metallic Corrosion*, pp. 807–811.
26. Shimada, H., et al. (1976). "Corrosion Behavior of Steels in Various Marine Environments (A Joint Study with Maritime Safety Agency)." *Nippon Steel Technical Report* (8), 31–42.
27. Efird, K. D. (1976). "Inter-relation of Corrosion and Fouling for Metals in Seawater." *Materials Performance* **15** (4), 16–25.
28. Chandler, H. E., (1979). "Corrosion-Biofouling Relationship of Metals in Seawater." *Metal Progress* **115** (6), 47–49, 54.
29. Traverso, E., et al. (1979). "Effects of Rolling Direction and Tensile Stress on the Corrosion of Naval Carbon Steel in Natural Marine Environments—I: Natural Seawater." *British Corrosion Journal* **14** (2), 97–102.
30. Sasaki, H., et al. (1977). "Corrosion Behavior of Mild Steel in Artificial Seawater Containing Hydrogen Sulfide." *Corrosion Engineering* (Tokyo) **26** (5), 229–236.
31. Sasaki, H., et al. (1977). "Corrosion Behavior of Mild Steel in Artificial Seawater Containing Hydrogen Sulfide." *Corrosion Engineering* (Tokyo) **26** (3), 125–132.
32. Larrabee, C. P. (1962). "Steel Has Low Corrosion Rate During Long Seawater Exposure." *Materials Protection* **1** (12), 95–96.
33. Larrabee, C. P. (1945). "Corrosion of Steels in Marine Atmospheres and in Seawater." *Transactions of the Electrochemical Society* **87**, 171–182.
34. Sandoz, G. (1972). "High Strength Steels." *Stress Corrosion Cracking in High Strength Steel and in Titanium and Aluminum Alloys* (B. F. Brown, ed). NRL, Washington, DC.
35. Sandoz, G. (1970). "ARPA Coupling Program on Stress-Corrosion Cracking." *Quarterly Report* 13, *NRL-MR*-2101, 68.
36. Sandoz, G., (1971). "The Effects of Alloying Elements on the Susceptibility to SCC of Martensitic Steels in Salt Water." *Metallurgical Transactions* **2**, 1055.

37. Carter, C. S. (1969). "The Effect of Silicon on the Stress Corrosion Resistance of Low Alloy High Strength Steels." *Corrosion* **25** (10), 423.
38. Benjamin, W. D., and Steigerwald, E. A. (1968). *Environmentally Induced Delayed Fractures in Martensitic High-Strength Steels.* Air Force Materials Laboratory Technical Report AFML-TR-68-80.
39. Puzak, P. P., and Lange, E. A. (1969). *Fracture Toughness Characteristics of the New Weldable Steels of 180- to 210-ksi Yield Strengths.* Naval Research Laboratory Report NRL-6931.
40. Gross, J. H. (1968). "The New Development of Steel Weldments." 1968 Adams Lecture, *Welding Journal* **47**, 241s.
41. Webster, D., (1969). *The Stress Corrosion Resistance and Fatigue Crack Growth Rate of a High Strength Martensitic Stainless Steel, AFC 77.* Boeing Document D6-23973, The Boeing Company.
42. Kennedy, J. W., and Whittaker, J. A. (1968). "Stress Corrosion Cracking of High Strength Steels." *Corrosion Science* **8** (6), 359.
43. Tiner, N. A., and Gilpin, C. B. (1966). "Microprocesses in Stress Corrosion of Martensitic Steels." *Corrosion* **22**, 271.
44. Slunder, C. J., and Boyd, W. K. (1961). *Environmental and Metallurgical Factors of Stress Corrosion Cracking in High Strength Steels.* Battelle Memorial Institute, DMIC Report 151.
45. Davis, R. A., et al. (1964). "Stress Corrosion Cracking Study of Several High Strength Steels." *Corrosion* **20** (3), 93t.
46. Benjamin, W. D., and Steigerwald, E. A. (1967). *Stress Corrosion Cracking Mechanisms in Martensitic High Strength Steels.* Air Force Materials Laboratory Technical Report AFML-TR-67-98.
47. Taniguchi, N., and Troiano, A. R. (1969). "Stress Corrosion Cracking of 4340 Steel in Different Environments." *Transactions of the Iron Steel Institute of Japan* **9** (4), 306.
48. Sandoz, G., and Newbegin, R. L. (1969). "A Comparison of Laboratory Salt Water and Flowing Natural Seawater as an Environment for Tests of Stress Corrosion Cracking Susceptibility." *Naval Research Laboratory Progress Report*, p. 29.
49. Phelps, E. H., (1969). "A Review of the Stress Corrosion Behavior of Steels with High Yield Strength." In: *Proceedings of Conference on Fundamental Aspects of SCC*, (R. W. Staehle, ed). NACE, p. 398.
50. Kennedy, J. W., and Whittaker, J. A. (1968). "Stress Corrosion Cracking of High Strength Steels." *Corrosion Science* **8** (6), 359.
51. Leckie, H. P. (1969). "Effects of Environments on Stress Induced Failure of High Strength Maraging Steels." *Proceedings of Conference on Fundamental Aspects of Stress Corrosion Cracking*, NACE, Houston, TX, p. 411.
52. Logan, H. L. (1965). "The Stress-Corrosion Cracking of Metals." *Metals Engineering Quarterly* **5** (2), 32.
53. Van Der Sluys, W. A. (1969). "Mechanisms of Environment Induced Subcritical Flaw Growth in AISI 4340 Steel." *Engineering Fracture Mechanics* **1**, 447.
54. Uhlig, H. H., (1977). "Effect of Aqueous Sulfides and Their Oxidation Products on Hydrogen Cracking of Steel." *Materials Performance* **16** (1) 22–25.
55. Herbsleb, G., et al. (1981). "Occurrence and Prevention of Hydrogen Induced Stepwise Cracking and Stress Corrosion Cracking of Low Alloy Pipeline Steels." *Corrosion* **37** (5), 246–256.
56. Moore, E. M., and Warga, J. J. (1976). "Factors Influencing the Hydrogen Cracking Sensitivity of Pipeline Steels." *Materials Performance* **15** (6), 17.
57. Southwell, C. R., et al. (1976). "Corrosion of Metals in Tropical Environments—Final Report of 16-Year Exposures." *Materials Performance* **15** (7), 9–25.

58. Wesley, W. A. (1940). "Controlling Factors in Galvanic Corrosion." *ASTM Proceedings* **40**, 690.
59. Evans, U. R. (1938). *Metallic Corrosion, Passivity and Protection.* Edward Arnold and Company, London.
60. Hack, H. P. (1976). "Seawater Corrosion of Fasteners in Various Structural Materials." *Naval Ship R & D Center Report DTNSRDC*-76-0034, p. 34.
61. Perkins, J., et al. (1979). "Flow Effects on Corrosion of Galvanic Couples in Seawater." *Corrosion* **35** (1), 23–33.
62. Kinson, K., et al. (1981). "Steel Piling in Materials." *BHP Technical Bulletin* **25** (2), 8–15.
63. Escalante, E., and Iverson, W. P. (1978). "Protection of Steel Piles by Nonmetallic Coatings in Seawater." *Materials Performance* **17** (10), 9–15.
64. Mor, E. D., et al. (1979). "Effects of Rolling Direction and Tensile Stress on the Corrosion of Naval Carbon Steel in Natural Marine Environments—II. Harbour Anaerobic Sediment." *British Corrosion Journal* **14** (2), 103–108.
65. Johnson, M. J., and Pavlik, P. J. (1980). "Atmospheric Corrosion of Stainless Steel." *Atmospheric Corrosion, Hollywood, FL*, pp. 461–473.
66. Southwell, C. R., and Bultman, J. D. (1980). "Atmospheric Corrosion Testing in the Tropics." *Atmospheric Corrosion, Hollywood, FL*, pp. 943–967.
67. Johnson, K. E., and Abbott, J. S. (1974). "Bimetallic Corrosion Effects on Mild Steel in Natural Environments." *British Corrosion Journal* **9** (3), 171–176.
68. Prince, D. E. (1975). *Corrosion Behavior of Metal Fasteners in Graphite-Epoxy Composites.* AFML, Wright-Patterson AFB, OH AFML-TR-75-53, Final, ADB006394-Limited.
69. Morris, A. W. (1980). "Electrochemical Screening of Fastener Materials for Carbon Composite Joint Applications." *Corrosion '80, Chicago, IL*, Paper 23.
70. Stanley, J. K. (1972). "The Current Situation on the Stress Corrosion Cracking and Hydrogen Embrittlement of High Strength Fasteners." *13th Structures, Structural Dynamics, Materials Conference, TX.*
71. Money, K. L., and Kirk, W. W., (1978). "Stress Corrosion Cracking Behavior of Wrought Fe-Cr-Ni Alloys in Marine Atmosphere." *Materials Performance* **17** (7), 28–36.
72. Phelps, E. H. (1967). "Stress Corrosion Behavior of High Yield-Strength Steels." *Seventh World Petroleum Congress* **9**, 201–209.
73. Slunder, C. J. (1961). "Stress-Corrosion Cracking of High-Strength Stainless Steels in Atmospheric Environments." *DMIC Report 158.*
74. Sprowls, D. O., Shumaker, M. B., and Walsh, J. D. (1973). *Evaluation of Stress-Corrosion Cracking Susceptibility Using Fracture Mechanics Techniques.* Alcoa, PA, Final-Part I, NACA, NAS8-21487.
75. Humphries, T. S., and Nelson, E. E. (1980). *Stress Corrosion Cracking Evaluation of Martensitic Precipitation Hardening Stainless Steels.* NASA, Marshall Space Flight Center, AL, N80-16142, NASA-TM-78257.
76. Southwell, C. R., Bultman, J. D., and Alexander, A. L. (1976). "Corrosion of Metals in Tropical Environments—Final Report of 16-Year Exposures." *Materials Performance* **15** (7), 9–26.
77. Larrabee, C. P. (1945). "Corrosion of Steels in Marine Atmospheres and in Seawater." *Transactions of the Electrochemical Society* **87**, 171–182.
78. Alexander, A. L., Southwell, C. R., and Forgeson, B. W. (1961). "Corrosion of Metals in Tropical Environments, Part 5—Stainless Steel." *Corrosion* **17** (7), 345t–353t.
79. Reinhart, F. M. (1966). *Corrosion of Materials in Hydrospace.* U.S. Naval Civil Engineering Laboratory, R-504, AD644473.
80. Lawson, H. H., Doughty, S. E., and Jones, R. T. (1974). "Evaluating Material Performance in a 3000-GPD Stainless Steel Desalination Test Plant." *Materials Performance* **13** (3), 11–16.

81. Kowaka, M., Nagano, H., and Suzuki, E. (1982). "Crevice Corrosion Test Method for Stainless Steel." *The Sumitomo Search* **27**, 9–21.

82. Baker, D. W. C., Heaton, W. E., and Patient, B. C. (1972). "Evaluation of Stainless Steels for Condenser Tube Nest Construction." *Corrosion Science* **12** (3), 247–264.

83. Hoskinson, D. W., and Kuester, C. W. (1965). "Operating Experiences with Stainless Steel Condenser Tubes in Central Stations." *Journal of Engineering for Power* **87** (4); *Transactions of the ASME*, Series A, pp. 401–404.

84. Hack, H. P. (1982). "Crevice Corrosion Behavior of 45 Molybdenum-Containing Stainless Steels in Seawater." *Corrosion '82, Houston, TX*, Paper 65.

85. Lee, T. S. (1981). "A Method for Quantifying the Initiation and Propagation Stages of Crevice Corrosion." *Electrochemical Corrosion Testing*, ASTM-STP-727, pp. 43–68.

86. Asphahani, A. I., Manning, P. E., Silence, W. L., and Hodge, F. G. (1980). "Highly Alloyed Stainless Materials for Seawater Applications." *Corrosion '80, Chicago, IL*, Paper 29.

87. Ulanovskii, I. B., Egorova, V. A., and Kakasheva, T. A. (1976). "Corrosion of Steel, Titanium, and Copper at Various Depths in the Black Sea." *Protection of Metals* **12** (4), 394–395.

88. Garner, A. (1980). "Crevice Corrosion of Stainless Steel in Seawater: Correlation of Field and Laboratory Tests." *Corrosion '80, Chicago, IL*, Paper 35.

89. Kovach, C. W., Redmerski, L. S., and Kurtz, H. D. (1980). "Crevice Corrosion Performance of a Ferritic Stainless Steel Designed for Saline Water Condenser and Heat Exchanger Applications." *Corrosion '80, Chicago, IL*, Paper 95.

90. Eden, A. (1979). "Recent Advances in the Use of Stainless Steel for Offshore and Seawater Applications." *Anti-Corrosion Methods and Materials* **26** (11), 7, 9–13.

91. Liebert, B. E., Sethuramalingam, K., and Larsen-Basse, J. (1981). "Corrosion Effects in OTEC Heat Exchanger Materials." *Materials Performance* **20** (8), 22–28.

92. Willingham, C. A. (1971). *Anaerobic Bacterial Corrosion of Metals in Seawater in Elevated Hydrostatic Pressures.* Clapp Laboratories, Duxbury, MA, AD 727221, Final, N00014-68C0365.

93. Lendvai-Lintnor, E. et al. (1978). "An Innovative Approach to Controlling Salt Water Cooling Tower System Problems." *Materials Performance* **17** (12), 16–22.

94. Walston, K. R. (1975). "Materials Problems in Salt Water Cooling Towers." *Materials Performance* **14** (6), 22–26.

95. LaQue, F. L. (1982). "Topics for Research in Marine Corrosion." *Materials Performance* **21** (4) 13–18.

96. Brennert, S., and Lindh, B. (1975). "Resistance to Corrosion of Metallic Material for Seawater Bearing Pipes." *7th Scandinavian Corrosion Congress, Trandheim, Norway*, pp. 346–355.

97. Little, B., Morse, J., Loeb, G., and Spiehler, F. (1981). "Gulf of Mexico Study of Biofouling on OTEC Heat Exchanger Candidate Alloys." *Materials Performance* **20** (8), 16–21.

98. Mankowski, J., Szklarska-Smialowska, Z. (1977). "The Effects of Specimen Position on the Shape of Corrosion Pits in an Austenitic Stainless Steel." *Corrosion Science* **17** (9), 725–735.

99. Scotto, V., Ventura, G., and Traverso, E. (1970). "The Influence of Non metallic Inclusion Nature and Shape on the Pitting Corrosion Susceptibility of 18Cr9Ni and 17Cr11Ni2Mo Austenitic Stainless Steels." *Corrosion Science* **19** (4), 237–250.

100. Eklund, G. S. (1976). "On the Initiation of Crevice Corrosion on Stainless Steel." *Journal of Electrochemical Society* **123** (2), 170–173.

101. Brennert, S., and Eklund, G. (1976). "On the Mechanism of Pitting Corrosion of Stainless Steels in Chloride Solutions—Initiation." *Scandinavian Journal of Metallurgy* **5** (2), 16–20.

102. Kain, R. M. (1981). "Crevice Corrosion of Stainless Steels in Seawater and Related Environments." *Corrosion '81, Toronto, Ontario, Canada*, Paper 200.

103. Wheatfall, W. L. (1967). *Metal Corrosion in Deep-Ocean Environments.* U.S. Navy Marine Engineering Laboratory, Annapolis, MD, R-429/66.

104. Fukuzuka, T., Shimogori, K., Fujiwara, K., and Sugie, K. (1980). "Corrosion Resistance of High Molybdenum Stainless Steel (20Cr-22Ni-5Mo-0.2N) to Seawater." *Transactions of Iron and Steel Institute of Japan* **20** (9), B403.

105. Weber, J. (1981). "Materials for Handling Seawater at High Flow Velocities." In: *Materials to Supply Energy Demand, Harrison, BC, ASM*, pp. 49–80.

106. Todd, B. (1978). "Nickel Containing Materials for Marine Applications." *Anti-Corrosion* **25** (10), 4–7.

107. Hack, H. P. (1980). "Galvanic Corrosion of Piping and Fitting Alloys in Sulfide Modified Seawater." *Journal of Testing and Evaluation* JTEVA **8** (2), 74–79.

108. Deverell, H. E., and Davis, J. A. (1978). "Effect of Sulfides and Other Pollutants on the Seawater Corrosion of Stainless Steels and Copper Base Alloys." *Corrosion '78, Houston, Texas*, Paper 27.

109. Hodgkiess, T., and Rigas, S. (1983). "A Comparison of the Corrosion Resistance of Some Higher-Alloy Stainless Steels in Seawater at 20–100°C." *Desalination* **44**, 283–294.

110. Kain, R. M. (1979). "Crevice Corrosion Resistance of Austenitic Stainless Steels in Ambient and Elevated Temperature Seawater." *Corrosion '79, Atlanta, GA*, Paper 230.

111. Reinhart, F. M., and Jenkins, J. F. (1971). *The Relationship Between the Concentration of Oxygen in Seawater and the Corrosion of Metals.* U.S. Naval Civil Engineering Laboratory, Port Hueneme, CA, pp. 562–577.

112. Veingarten, A. M., Grinvald, I. I., and Laanson, G. A. (1979). "Stress Corrosion Resistance of High-Strength Dispersion-Hardening Stainless Steels." *Soviet Materials Science* **6** (4), 445–449.

113. Dundas, J. F., and Bond, A. P. (1980). "Relation of Composition to Stress Corrosion Cracking of Austenitic Stainless Steels." *Corrosion '80, Chicago, IL*, Paper 89.

114. Redmond, J. D. and Miska, K. H. (1983). "High-Performance Stainless Steels For High-Chloride Service." *Chemical Engineering* **90** (15), 93–96.

115. Speidel, M. O. (1981). "Stress Corrosion Cracking of Stainless Steels in NaCl Solutions." *Metallurgical Transactions* **12A** (5), 779–789.

116. Cramer, S. D., and Carter, J. P. (1980). "Laboratory Corrosion Studies in Low- and High-Salinity Geobrines of the Imperial Valley, California." U.S. Bureau of Mines, Washington, DC, Report of Investigation 8415.

117. Truman, J. E. (1977). "Influence of Chloride Content, pH, and Temperature of Test Solution on the Occurrence of Stress Corrosion Cracking (SCC) with Austenitic Stainless Steel." *Corrosion Science* **17** (9), 737–746.

118. Herbsleb, G., and Popperling, R. K. (1980). "Corrosion Properties of Austenitic-Ferritic Duplex Steel AF 22 in Chloride and Sulfide Containing Environments." *Corrosion '80, Chicago, IL*, Paper 13.

119. Lin, L. F., Cragnolino, G., Szklarska-Smialowska, A., and MacDonald D. D. (1981). "Stress Corrosion Cracking of Sensitized Type 304 Stainless Steel in High Temperature Chloride Solutions." *Corrosion* **37** (11), 616–626.

120. Radhakrishnan, V. R., and Iyer, S. R. (1979). "Stress Corrosion Testing and Data Analysis." *Werkstoffe und Korrosion* **30** (3), 185–189.

121. Gurevich, S. E. (1979). "Investigation of the Stress Corrosion Crack Growth Rates of Steel and Titanium Alloy in Artificial Sea Water." *Mechanical Behavior of Materials* **2**, 457–462.

122. McGuire, M. F., Troiano, A. R., and Hehemann, R. F. (1973). "Stress Corrosion of Ferritic and Martensitic Stainless Steels in Saline Solutions." Corrosion **29** (7), 268–271.

124. Werchniak, W., and Gudas, J. P. (1981). "Seawater Corrosion and Corrosion Fatigue of High-Strength Cast Alloys for Propellers and Impellers." *Corrosion '81, Toronto, Ontario, Canada*, Paper 199.

125. Pini, G., and Weber, J. (1979). "Materials for Pumping Seawater and Media with High Chloride Content." *Sulzer Technical Review* **2**, 69–80.
126. Lee, T. S. (1983). "Seawater Velocity Effects on Corrosion Behavior of Materials." *Sea Technology* **24** (11), 51, 53, 55–59.
127. Hohman, A. E., and Kennedy, K. W. (1963). "Corrosion and Materials Selection Problems on Hydrofoil Craft." *Materials Protection* **2** (9), 56–68.
128. Simoneau, R., Fihey, J. L., and Roberg, R. (1981). "The Effect of Corrosion in Low-Intensity Cavitation Erosion." In: *Cavitation Erosion in Fluid Systems*, 71–81.
129. Samans, C. H. (1964). "Stress Corrosion Cracking Susceptibility of Stainless Steels and Nickel-Base Alloys in Polythionic Acids and Acid Copper Sulfate Solution." *Corrosion* **20** (8), 256t–261t.
130. Steigerwald, R. F. (1974). "New Molybdenum Stainless Steels For Corrosion Resistance: A Review of Recent Developments." *Materials Performance* **13** (9), 9–16.
131. Maurer, J. R. (1979). "Controlling Corrosion Problems with the New High Technology Stainless Steels." *Journal Materials for Energy Systems* **1** (1), 41–51.
132. Hodgkiess, T., Eid, N., and Hanbury, W. T. (1978). "Corrosion of Welds in Seawater." *Desalination* **27** (2), 129–136.
133. Henrikson, S., Olsson, J., and Arvesen, J. (1981). "Corrosion Tests of Stainless Steels in Flowing Baltic Sea Water." *6th Scandinavian Corrosion Congress, Gothenburg, Sweden* CONF-710520-1.
134. Lee, T. S., and Tuthill, A. H. (1983). "Use of Carbon Steel to Mitigate Crevice Corrosion of Stainless Steels in Seawater." *Materials Performance* **22** (1), 48–53.
135. Henrikson, S., and Asberg, M. (1975). "Corrosion Potentials and Galvanic Corrosion Currents for Various Material Combinations in Baltic Sea-water." *7th Scandinavian Corrosion Congress, Trandheim, Norway*, pp. 371–384.
136. Smith, C. A., and Compton, K. G. (1975). "Potentials of Selected Metal Alloys in Seawater at Elevated Temperatures." *Corrosion* **31** (9), 320–326.
137. Ebihara, W. T. (1979). *Tropical Exposure of Galvanically Coupled Metal Systems.* Army Armament R&D, Dover, NJ, ADB039047-Limited, ARSCD-TR-79004, p. 18.
138. Waldron, L. J., Forgeson, B. W., Peterson, M. H., and Brown, B. F. (1963). *Performance of Stainless Steel Galvanically Coupled to Other Metals.* NRL, Washington, DC, NRL-MR-1388, AD679659.
139. Gehring, G. A., Jr., and Maurer, J. R. (1981). "Galvanic Corrosion of Selected Tubesheet/Tube Couples Under Simulatd Seawater Condenser Conditions." *Corrosion '81, Toronto, Ontario, Canada*, Paper 202.
140. Gehring, G. A., Jr., and Kyle, R. J. (1982). "Galvanic Corrosion in Steam Surface Condensers Tubed with either Stainless Steel or Titanium." *Corrosion '82, Houston, TX*, Paper 60.
141. Pelensky, M. A., Jaworski, J. J., and Gallaccio, A. (1976). "Air, Soil, and Sea Galvanic Corrosion Investigation at Panama Canal Zone." In: *Galvanic and Pitting Corrosion*, ASTM-STP-576, pp. 94–113.
142. Charlot, L. A., Wong, C. M., and Gillam, W. S. (1970). *Investigation of Galvanically Induced Hydriding of Titanium in Saline Solutions.* Battelle Northwest Laboratories, Richland, WA, RDPR-70-624, PB198639.
143. Scholes, I. R., Astley, D. J., and Rowlands, J. C. (1977). "Bimetallic Corrosion in Seawater." EROCOR '77, *European Congress on Metallic Corrosion, London, England*, pp. 161–169.
144. Mansfeld, F. (1973). "Galvanic Corrosion of Al Alloys Coupled to Coated PH 13-8 Mo Stainless Steel." *Corrosion* **29** (7), 276–281.
145. Davis, J. A., and Gehring, G. A. (1976). "Galvanic Corrosion as a Function of Applied Potential in High Velocity Seawater." *Corrosion '76, Houston, TX*, Paper 75.

146. Lennox, T. J., Jr., Peterson, M. H., Smith, J. A., and Groover, R. E. (1974). "Corrosion and Cathodic Protection of 5086-H32 Al Coupled to Dissimilar Metals." *Materials Performance* **13** (2), 31–36.

147. Miller, B. A., and Lee, S. G. (1976). *The Effect of Graphite-Epoxy Composites on the Galvanic Corrosion of Aerospace Alloys.* Wright-Patterson AFB, Final, AFML-TR-76-121, ADA035029.

148. Southwell, C. R., and Alexander, A. L. (1969). "Corrosion of Metals in Tropical Environments—Nickel and Nickel-Copper Alloys." *Materials Protection* **8** (3), 39–44.

149. Miska, K. H. (1974). "Most Nonferrous Metals and Alloys Weather Well." *Materials Engineering* **79** (4), 64–66.

150. Johnson, K. E., and Abbott, J. S. (1974). "Bimetallic Corrosion Effects on Mild Steel in Natural Environments." *British Corrosion Journal* **9** (3), 171–176.

151. Bomberger, H. B., et al. (1954). "Corrosion Properties of Titanium in Marine Environments." *Journal of the Electrochemical Society* **101** (9), 442–447.

152. Ballantine, W. (1976). "Choosing Corrosion-Resistant Fasteners for Marine Applications." Sea Technology **17** (2), 8–10.

153. Anderson, D. B., and Ross, R. W., Jr. (1976). "Protection of Steel Piling in Marine Splash and Spray Zones—Metallic Sheathing Concept." *4th International Congress on Marine Corrosion and Fouling, Boulogne, France*, pp. 461–473.

154. Davis, J. W. (1972). "Corrosion and Stress Corrosion Susceptibility of Several High Temperature Materials." *McDonnell Douglas, 3rd Quarterly*, NAS8-27270, 34 pp.

155. Montano, J. W. (1971). *A Mechanical Property and Stress Corrosion Evaluation of MP35N Multiphase Alloy.* NASA Report NASA-TM-X-64591, 25.

156. Deel, O. L., and Mindlin, H. (1972). *Engineering Data on New Aerospace Structural Materials.* Battelle, Columbus, OH, Final Report AFML-TR-72-196-VI, 286.

157. Lennox, T. J., Jr., et al. (1982). "Corrosion Resistance and Response to Cathodic Protection of Advanced Alloys in Seawater." *Corrosion '82, Houston, TX*, (64).

158. Reinhart, F. M. (1967). "Corrosion of Materials in Hydrospace. Part II: Nickel and Nickel Alloys." *Naval Civil Engineering Lab Report* NCEL-TN-915, p. 65.

159. Tipton, D. G., and Kain, R. M. (1980). "Effect of Temperature on the Resistance to Pitting of Monel Alloy 400 in Seawater." *Corrosion '80, Chicago, IL*, Paper 36, p. 13.

160. Asphahani, A. I., et al. (1980). "Highly Alloyed Stainless Materials for Seawater Applications." *Corrosion '80, Chicago, IL*, Paper 29, p. 42.

161. Anderson, D. B. (1976). "Statistical Aspects of Crevice Corrosion in Seawater." *Galvanic and Pit Corr-Field and Lab Studies* ASTM-STP-576, pp. 231–242.

162. Kain, R. M. (1982). "Crevice Corrosion Resistance of Several Iron and Nickel–Base Cast Alloys in Seawater." *Corrosion '82, Houston, TX* (66), p. 25.

164. Vreeland, D. C., *Galvanic Corrosion Behavior of Wear-Resistant Materials for Mechanical Shaft Seals.* Navy Marine Engineering Laboratory, MEL-242/66, p. 13.

165. Ailor, W. H. (1966). "Five-Year Corrosion of Aluminum Alloys at Several Marine Sites." *British Corrosion Journal* **18** (6) 237–243.

166. Ailor, W. H., (1974). "Metal Exposures at Tropical and Marine Sites." *6th Annual Offshore Technology Conference* **1**, 293–312.

167. Pelensky, M. A., et al. (1978). "Corrosion Investigations at Panama Canal Zone." *Atmospheric Exposure Testing Symposium*, pp. 48–73.

168. Lage, K., and Sune, L. (1976). "Galvanic Corrosion of Seven Different Al Alloys in Contact with Copper, Steel, Stainless Steel, and Zinc—Field Tests in Outdoor Atmospheres." In: *Corrosion and Corrosion Protection of Aluminum*, pp. 95–106.

169. Sprowls, D. O., et al. (1973). *Evaluation of Stress-Corrosion Cracking Susceptibility Using Fracture Mechanics Techniques.* Alcoa, PA, NAS8-21487, p. 117.

170. Dorward, R. C., et al. (1978). "Marine Atmosphere Stress Corrosion Tests on Precracked Specimens From High Strength Aluminum Alloys: Effect of Corrosion-Product Wedging." *Journal of Testing and Evaluation* **6** (4), 268–275.

171. Sprowls, D. O. (1976). "High Strength Aluminum Alloys with Improved Resistance to Corrosion and Stress Corrosion Cracking." *1976 Tri-Service Conference on Corrosion*, MCIC-77-33, pp. 89–120.

172. Jankowsky, E. J., et al. (1979). *Aircraft Carrier Exposure Tests of Aluminum Alloys.* Naval Air Development Center, NADC-79251-60, p. 34.

173. Ketcham, S. J., and Jankowsky, E. J. (1980). "Shipboard Exposure Tests of Aluminum Alloys." *Proceedings of the 1980 Tri-Service Conference on Corrosion*, AFWAL-TR-81-4019.

174. Brooks, C. L. (1970). "Aluminum-Magnesium Alloys 5086 and 5456-H116." *Naval Engineers Journal* **82** (4), 29–32.

175. Czyryca, E. J., and Hack, H. P. (1975). "Corrosion of Aluminum Alloys in Exfoliation-Resistant Tempers Exposed to Marine Environments for 2 Years." *Proceedings of the 1974* Tri-Service Corrosion of Military Equipment Conference, AFML-Tr-75-42.

176. Southwell, C. R., et al. (1965). "Corrosion of Metals in Tropical Environments—Aluminum and Magnesium." *Materials Protection* **4** (12), 30.

177. Reinhart, F. M. (1969). *Corrosion of Materials in Hydrospace. Part V: Aluminum Alloys.* Naval Civil Engineering Laboratory, NCEL-TN-N-1008, p. 89.

178. Beberdick, L. E. (1981). *A Preliminary Investigation of the Corrosion and Stress Corrosion Susceptibility of Thermomechanically Processed High Magnesium, Aluminum-Mg Alloys.* Naval Postgrad School, Monterey, CA, p. 85.

179. Bonora, P. L. (1977). "The Anodic Dissolution of Aluminum Zinc Alloys in Seawater." *Materials Chemistry* **2** (1), 35–49.

180. Zamin, M. (1981). "The Role of Manganese in the Corrosion Behavior of Al-Mn Alloys." *Corrosion* **37** (11), 627–632.

181. Klimko, A. F., and Mochalova, L. M. (1973). "Effect of Microalloying on Aging and Corrosion Resistance of AMg6 Alloys." *Soviet Materials Science* **9** (1), 103–104.

182. Dexter, S. C. (1980). "Effect of Variations in Seawater Upon the Corrosion of Aluminum." *Corrosion* **36** (8), 423–432.

183. Larsen-Basse, J. (1984). "Corrosion of Aluminum Alloys in Ocean Thermal Energy Conversion Seawaters." *Materials Performance* **23** (7), 16–21.

184. Godard, H. P., and Booth, F. F. (1964). "Corrosion Behavior of Aluminum Alloys in Seawater." *International Congress of Marine Corrosion and Protection, Cannes, France*, pp. 37–52.

185. Johnson, W. K. (1971). "Recent Developments in Pitting Corrosion of Aluminum." *British Corrosion Journal* **6**, 782.

186. Zossi, A. M., et al. (1976). "Effect of Solidification Microstructure on Corrosion Behavior of the Chill Zone in Aluminum Binary Alloys." *Metallurgical Transactions*, **7A** (10), 1489–1496.

187. Godard, H. P., and Booth, F. F. (1964). "Corrosion Behavior of Aluminum Alloys in Seawater" International Congress of Marine Corrosion and Protection," Cannes, France, 37–52.

188. Sujimoto, et al. (1970). *Journal of Japanese Institute of Metals* **34**, 312.

189. Muller, I. L., and Galvele, J. R. (1977). "Pitting Potential of High Purity Binary Aluminum Alloys. II: Al-Mg and Al-Zn Alloys." *Corrosion Science* **17** (12), 995–1007.

190. Muller, I. L., and Galvele, J. R. (1977). "Pitting Potential of High Purity Binary Aluminum Alloys—Al-Cu Alloys Pitting and Intergranular Corrosion." *Corrosion Science* **17** (3), 179–189.

191. Groover, R. E. et al. (1969). "Characterization of the corrosion Behavior and the Response to Cathodic Protection of Nineteen Aluminum Alloys in Seawater." Naval Research Laboratory, NRL Memo 1961.
192. Holtan, H., and Sigurdsson, H. (1977). "Pitting Potentials of Al Alloys." *Werkstoffe und Korrosion* **28** (7), 475–477.
193. Niskanen, P., et al. (1982). "Corrosion of Aluminum Alloys Containing Lithium." *Corrosion Science* **22** (4), 283–304 (1982).
194. Galvele, J. R., and DeMicheli, S. M. (1970). *Corrosion Science* **10**, 795.
195. Sugimoto, K. et al. (1975). "Stress Corrosion Cracking of Aged Al-4 Percent Cu Alloy in NaCl Solution." *Corrosion Science* **15** (11/12), 709–720.
196. Rynewicz, J. (1974). "Corrosion in Deep Oceans." *Ocean Engineering*—Petroleum Engineer **46** (13), 57–66.
198. Dexter, S. C. (1979). "Oxygen, Temperature, and pH Effects on Corrosion of Aluminum in Seawater." *Sixth Ocean Thermal Energy Conversion Conference*, 2 (CONF-790631), pp. 12.4 to 1-11.
199. Ahmed, Z. (1981). "Corrosion and Corrosion Prevention of Aluminum Alloys in Desalination Plants—II." *Anti-Corrosion Methods and Materials* **28** (7), 4–10.
200. Alier, A. F., and Freiman, L. I. (1976). "Electrochemical Behavior of Aluminum Alloys in Connection with the Operating Conditions of a Seawater Thermal Desalination Plant." *Protection of Metals* **12** (4), 351–356.
201. Herrigel, H. R. (1976). Materials Performance **15**, 43–45.
202. Ahmad, Z. (1982). "Effect of Velocity on the Corrosion Behavior of Al-Mg 3 in North Sea Water." *Metal Corrosion Industry* **57** (677), 11–26.
203. Davis, J. A., and Gehring, G. A., Jr. (1975). "Effect of Velocity on the Seawater Corrosion Behavior of High Performance Ship Materials." *Materials Performance* **14** (4), 32–39.
204. Symonds, J. (1978). "Effect of Velocity on the Seawater Corrosion Resistance of Two Aluminum Alloys." *5th Ocean Thermal Energy Conversion Conference* **4**, viii/222–230.
205. Gehring, G. A., Jr., and Peterson, M. H. (1981). "Corrosion of 5456-H1117 Aluminum in High Velocity Seawater." *Corrosion* **37** (4), 232–242.
206. Mansfeld, F., et al. (1974). "Galvanic Corrosion of Aluminum Alloys. I: Effect of Dissimilar Metal." *Corrosion* **30** (10), 343–353.
207. Southwell, C. R., and Bultman, J. D. (1980). "Atmospheric Corrosion Testing in the Tropics." *Atmospheric Corrosion* Nov, 943–967.
208. Scholes, I. R., et al. (1977). "Bimetallic Corrosion in Seawater." *European Congress on Metallic Corrosion, London, England*, pp. 161–169.
209. Speidel, M. O. (1975). "Stress Corrosion Cracking of Aluminum Alloys." *Metallurgical Transactions* **6A** (4), 631–651.
210. Holtyn, C. H. (1966). *The Age of Ships*. Society of Naval Architects and Marine Engineers, Preprint 9.
211. Ohnishi, T., et al. (1976). "Effect of Microstructure of Stress Corrosion Susceptibility." *Journal of Japan Institute of Light Metals* **26** (1), 8–17.
212. Baba, Y., et al. (1973). "Effect of Addition of Bi on Stress-Corrosion Cracking of Al-Mg Alloy." *Sumitomo Light Metal Technical Reports*, **14** (2), 1–6.
213. Chandrashekar, R., and Vasu, K. I. (1973). "Morphology of Stress Corrosion Cracks and Crack Branching in Aluminum-Magnesium Alloys." *Metallography* **6** (5), 377–392.
214. Baba, Y. (1972). "Effect of Additions of Cu and Zr on Stress Corrosion Cracking of Al-Mg Alloys." *Journal of Japan Institute of Metals* **36** (4), 341–346.
215. Ford, J. A., and Pryor, M. J. (1970). *Corrosion Performance of MRL-ADJ*. Olin Metals Research Laboratories Report MRL-70-PR-11.

216. McMillan, J. C., and Hyatt, M. V. (1968). *Development of High-Strength Aluminum Alloys with Improved Stress-Corrosion Resistance. Phase II.* Boeing Report AFML-TR-68-148, pp. 142.
217. Bhat, M. V., and Vasu, K. I. (1983). "Electrochemical Aspects of Stress Corrosion Cracking in Aluminum-5 Percent Zinc and Aluminum-5 Percent Copper Alloys Containing Added Titanium." *Journal of the Electrochemical Society of India* **32** (3), 245–251.
218. Bhat, M. V., and Vasu, K. I., (1983). "Relevance of Microstructure to Stress Corrosion Cracking in Al-5 Percent Zn and Al-5 Percent Cu Alloys Containing Titanium." *Practical Metallography* **20** (3), 130–137.
219. Speidel, M. O., and Hyatt, M. V. (1972). *Advances in Corrosion Science of Technology*, Vol. 2. Plenum Press, New York, NY, pp. 115–335.
220. Miller, W. S., and Scott, V. D. (1978). "Effect of Composition on the Mechanical and Stress-Corrosion Behaviour of Al-Zn-Mg Alloys." *Metal Science* **12** (2), 95–101.
221. Chen, C., and Judd, G., (1973). *Stress Corrosion Cracking Susceptibility and Aging Characteristics of Al-Zn-Mg-Ti Alloys.* Rensselaer Polytechnic Institute, Report TR-5, p. 22.
222. Chen, C., and Judd, G. (1974). *Microstructural Characterization and Stress Corrosion Cracking Susceptibility of Al-Zn-Mg-Ti Alloys.* Rensselaer Polytechnic Institute, Report TR-6, p. 19.
223. Evans, G. B. (1969). "The Choice of Materials." *29th Meeting of Structures and Materials Panel, AGARD, Turkey*, p. 10.
224. Polmaer, I. J. (1960–61). "The Properties of Commercial Al-Zn-Mg Alloys, Practical Implications of Trace Additions of Silver." *Journal of the Japan Institute of Metals* **89**, 51–59.
225. Rosenkranz, W. (1963). "Development of a High Strength, Stress-Corrosion Resistant Alloy of the Al-Zn-Mg Type. Part II: the Effect of the Chemical Composition of the Artificial Aging and Stress-Corrosion of High Strength Al-Zn-Mg-Cu Alloys." *Aluminum* **39**, 741–752.
226. Sprowls, D. O., and Brown, R. H. (1969). "Stress Corrosion Mechanisms for Aluminum Alloys." *Proceeding of Conference on Fundamental Aspects of Stress Corrosion Cracking*, NACE, p. 466.
227. Staley, J. T. (1969). *Investigation to Improve the Stress Corrosion Resistance of Aluminum Aircraft Alloys Through Alloy Additions and Specialized Heat Treatment.* Alcoa Final Report N00019-68C0146, p. 199.
228. Pizzo, P. P. (1982). *The Relative Stress Corrosion Cracking Susceptibility of Candidate Aluminum-Lithium Alloys for Aerospace Applications.* Advanced Research and Applications Corporation Report NASA-CR-3578, p. 126.
229. Enjo, T., et al. (1979). "Effects of Relatively Insoluble Compounds and Beta Phase on Stress Corrosion Cracking in 5083 Aluminum Alloy." *Transactions of JWRI* **8** (1) 67–75.
230. Niederberger, R. B., et al. (1966). "Corrosion and Stress Corrosion of 5000 Series Al Alloys in Marine Environments." *Corrosion* **22** (3), 68–73.
231. Hayne, F. H., and Boyd, W. K. (1966). *Stress Corrosion Cracking of Aluminum Alloys.* DMIC Report 228, Battelle Memorial Institute.
232. Collins, J. F., and Maduell, C. L. (1977). "Polyalkylene Glycol Quenching of Aluminum Alloys." *Materials Performance* **16** (7), 20–23.
233. Sprowls, D. O., and Nock, J. A., Jr. (1965). *Thermal Treatment of Aluminum Base Alloy Article*, U.S. Patent 3,198,676.
234. Doig, P., et al. (1977). "The Stress Corrosion Susceptibility of 7075 Al-Zn-Mg-Cu Alloys Tempered from T6 to an Overaged T7X." *Corrosion* **33** (6), 217–221.

235. Koch, G. H., and Kolijn, D. T. (1979). "The Heat Treatment of Commercial Aluminum Alloy 7075." *Journal of Heat Treating*: **1** (3), 3–14.
236. Biggs, L. A., (1980). *Stress Corrosion Resistance of* 7010-*T*73651. Naval Air Development Center, NADC-80038-60, p. 13.
237. Dorward, R. C., and Hasse, K. R. (1976). *Flaw Growth of 7075, 7050, and 7049 Aluminum Plate in Stress Corrosion Environments*. Kaiser Aluminum, NAS8-30890, p. 212.
238. Shaffer, I. S. (1970). *Stress Corrosion Resistance of Die Forged 7175–T736 Aluminum Alloy*. Naval Air Development Center, NADC-MA-7055, p. 18.
239. Staley, J. T., and Hunsicker, H. Y. (1970). *Investigation to Develop a High Strength Stress-Corrosion Resistant Aluminum Aircraft Alloy*. Alcoa Final Report, p. 182.
240. Reimann, W. H., and Brisbane, A. W. (1973). "Improved Fracture Resistance of 7075 Through Thermomechanical Processing." *Engineering Fracture Mechanics* **5** (1), 67–78.
241. Hyatt, M. V., and Speidel, M. O. (1972). "High Strength Aluminum Alloys" In: *Stress Corrosion Cracking in High Strength Steels and in Titanium and Aluminum Alloys* (B. F. Brown, ed). Naval Research Laboratory, Chapter 4.
242. Hyatt, M. V. (1969). *Effects of Residual Stresses on Stress-Corrosion Crack Growth Rates in Aluminum Alloys*. Boeing Company, D6-24469, p. 23.
243. Ratke, L. (1982). "Influence of Surface Deformation on the Stress Corrosion Cracking Behavior of Age-Hardened AlZn5Mg3." *Aluminum and Supplement in English* **58** (12), 715–718.
244. Kohler, W. (1982). "Improvement of Stress Corrosion Resistance of Aluminum Weldments by Shot Peening." *2nd International Conference on Aluminum Weldments, Munich, FRG*, pp. II: 7.1-7.8.
245. Hanisch, A. H., and Burck, L. H. (1982). "Effects of Overloads on the Incubation Time for Stress Corrosion Cracking of 7075 Aluminum." Corrosion **38** (6), 330–335.
246. Helfrich, W. F. (1967). "Influence of Stress and Temperature on Short Transverse Stress Corrosion Cracking of an Al-4.2Zn-2.5Mg Alloy." ASTM-STP-425.
247. Hyatt, M. V., and Quist, W. E. (1970). *Effect of Exposure Time at 250°F on Stress Corrosion Crack Growth Rates in 2024-T351 Aluminum*. Boeing Company, D6-25218, p. 20.
248. Lifka, B. W., and Sprowls, D. O. (1972). "Significance of Intergranular Corrosion in High Strength Aluminum Alloy Products, Localized Corrosion—Cause of Metal Failure. *ASTM-STP-516*, pp. 120–144.
249. Rhodes, D., and Radon, J. C. (1978). "Fracture Analysis of Exfoliation in an Aluminum Alloy." *Engineering Fracture Mechanics* **10** (4) 843–852.
250. Tohma, K., et al. (1983). "Effect of Cu Addition on Intergranular Corrosion of Al-5.5Zn-2.5Mg Alloy." *Journal of the Japan Institute of Light Metals* **33** (7), 377–385.
251. Staley, J. T. (1970). *Investigation to Develop a High Strength Stress-Corrosion Resistant Naval Aircraft Aluminum Alloy*. Alcoa Final Report, p. 114.
252. Reinhart, F. M. (1967). *Corrosion of Materials in Hydrospace. III: Titanium and Titanium Alloys*. U.S. Naval Civil Engineering Laboratory Report NCEL-TN-921, p. 30.
253. Bomberger, H. B., et al. (1954). "Corrosion Properties of Titanium in Marine Environment." *Journal of the Electrochemical Society* **101** (9), 442–447
254. Cotton, J. B., and Downing, B. P. (1957). "Corrosion Resistance of Titanium to Seawater." *Transactions of the Institute of Marine Engineers* **69**, 311.
255. Feige, N. G., and Murphy, T. J. (1966). "Environmental Effects on Titanium Alloys." *Corrosion* **22** (11), 320–324.
256. Griess, J. C., Jr. (1968). "Crevice Corrosion of Titanium in Aqueous Salt Solutions." *Corrosion*. **24** (4), 96–109.
257. Minkler, W. W. (1978). "Titanium for Chemical Processing Equipment." *Metal Progress* **113**, 27–31.

258. Covington, L. C., et al. (1977). "Selection of Titanium Alloys for Corrosion Resistance." *Symposium on Selection Materials, AIChE Annual Meeting*, p. 42.

259. Adamson, W. L. (1976). *Marine Fouling of Titanium Heat Exchangers*. Report PAS-75-29, David W. Taylor Naval Ship R & D Center.

260. Franson, I. A., and Covington, L. C. (1977). "Titanium and the OTEC Environment." *Ocean Thermal Energy Conversion (OTEC) Symposium*, pp. 293–304.

261. "Properties and Applications of Titanium for Tubes." *Pipes and Pipelines International* **22** (5), 11–15.

262. Henrikson, S., and Knutsson, L. (1975). "Corrosion Tests in Baltic Sea Water on Heat Exchanger Tubes of Various Metallic Materials." *British Corrosion Journal* **10** (3), 128–135.

263. Davis, J. A., and Gehring, G. A., Jr., (1975). "Effect of Velocity on the Seawater Corrosion Behavior of High Performance Ship Materials." *Materials Performance* **14** (4), 32–39.

264. Basil, J. L. (1960). *Corrosion of Materials in High Velocity Seawater*. U.S. Naval Engineering Exp. Station Report 910160A, p. 9.

265. Crowe, C. R., and Hasson, D. F. (1981). "The Use of Titanium in Deep Sea Exploration." *Titanium for Energy and Industrial Applications*, pp. 93–110.

266. Anon (1970). "Titanium Heat Exchangers for Service in Seawater, Brine, and Other Natural Environments: The Corrosion Erosion and Galvanic Corrosion Characteristics of Titanium in Seawater, Polluted Inland Waters, and Brines." *Titanium Information Bulletin*, Imperial Metal Industries (Kynoch) Limited.

267. Hohman, A. E., and Kennedy, W. L. (1963). "Corrosion and Materials Selection Problems on Hydrofoil Craft," Materials Protection **2**, (9), 56–68.

268. Cox, B. (1981). "Titanium for Desalting Plants." *Titanium for Energy and Industrial Applications*, pp. 123–142.

269. Morris, A. W. (1980). "Electrochemical Screening of Fastener Materials for Carbon Composite Joint Applications." *Corrosion '80, Chicago, IL*, Paper 23, p. 9.

270. Brown, B. F. (1966). "A New Stress Corrosion Cracking Test for High-Strength Alloys." *Materials Research and Standards* **6** (3), 129–133.

271. Reinhart, F. M., and Jenkins, J. F. (1972). *Corrosion of Materials in Surface Seawater After 12 and 18 Months of Exposure*. U.S. Naval Civil Engineering Laboratory Final Report NCEL-TN-1213, 106 pp.

272. Blackburn, M. J., and Williams, J. C. (1969). "Strength, Deformation Modes and Fracture in Titanium-Aluminum Alloy." *Trans. Amer. Soc. Metals* **62**, 398.

273. Macky, T. L., and Tiner, N. A. (1969). "Stress Corrosion Susceptibility of Titanium Alloys in Aqueous Environments." *Metals Engineering Quarterly* **9** (4), 59–60.

274. Boyd, J. D., et al. (1970, 1971). *The Effect of Composition on the Mechanism of SCC of Titanium Alloys in Nitrogen Dioxide and Aqueous and Hot Salt Environments*. NASA CR-1525 (1970) and NASA CR-1846 (1971).

275. Cavallaro, J. L., and Wilcox, R. C. (1971). "Embrittlement of Ti-7Al Binary Alloys in Seawater." *Corrosion* **27** (4), 157–163.

276. Lane, I. R., and Cavallaro, J. L. (1968). "Metallurgical and Mechanical Aspects of the Seawater Stress Corrosion of Titanium." In: *Application Related Phenomena in Titanium Alloys*. ASTM STP 432, pp. 143–169.

277. Blackburn, M. J., and Cavallaro, J. L. (1969). "Metallurgical and Mechanical Aspects of the Stress Corrosion Cracking of Titanium Alloys." In: *Fundamental Aspects of Stress Corrosion Cracking*, R. W. Staehle, et al. ed). NACE, p. 620.

278. Crossley, F. A., and Carew, W. F. (1957). "Embrittlement of Ti-Al Alloys in the 6 to 10 Percent Aluminum Range." *Journal of Metals* **209** (1).

279. Crossley, F. A. (1981). "Aircraft Application of Titanium: A Review of the Past and Potential for the Future." *Journal of Aircraft* **18** (12), 993–1002.

280. Williams, W. L. (1969). "Development of Structural Titanium Alloys for Marine Applications." *Ocean Engineering* **1**, 375–384.
281. Wood, R. A., et al. (1973). "The Effects of Alloy Composition on the Salt Water Stress Corrosion Susceptibility of Titanium-Aluminum-Base Alloys." *Titanium Science and Technology*, Vol. 4. Plenum Press, New York, pp. 2639–2654.
282. Seagle, S. R., et al. (1968). "The Influence of Composition and Heat Treatment on the Aqueous Stress Corrosion of Titanium." In: *Application Related Phenomena in Titanium Alloys*. ASTM STP 432, pp. 170–188.
283. Curtis, R. E., et al. (1969) "Relationship Between Composition, Microstructure, and Stress-Corrosion Cracking (in Salt Solution) in Titanium Alloys." *Transactions of ASM* **62**, 457.
284. Sorkin, G., et al. (1976). "Ti-6Al-4V for Marine Uses." *Titanium and Titanium Alloys: Scientific and Technological Aspects* (Moscow, USSR) **3**, pp. 2139–2147.
285. Gastelow, C. R., and Weaver, M. J. (1977). *The Behavior of Titanium Alloys in Simulated Marine Gas Turbine Environments*. National Gas Turbine Establishment Report 344.
286. Shannon, J. L., and Brown, W. F. (1968). "A Review of Factors Influencing the Crack Tolerance of Titanium Alloys." In: *Application Related Phenomenon Titanium Alloys*. ASTM STP 432, pp. 33–63.
287. Piper, D. E., et al. (1968). "Corrosion Fatigue and Stress Corrosion Cracking in Aqueous Environments." *Metal Engineering Quarterly* **8** (3), 50–63.
288. Wanhill, R. J. (1973). "Fractographic Interpretation of Subcritical Cracking in a High Strength Titanium Alloy (IMI 550)." *Corrosion* **29** (11), 435–441.
289. Onman, S., and Picton, G. (1974). "Role of Hydrogen in the Stress Corrosion Cracking of Titanium Alloys." *Corrosion Science* **14**, 451–459.
290. Howe, D. G., and Goode, R. J. (1968). "Effect of Heat Treating Environmental Conditions on the Stress Corrosion Cracking Resistance of Several Titanium Alloys." In: *Applications Related Phenomena in Titanium Alloys*. ASTM STP 432, pp. 189–201.
291. Sandoz, G. (1969). "Subcritical Crack Propagation in Ti-8Al-1Mo-1V Alloy in Organic Environments, Salt Water, and Inert Environments." In: *Fundamental Aspects of Stress Corrosion Cracking* (R. W. Staehle, et al., ed). NACE, pp. 684–690.
292. Meyn, D. A. (1974). Effect of Hydrogen on Fracture and Inert Environment Sustained Load Cracking Resistance of Titanium Alloys." *Metallurgical Transactions* **5** (11), 2405–2414.
293. Blackburn, M. J. (1967). *Trans AIME* **239**, 1200.
294. Koch, G. H., et al. (1981). "A Comparison of Gaseous Hydrogen Embrittlement, Slow-Strain-Rate Hydrogen Embrittlement, and Stress Corrosion Cracking in Ti-8Al-1Mo-1V." *Metallurgical Transactions* **12A**, 1833–1843.
295. Lane, I. R., and Cavallaro, J. L. (1969). "Metallurgical Aspects of Stress Corrosion Cracking of Titanium Alloys." In: *Fundamental Aspects of Stress-Corrosion Cracking* (R. W. Staehle, et al. ed)., NACE, p. 620.
296. Mackay, T. L. (1969). *Stress Corrosion Cracking of Titanium Alloys at Ambient Temperature in Aqueous Solutions*. NASA CR-1464.
297. Hickman, B. S. (1969). "The Formation of Omega Phase in Titanium and Zirconium: A Review." *Journal of Material Science* **4**, 554.
298. Fager, D. N., and Spurr, W. F. (1968). "Some Characteristics of Aqueous Stress-Corrosion in Titanium Alloys." *Transactions ASM* **61** 283–292.
299. Meyn, D. A. (1965). "*A Study of the Crystallographic Orientation of Cleavage Facets Produced by Stress Corrosion Cracking of Ti-7Al-2Nb-1Ta in Water*." NRL Progress Report.
300. Koch, G. H., et al. (1978). "Comparison and Interpretation of Fracture Surfaces Produced in Ti-8Al-1V-1Mo by Stress Corrosion Cracking and Slow Strain Rate Hydrogen Embrittlement." *Metallurgical Transactions A*, 129.

301. May, T. P., Holmberg, E. G., and Hinde, J. (1962). "Seawater Corrosion and Elevated Temperatures." *Deckema-Monograhian Bank*: **47**, 253–274. (Inco Reprint).

302. Osborn, D. H. (1963). "The Truth About Miss Liberty" (Analysis of Patina on Statue of Liberty). *Materials Design Engineering* **57** 80–82.

303. Hummer, C. W., Southwell, C. R., and Alexander, A. L. (1968). "Corrosion of Metals in Tropical Environments-Copper and Wrought Copper Alloys." *Materials Protection* **7** (1), 41–47,

304. Landegren, W., and Mattsson, E. (1970). "Stress Corrosion of Brass; Field Tests in Different Types of Atmosphere." *British Corrosion Journal* **II** (21), 291.

305. Reinhart, F. M. (1968). *Corrosion of Materials in Hydrospace, Part IV, Copper and Copper Alloys.* Naval Civil Engineering Laboratory, Technical Note N-961.

306. Beavers, J. A., Agrawal, A. K., and Berry, W. E. (1980). *Corrosion Related Failures in Power Plant Condensers.* EPRI NP-1468.

307. Mattson, M. E., and Fuller, R. M. (1965). *A Study of Materials of Construction in Distillation Plants.* Joint Report by Office of Saline Water and the International Nickel Company, Inc., R & D Progress Report No. 163.

308. Schreiber, C. F., Osborn, O., and Coley, F. H. (1968). "Corrosion of Metals in Desalination Environments." *Materials Protection* **7** (10) 20–25.

309. Rogers, T. H. (1948–1949). "The Promotion and Acceleration of Metallic Corrosion by Microorganisms." *Journal of the Institute of Metals* **75**, 19–38.

310. LaQue, F. L., and Clapp, W. F. (1945). "Relationships Between Corrosion and Fouling of Copper-Nickel Alloys in Seawater." *Transactions of the Electrochemical Society* **87**, 103–125.

311. Clapp, W. F. (1948). "Microorganisms in Seawater and their Effect on Corrosion." In: *Corrosion Handbook* (H. H. S. Uhlig, ed). Wiley and Sons, p. 443.

312. May, T. P., and Weldon, B. A. (1964). "Copper-Nickel Alloys for Service in Seawater." *Congress International de la Corrosion Marine et des Salissures, Cannes, France*, pp. 141–156.

313. Gudas, J. P., and Hack, H. P. (1979). "Parametric Evaluation of Susceptibility of Copper-Nickel Alloys to Sulfide Induced Corrosion in Seawater." *Corrosion* **35** (6), 259–264.

314. Syrett, B. C. (1981). "The Mechanism of Accelerated Corrosion of Copper Nickel Alloys in Sulfide-Polluted Seawater." *Corrosion Science* **21** (3), 187–209.

315. Syrett, B. C., and Wing, S. S. (1980). "Effect of Flow on Corrosion of Copper-Nickel Alloys in Aerated Seawater Polluted with Sulfide and Sulfide Oxidation Products." *Corrosion* **35** (2), 73–85.

316. Syrett, B. C., MacDonald, D. D., and Wing, S. S. (1979). "Corrosion of Copper-Nickel Alloys in Seawater Polluted with Sulfide and Sulfide Oxidation Products." *Corrosion* **35** (9), 409–422.

317. Eiselstein, L. E., Syrett, B. C., Wing, S. S., and Caligiuri, R. D. (1983). "The Accelerated Corrosion of Copper-Nickel Alloys in Sulfide-Polluted Seawater: Mechanism No. 2." *Corrosion Science* **23** (3), 223–239.

318. Eiselstein, L. E., Caligiuri, R. D., Wing, S. S., and Syrett, B. C. (1981). *Mechanisms of Corrosion of Copper-Nickel Alloys in Sulfide-Polluted Seawater.* SRI Int., Menlo Park, CA, ADA 096865, Final, N00014–77C0046, p. 71.

319. Schrader, M. E. (1980). *Mechanism of Sulfide-Accelerated Corrosion of Copper-Nickel* (90–10) *Alloys in Seawater.* Naval Ship R & D Center, Bethesda, MD, DTNSRDC-80-127, ADA093951, p. 27.

320. Sanchez, S. R., and Schiffrin, D. J. (1982). "The Flow Corrosion Mechanism of Copper Base Alloys in Seawater in the Presence of Sulfide Contamination." *Corrosion Science* **22** (6), 585.

321. Freeman, J. R., Jr., and Tracy, A. W. (1949). "Comparative Corrosion Resistance of Some Copper Alloy Condenser Tubes." *Corrosion* **5** (8) 245–248.

322. Otsu, T., and Sato, S., (1961). "Comparative Corrosion Test of Condenser Tube Materials by Model Condenser at Meiko Power Station." *Sumitomo Light Metal Technical Reports* **2** (4) 299–322.

323. Pruvot, C. (1976). "Service Life of Condenser Tubes at the Dunkerque Station." *Rev. Generale de Thermique* **174–175**, 505–519.

324. Tracy, A. W., and Hungerford, R. L. (1945). "The Effect of the Iron Content of Cupro-Nickel on its Corrosion Resistance in Seawater." *Proceedings of the American Society of Testing and Materials* **45**, 591.

325. LaQue, F. L., and Mason, J. F., Jr., (1950). "The Behavior of Iron Modified 70–30 Cupro-Nickel Alloy in Salt Water and in Some Petroleum-Industry Environments." Paper presented to the *15th Mid-Year Meeting of the American Petroleum Institute's Division of Refining, Cleveland, OH*, **3M** (3).

326. Stewart, W. C., and LaQue, F. L. (1952). "Corrosion Resisting Characteristics of Iron Modified 90:10 Cupro Nickel Alloy." *Corrosion* **8** (8), 259–277.

327. Syrett, B. C. (1976). "Erosion Corrosion of Copper-Nickel Alloys in Seawater and Other Aqueous Environments—A Literature Review." *Corrosion* **32** (6), 242–252.

328. Heaton, W. E. (1977). "Impingement Corrosion of Condenser Tubes." *Chemistry and Industry* (13), 29–36.

329. Anderson, D. B., and Badia, F. A., (1973). "Chromium Modified Copper-Nickel Alloys for Improved Seawater Impingement Resistance." *Transactions of the ASME*, Series A; *Journal of Engineering for Power* **95** (2), 132–135.

330. Sato, S., and Nagata, K. (1978). "Factors Affecting Corrosion and Fouling of Condenser Tubes of Copper Alloys and Titanium." *Sumitomo Light Metal Technical Reports* **19** (3–4), 83–94.

331. Michels, H. T., Kirk, W. W., and Tuthill, A. H. (1978). "The Influence of Corrosion and Fouling on Steam Condenser Performance." Paper presented at the *16th INCO Power Conference*, Wrightsville Beach, NC, p. 54.

332. Efird, K. D., and Anderson, B. D. (1975). "Seawater Corrosion of 90-10 and 70-30 Cu-Ni: 14 Year Exposures." *Materials Performance* **14** (11), 37–40.

333. Sato, S., and Nosetani, T. (1970). "Allowable Water Velocity and Cleanliness Factor of Aluminum Brass Condenser Tube with Ferrous Ion Addition into Seawater." *Sumitomo Light Metal Technical Reports* **11** (4), 271–280.

334. Efird, K. D. (1977). "Effect of Fluid Dynamics on the Corrosion of Copper-Base Alloys in Seawater." *Corrosion* **33** (1), 3–8.

335. Ferrara, R. J., and Gudas, J. P. (1977). "Corrosion Behavior of Copper-Base Alloys with Respect to Seawater Velocity." In: *Proceedings of the 3rd International Congress on Marine Corrosion and Fouling*, pp. 285–298.

336. Davis, J. A. (1977). "Review of High Velocity Seawater Corrosion: NACE T-7C-5 Task Group Report." *Corrosion/77, San Francisco, CA (March 14-18, 1977).*

337. Beavers, J. A., and Boyd, W. K. (1977). *Evaluation of Copper-Nickel Alloys for OTEC Application.* Final Report to Battelle's Pacific Northwest Laboratories, Battelle's Columbus Laboratories.

338. Anderson, D. B. (1971). "In Desalination Service Copper Alloys Provide Optimum Corrosion Resistance at Temperatures Above 200°F" *Materials Performance* **10** (11), 26.

339. Fontana, M. G., and Greene, N. D. (1978). *Corrosion Engineering.* McGraw-Hill.

340. Sato, S. (1962). "Effects of Chlorination of Seawater on the Corrosion of Condenser Tubes." *Sumitomo Light Metal Technical Reports* **3** (3), 106–116.

341. Tanabe, Z. I. (1968). "On The Erosion of Condenser Tube Alloys by Sand in Water." *Sumitomo Light Metal Technical Reports* **9** (3), 169–175.

342. Tanabe, Z. I. (1969). "On the Sand Erosion of Aluminum Brass." *Journal of the Japan Institute of Metals* **33** (4), 476–480.

343. Sato, S., Nosetani, T., Yamaguchi, Y., and Onda, K. (1975). "Factors Affecting the Sand Erosion of Aluminum Brass Condenser Tubes.", *Sumitomo Light Metal Technical Reports* **16** (1-2), 23-37.

344. Lockhart, A. M. (1965). "Curing Condenser Tube Corrosion at Kincardine Power Station." *The Engineer* **219**, 499-500.

345. Nosetani, T., Shimono, M., Tanabe, Z., Sato, S., and Hirose, H. (1974). "Service Experience of Welded Titanium Tubes in Air Cooling Zone of Surface Condensers." *Sumitomo Light Metal Technical Reports* **15** (3), 163-173.

346. Skulte, Peteris (July 1978). Private Communication.

347. Gehring, G. A., Jr., and Maurer, J. R., (1981). "Galvanic Corrosion of Selected Tubesheet/Tube Couples Under Simulated Seawater Condenser Condition." *Corrosion/81*, Paper 202.

348. Bulow, C. L. (1965). "Use of Copper Base Alloys in Marine Service." *Naval Engineers Journal* **77** (3), 470-482.

349. Speidel, M. O., *ARPA Handbook on Stress Corrosion Cracking*, to be published.

5

Corrosion Control

R. HEIDERSBACH

California Polytechnic State University
San Luis Obispo, California

and

J. SMART

Welchem, Inc.
Houston, Texas

ISBN 0-12-341828-3

I. Introduction

Most marine corrosion control depends on three techniques—materials selection, protective coatings, and cathodic protection. Combinations of these techniques are sometimes used. The selection of a protective method depends on economic considerations. Because of economics, carbon steel, a material with little natural corrosion resistance, is the most commonly used material for marine applications. Corrosion allowances, based on the assumption that corrosion will occur, are required on many marine structures and vessels. Depending on the application, excess metal thicknesses of 50% or more may be required.

II. Materials Selection

In the following presentation, the Unified Numbering System (UNS) is not used. This system for identifying metals and alloys was developed to correlate the many numbering systems currently used by trade associations, technical societies, government organizations, and producers (*1*). The UNS covers all major alloy systems. Unfortunately it has not been widely accepted.

A. Carbon Steel

Because carbon steels are the lowest cost metals available for use in marine applications, all material selections based on corrosion resistance must be compared to carbon steel. If carbon steel can be made to perform adequately through some economical combination of protective coatings, cathodic protection, or a combination of the two, then the selection of a more expensive material cannot be justified.

Commercial carbon steels are iron-based alloys with carbon additions from 0.04% to 1.2%. The majority of steels used in the marine environment have less than 0.5% carbon and would be considered to be low or medium carbon steels. The binary iron-carbon quasi-equilibrium phase diagram is shown in Figure 1. The most common phases found in steels are:

(a) Austenite—face centered cubic gamma iron. This ductile phase of iron is stable above approximately 727°C (1341°F); but depending on alloying elements in the steel, it can be retained at room temperature. Because austenite has the ductile face centered cubic crystal structure, steels are heated

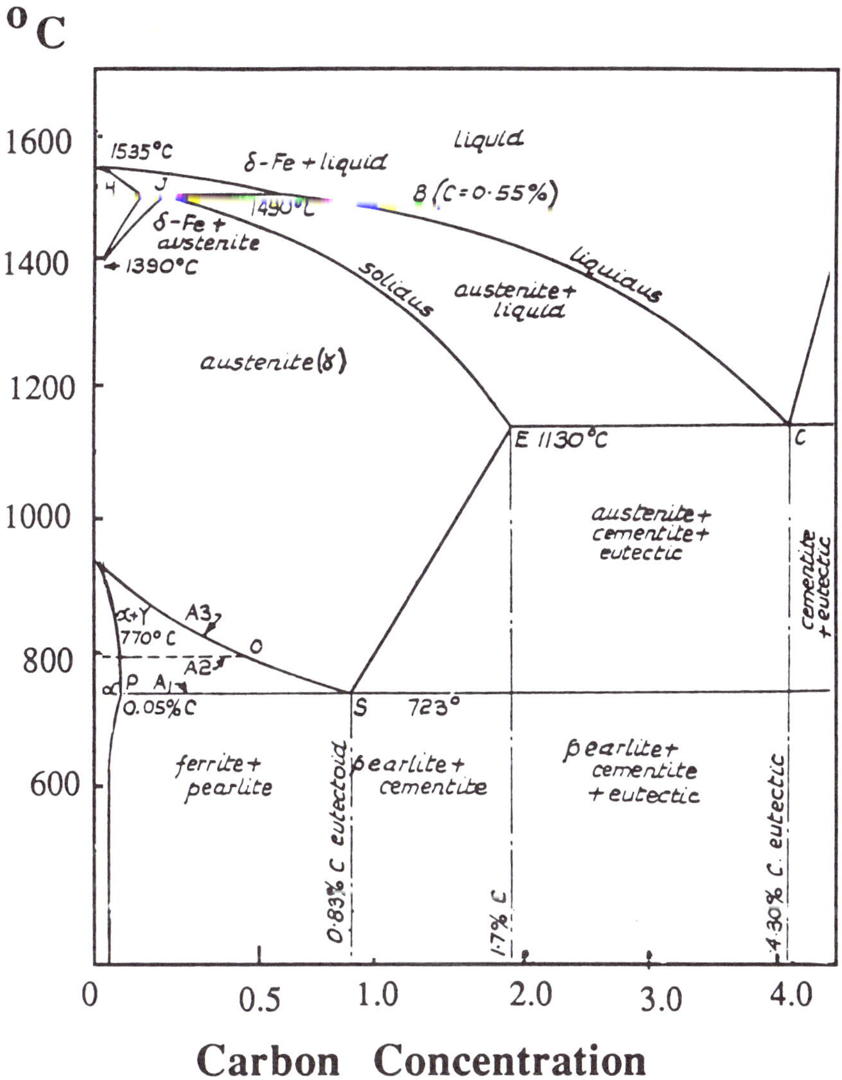

A_1—the eutectoid line, 723°C;
A_2—the ferrite Curie point;
A_3—the $\alpha \rightarrow \gamma$ transformation line, *GS*;
A_4—the $\gamma \rightarrow \delta$ transformation, 1390°C.

Fig. 1. Iron–carbon quasi-equilibrium phase diagram.

to the austenite temperature range for hot working (rolling, forging, upsetting, etc.).

(b) Ferrite—body centerd cubic alpha iron. This ductile, magnetic form of iron coexists with iron carbide, Fe_3C or cementite, in all commercial carbon steels. The varying forms of this two-phase structure account for the wide variability of mechanical properties which can be achieved in commercial carbon steels.

(c) Martensite—a body centered tetragonal form of iron with carbon, achieved by supercooling high temperature austenite to a temperature below approximately 300°C (572°F). The resulting metal is so brittle that it must be tempered to be of any practical use. Tempering is done by heating the steel at temperatures below the austenite region. Complete discussions of this process are available (*2–4*).

(d) Pearlite—a combination of two phases, ferrite and cementite, in alternating plates which are so fine that they appear as a separate phase under moderate magnification in a microscope. Figure 2 shows a pearlite colony growing at a grain boundary between austenite grains.

Most alloying additions found in carbon steels are intended to refine the crystal (grain) size, to alter the response to heat treatment, or to enhance other properties.

Figure 3 shows an S curve or a time-temperature-transformation curve (T-T-T) for carbon steel. Diagrams of this type are used to determine heat treatment programs for various steels. This diagram shows the effect of adding molybdenum to carbon steel. Molybdenum is one of several elements

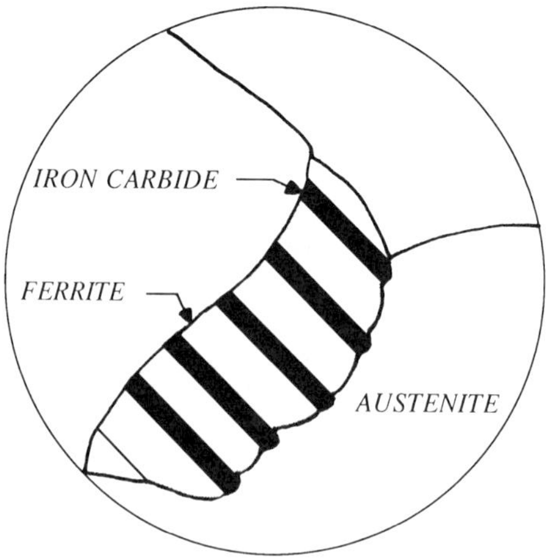

Fig. 2. Pearlite colony growing at an austenite grain boundary.

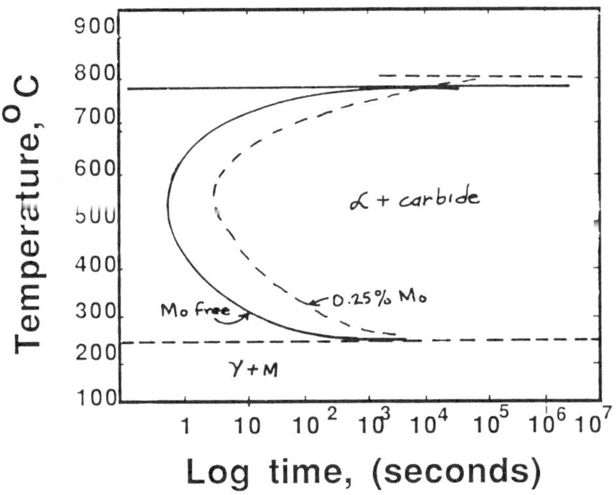

Fig. 3. Time–temperature–transformation (T–T–T) curve showing the effect of molybdenum additions on the time to transformation.

added to steels to improve their thick section heat treatment characteristics. Corrosion problems occur near welds in these thick section steels when differential cooling rates create different microstructures in and near the weldment.

Another problem associated with welding is the formation of cooling or quenching cracks when the outer surface of the metal cools faster than the hot inside. Figure 4 illustrates this phenomena. These cracks can serve as initiation sites for corrosion fatigue failures.

B. Cast Irons

Most cast irons are iron–carbon alloys with from 2% to 4% carbon additions. Almost all cast irons are brittle and subject to damage from thermal and mechanical shock. Four types of cast irons are commonly available: grey, white, nodular, and malleable.

Grey cast iron contains approximately 2% silicon. This lowers the melting point and decreases the viscosity of the liquid metal. The carbon–rich phase in carbon steels is usually Fe_3C, but in grey cast iron the silicon promotes precipitation of long flaky graphite particles. These particles are easy to machine and provide good vibration damping properties. Unfortunately their shape makes grey cast iron subject to mechanical shock damage and to corrosion.

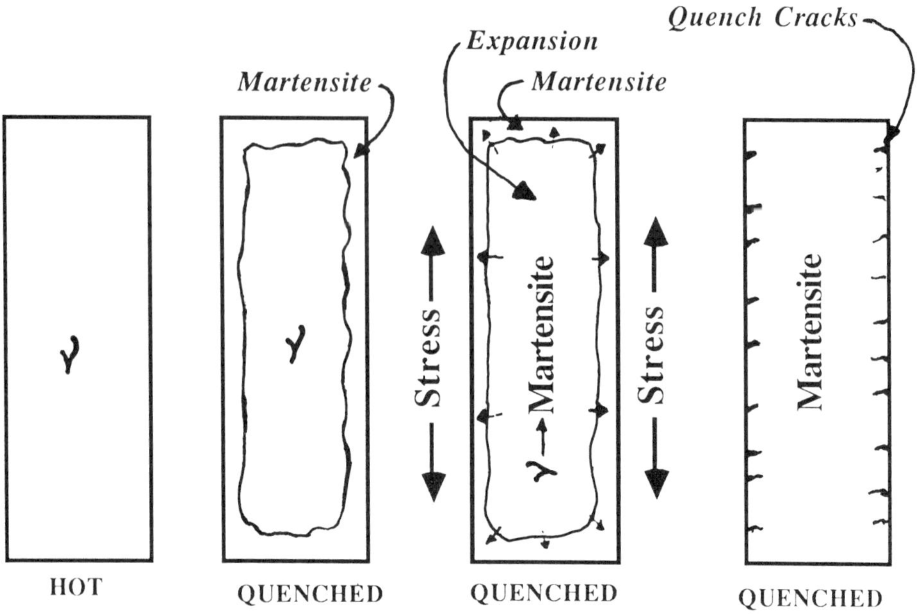

Fig. 4. Formation of quench cracks due to residual stresses produced during quenching.

White cast iron has less silicon or is cooled quickly from the melt. This produces a Fe_3C second phase which is much harder than the graphite found in grey cast iron. White cast iron is very hard and wear resistant. The poor mechanical properties of grey cast iron are caused by the shape of the graphite flakes. If white cast iron is heated, some of the Fe_3C particles become rounded graphite clusters. The resulting malleable iron has better mechanical properties than grey cast iron and is easier to machine than is white cast iron. Malleable cast iron is sometimes used as deck piping.

A similar product can be created by adding magnesium or cerium to molten grey cast iron. The resulting casting has nodular graphite instead of the undesirable graphite flakes. Nodular cast iron piping is sold as "ductile cast iron" in the United States.

Alloy cast irons high in chromium or silicon are available for corrosion resistant applications.

C. Low Alloy Steels

Steels containing up to 5% alloying elements other than carbon are called low-alloy steels. These steels have phase equilibria related to the iron-carbon phase diagram shown in Figure 1. Table I lists several common alloy series.

TABLE I

CARBON STEELS AND LOW-ALLOY STEELS

AISI or SAE Number	Composition
10XX	Plain-carbon steels*
11XX	Plain-carbon (resulfurized for machineability)
15XX	Manganese (1.0–2.0%)
41XX	Chromium (0.4–1.2%), molybdenum (0.08–0.25%)
43XX	Nickel (1.6–2.0%), chromium (0.4–0.9%), molybdenum (0.2–0.3%).

* The XX numbers show carbon content, 0.XX%, e.g., 1040 steel and 4340 steels are carbon steel and Cr-Ni-Mo steels having 0.40% carbon.

The alloying elements shown in Table I are added to improve mechanical properties, forming characteristics, weldability, and heat treatment response of the alloy. None of the alloy steels in Table I has enough alloying additions to make them corrosion resistant for immersion service. While some improved corrosion resistance may be noted, this improvement is not enough to justify the alloy cost premium. All these alloys require additional means of corrosion control, usually coatings, cathodic protection, or combinations of the two.

Low alloy steels do have sufficient alloying additions to make them behave as electrodes, normally cathodes, in galvanic couples with carbon steels. Direct metal contacts between these low alloy steels and carbon steels will normally result in accelerated corrosion of the carbon steel.

D. *Weathering Steels*

Weathering steels are a class of carbon steels which form a protective rust coating and can be used unpainted for atmospheric exposures. These steels require alternate wetting and drying to perform adequately. Inadequate drainage, immersion service, or buried-in-soil applications will lead to rapid corrosion. Weathering steels should never be specified for immersion or buried service unless they are painted or cathodically protected.

Weathering steels have been marketed in the United States under several different trade names. In the early 1970s, several producers developed "marine grade" weathering steels which were marketed under trade names like "Mariner" or "C-gard." These grades did not live up to expectations and

Fig. 5. Corroded weathering steel sheet piling along a tidal creek in Maryland.

are no longer marketed. Unfortunately, sales organizations still suggest weathering steels for immersion service and other improper applications. These misapplications have resulted in a poor reputation for weathering steels in some industries. Figure 5 shows a corroded sheet piling along a tidal creek in Maryland. Perforation of this piling occurred due to corrosion from the back (earth filled) side of the piling.

E. *Stainless Steels*

Chromium is added to iron to produce a variety of products called stainless steels. There are five general categories of stainless steels: martensitic, austenitic, ferritic, precipitation hardening, and duplex (*6*, *7*). Table II is a brief listing of some of the more common stainless steels. Many other specialty stainless steels exist and are used for unique applications. Like the properties of all metals and alloys, those listed in Table II can vary depending

TABLE II

COMMON STAINLESS STEELS

Alloy	Composition (Weight Percent)				Tensile Strength (ksi)	Condition
	Cr	Ni	C(max)	Others		
			Martensitic			
410	$11\frac{1}{2}$–$13\frac{1}{2}$	—	0.15	—	75	Annealed (soft)
			Ferritic			
430	16–18	—	0.12	—	75	Annealed
446	23–27	—	0.20	—	80	Annealed
			Austenitic			
304	18–20	8–10	0.08	—	85	Annealed
316	16–18	10–14	0.10	Mo 2–3	85	Annealed
			Precipitation-hardening			
17-7 pH	17	7	0.07	Al 1	130	Annealed
					235	Aged 950°F
17-4 pH	$16\frac{1}{2}$	$4\frac{1}{4}$	0.05	Cu 4	200	Aged 900°F
			Duplex			
Ferralium[a] 225	26	5	0.04	Mo 3.3	80–160 (Yield)	
VS-22[a]	22	5.5	0.03	Mo 3	65–160 (Yield)	
AF-22[b]	22	5.4	0.03	Mo 3	160	
AISI-326[c]	26	6.5	0.05	Mnl Si 1.6 Ti 0.25		

[a] CabVal product literature.
[b] Mannesmann product literature.
[c] Sedriks, J. (1979). *Corrosion of Stainless Steels*. Wiley, New York, p. 23.

on cross section, product form, and supplier. The mechanical properties listed are for illustrative purposes only. They should not be used for design without checking with industry standards or product supplier literature.

Martensitic stainless steels are iron-chromium alloys with up to approximately 16% chromium. They also have carbon additions for strength and/or wear resistance. They can be hardened by heat treatment to produce a material which will retain a hard cutting edge and are therefore used in stainless steel cutlery. They find limited use in the marine industries.

Ferritic stainless are also iron-chromium alloys, but they have higher chromium contents than do the martensitic stainless steels. The ferritic stainless steels are generally more resistant to hot-salt stress corrosion cracking than are the more common, austenitic stainless steels. In recent years, they have become popular condenser tube alloys for coastal power plants.

Austenitic stainless steels are based on iron containing 18% chromium and 8% nickel. These alloys are the workhorses of the chemical process and refining industries. They are also used for many marine applications. For many applications, they have outstanding corrosion resistance. Unfortunately, they have two major limitations which prevent their wider use. Nickel gives these alloys greater corrosion resistance than they would have with only chromium additions. Nickel is also an austenite stabilizer, which means that these steels have essentially single-phased face centered cubic (fcc) structures even at low temperatures. This fcc structure limits the strengths that can be obtained, especially in thick sections which cannot be cold worked to produce strength. Austenitic stainless steel alloys also suffer from stress corrosion cracking susceptibility in elevated temperature brines. This is a major limitation in any application where elevated-temperature seawater is encountered.

Molybdenum additions to austenitic stainless steel greatly increase the resistance of these alloys to crevice corrosion. Since fouling and scales can produce deposits in salt water, the molybdenum containing alloys (Type 316 and similar grades) should be used for most marine applications instead of the more susceptible molybdenum-free grades (304, etc.). Figure 6 shows a

Fig. 6. Corroded stainless steel nut.

corroded 304 stainless steel nut recovered from a navigational buoy after six months. The extensive crevice corrosion has followed forming lines and inclusions in the metal. Use of 316 stainless steel would have reduced, but not eliminated, this corrosion. Cathodic protection of stainless steel nuts using sacrificial zinc anodes sometimes works for short-term deployments of instrument packages (*5*).

Precipitation-hardening stainless alloys are ternary alloys with less nickel. They have lower corrosion resistance than the austenitic grades but can be heat treated for improved mechanical properties. These alloys are popular with mechanical designers, who like their ability to be machined to shape in the soft condition and then hardened for final use. Many designers do not realize that these desirable mechanical properties are obtained at considerable sacrifice in corrosion resistance. Product literature from alloy suppliers may not warn the user of how susceptible these alloys are to pitting and crevice corrosion.

Duplex stainless steels are marketed by a number of suppliers. These alloys have chromium (20%–32%), nickel (5%), and molybdenum (3%) in a duplex structure, which has approximately equal portions of austenite and ferrite. They are marketed as being more resistant to stress corrosion cracking than austenitic stainless steels. Industry standards for these proprietary alloys are not generally available and most of the popular grades are proprietary.

F. Aluminum alloys

Aluminum has excellent atmospheric corrosion resistance and is often specified for marine atmospheric exposure by highway departments and other construction agencies. It is also used for deck housing on ships and offshore platforms.

Aluminum's chief advantage over carbon steel is its high strength-to-weight ratio. Unfortunately the same alloying additions which are used to increase aluminum's strength also lower its resistance to corrosion. Pure aluminum is the most corrosion resistant form of aluminum. Unfortunately it has very limited strength and is not a practical structural metal. Pure aluminum is used as a corrosion-resistant coating both for aluminum alloys and for steel. The Alclad aluminum alloys with pure aluminum exteriors are so common that aluminum design handbooks list the mechanical properties of these clad alloys along with those for their unclad counterparts.

Table III lists the commercial wrought aluminum alloy series. The age-hardenable aluminum-copper alloys (2xxx series) have the best fatigue versus weight characteristics of any aluminum alloys. Unfortunately, microscopic aluminum-copper galvanic cells make these alloys very susceptible to intergranular corrosion. Detailed discussions of how this occurs are available (*6*).

TABLE III

WROUGHT ALUMINUM ALLOY SYSTEMS

1xxx	Commercially pure Al (> 99% Al)	Not aged
2xxx	Al-Cu	Age-hardenable
3xxx	Al-Mn	Not aged
4xxx	Al-Si and Al-Mg-Si	Age hardenable if Mg is present
5xxx	Al-Mg	Not aged
6xxx	Al-Mg_2Si	Age hardenable
7xxx	Al-Mg-Zn	Age hardenable

The two most useful marine alloys series are the work-hardened 5xxx series (Al-Mg) alloys, commonly used for deck housings, work boat hulls, etc., and the 6xxx series (Al plus Mg and Si). Both of these alloys systems are extensively used in the marine environment.

Many designers are tempted to use the 7xxx series aluminum alloys for marine designs, because they have much higher strength-to-weight ratios than the competitive 6xxx series alloys. Unfortunately, these alloys are notorious for their corrosion susceptibility. The T7 condition (Table IV), which produces lower strengths than the more popular T6 condition, was developed specifically to limit the stress corrosion cracking problems in these alloys. If these alloys must be used in the marine environment, the T7 (maximum strength) condition must be avoided. Unfortunately many designers are unaware of these problems.

A designer unfamiliar with aluminum corrosion could be misled into thinking aluminum alloys can be used in marine environments with few

TABLE IV

TEMPER DESIGNATIONS FOR ALUMINUM ALLOYS

F	As-fabricated (hot worked, forged, cast, etc.).
O	Annealed (in softest possible condition).
H1X	Cold worked only. (X refers to how much cold working and strengthening, e.g. H12 cold worked to strength midway between 0 and 14 tempers; other cold-worked designations are available).
W	Solution treated.
T	Age-hardened
	T6 Solution treated and aged to maximum strength
	T7 Solution treated and overaged to reduce susceptibility to stress-corrosion cracking.

[a] A complete alloy designation requires an alloy number plus a temper designation, e.g., 6061T6, 7075T7, etc.

concerns about corrosion. This is not true. Aluminum alloys should be anodized before use in the marine atmosphere. Most authorities also recommend that they be painted with coatings specifically chosen for the application in question. Uncoated aluminum alloys must never be used for marine immersion service except as sacrificial anodes.

G. *Copper alloys*

Unlike steel and aluminum, which are normally specified for their strength, copper alloys are frequently specified for their corrosion resistance.

There are two major classes of copper-based alloys, brasses and bronzes. At one time the term "brass" referred to copper-zinc alloys and "bronzes" were copper-tin alloys. Both terms have been expanded, and "brass" is now used to refer to any copper-based wrought alloy whereas "bronze" refers to copper-based casting alloys. Thus there are both aluminum brasses (wrought copper-zinc alloys with aluminum additions) and aluminum bronzes (cast copper-aluminum alloys).

Wrought copper-zinc products are available as single-phase alpha brasses having up to approximately 30% zinc. The corrosion resistance generally decreases and the strength increases with these alloys. Duplex alpha-plus-beta brasses are also available as castings. Copper-zinc alloys were once widely used, but they have been largely replaced in corrosion-sensitive marine applications by more corrosion-resistant alloys. They are still common in valves and in chromium-plated fixtures.

Copper-tin alloys are widely used and are reasonably corrosion resistant. Many castings and threaded fasteners are made from copper-tin bronzes.

Copper-aluminum alloys are widely used in large castings such as pump housings, pump impellers, and ship propellers.

Copper-nickel alloys exhibit complete solid solubility from pure copper to pure nickel. Three alloy groupings are important in this series. The 90 copper-10 nickel and 70–30 cupronickels are widely used as heat exchanger alloys and as castings. They are also used as antifouling cladding for tidal zones and have received extensive promotion as nonfouling hull materials for intermediate-sized fishing vessels.

Nickel-based 70 nickel-30 copper alloys are marketed under the tradename Monel. These highly corrosion resistant metals are not copper based alloys, but they are frequently compared with cupronickels.

Copper beryllium alloys are high strength metals with good corrosion resistance. Beryllium is very toxic, so these alloys are only used for special applications. Marine applications include springs and nonmagnetic pressure housings for underwater cables. They are also used for nonsparking tools—wrenches, screwdrivers, etc.

TABLE V

TITANIUM TUBING ALLOYS[a]

Grade	Pd	Al	V	Mo	Zr	Sn	Ni
1			Commercially pure[b]				
7	0.2						
9		3	$2\frac{1}{2}$				
10				12	6	4	
12				0.03			0.8

[a] Condensed from ASTM Standard B 339.

[b] All titanium alloys have maximum limits for impurity elements—N, C, H, Fe, O, etc.

H. *Titanium alloys*

Titanium alloys have become, in recent years, the third group of structural alloys along with steel and aluminum. Unlike the other structural metals, titanium is virtually immune to ambient temperature seawater. Titanium is also popular for structural applications because its strength-to-weight characteristics are even better than aluminum. Table V is a brief listing of some titanium alloys.

One structural application is the pressure hull of the U.S. Navy's deep submergence research submersible, Alvin. The forming, shaping, and welding of Alvin's pressure hull is described in reference 8. Other applications of titanium and its alloys in offshore platform applications are also given in the volume which contains reference 8.

Development of titanium alloys for structural applications, especially for larger submersibles, was slowed in the 1960s when it was learned that titanium was not unlike high strength steels and aluminum alloys in that it was subject to stress corrosion cracking in ambient temperature seawater. A similar development, the discovery of "hot salt stress corrosion cracking" of titanium, was creating problems for American supersonic transport development at about the same time.

Titanium is subject to crevice corrosion at elevated temperature. This can be minimized by the use of molybenum additions. (Molybenum is also added to stainless steels for the same reason). Palladium additions have the same effect, but, for cost reasons, titanium-palladium alloys have not found widespread commercial acceptance. Figure 7 shows a titanium casting which suffered crevice corrosion in marine service. The part in question was made from a titanium–6 w/o aluminum–4 w/o vanadium aerospace alloy. It should

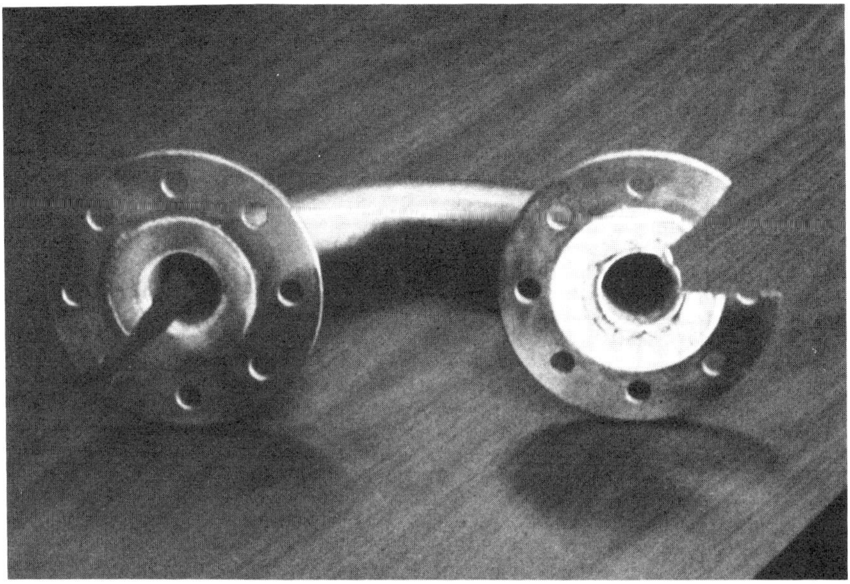

Fig. 7. Crevice corrosion of titanium.

not have been used for marine immersion service. A marine alloy with palladium or molybdenum additions would have been far more corrosion resistant.

Titanium is virtually immune to corrosion in ambient temperature seawater. This means, however, that it is susceptible to fouling and also acts as an efficient cathode in galvanic cells.

I. Amphoteric Coating Metals

Amphoteric metals have high corrosion rates in acids or bases but relatively low corrosion rates in neutral environments. Seawater has a pH of approximately 7.8, so it is a "neutral environment." The common amphoteric coating metals are zinc (galvanizing), cadmium, and aluminum.

Zinc is the most common of these materials, and galvanized steel is used for many marine applications. Typical uses include cables, fasteners, ladders, and walkways. Zinc corrodes relatively slowly, compared to carbon steel, in a marine air environment. Once the zinc is breached it still protects the steel substrate by sacrificial action. This is illustrated in Figure 8.

Cadmium is a metal that is chemically very similar to zinc. It is, however, somewhat harder than zinc and can be electroplated onto complex parts,

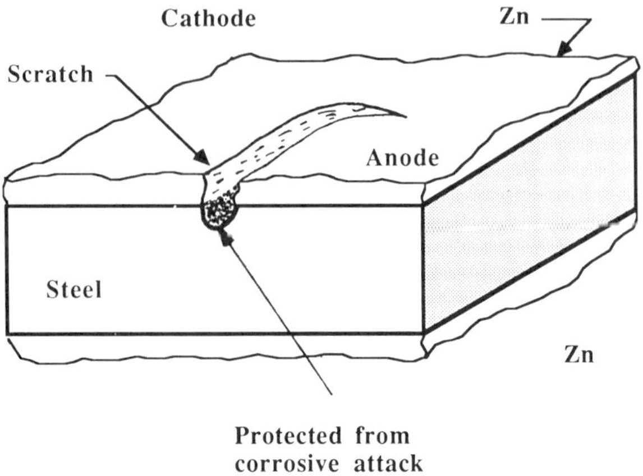

Fig. 8. The effect of a scratch (coating holiday) on the corrosion of galvanized steel.

such as threaded fasteners, to close dimensional tolerances. Unfortunately, cadmium is very toxic and is being removed from the marketplace for occupational safety and environmental pollution reasons. It should no longer be specified as an alternative to zinc or aluminum.

Aluminum is the third of the amphoteric coating metals. Unlike the other two, it cannot be electroplated but can be applied using physical vapor deposition techniques such as ion plating to small parts or by plasma spraying. Aluminum is somewhat more corrosion resistant than the other two metals in most environments. Onshore this can be a disadvantage when holidays open in an aluminum coating and the aluminum fails to protect the substrate alloy by sacrificial protection. This is less of a problem in the marine environment.

One interesting development in recent years is the use of flame-sprayed aluminum coatings for atmospheric corrosion protection. The U.S. Navy reports success for walkways and other topside applications. Highway bridges in coastal locations are also cited as another example of corrosion protection provided to steel by aluminum.

The use of amphoteric coating metals for marine immersion service is very controversial. Most authorities caution that these coatings cannot provide long-term protction. Conoco, however, has specified flame-sprayed aluminum as the corrosion control system for the tension legs of their Hutton tension-leg platform launched in the North Sea in 1984 (*9*, *10*). If flame-sprayed aluminum works in this application, it will open up new markets for these materials. Many "experts" are of the opinion that it will not work.

J. Noble Coatings

Noble coatings, or corrosion resistant alloys, can be applied to corrosion susceptible substrates by a number of means. Stainless steel, copper, Monel, nickel, and chromium are examples of materials used in this way. These coatings provide excellent corrosion resistance until they develop "holidays" or holes in the coating. When holidays develop, the large cathodic area of the coating causes accelerated corrosion of the small exposed anode substrate. These coatings should be used with caution.

III. Marine Coatings

Coatings are the only protection for steel in the marine environment above the water line. They are widely used on offshore platforms, ships, pipelines, buoys, and piers. Coastal installations are equally subject to corrosion, and marine coatings systems are used there also. In fact, the surf conditions on beaches often generate more corrosive conditions than exist offshore.

A. Surface Preparation

Proper surface preparation is the key to successful performance of marine coatings. The Steel Structures Painting Council (SSPC), National Association of Corrosion Engineers (NACE), and other groups have developed standards for surface preparation. Table VI lists some of these standards. Wet abrasive blast cleaning and water blasting are not yet included in the standards but are now being extensively used. Wet blasting is useful for dust control and to avoid electrical sparking in Class I (explosive) areas. Generally, a small amount of nitrate inhibitor is added to the water to prevent rerusting before priming.

Water blasting can be used around rotating equipment, such as pumps, turbines, and generators—and in Class I areas. No grit is used, thus few solids result which could harm equipment. Water blasting at 5000–20,000 psi will result in a white metal surface and, once blown dry, an excellent surface for painting. Waterblasting with detergent in the water, at lower pressures, can be as good as solvent washing to prepare oily and greasy surfaces for painting.

Grit blasting is usually used for surface preparation for marine coatings. The severe corrosion exposure conditions in offshore and coastal locations require as good a surface preparation as possible.

TABLE VI

SURFACE PREPARATION STANDARDS

Solvent Cleaning SSPC-SP1 NACE—none Visual Standards none	Solvent cleaning is used to remove oil, grease, dirt, soil, drawing compounds, and various other foreign matter by use of solvents, emulsions, cleaning compounds, or steam cleaning. Solvent cleaning does not remove rust or millscale.
Hand Tool Cleaning SSPC-SP2 NACE—none Visual Standards SSPC-Vis 1: BSt2, CSt2, DSt2	Hand tool cleaning is the method of removing loose rust and loose millscale or any other loose contaminants from the surface by hand tools, such as wire brushes. It does not require the removal of tight rust or millscale.
Power Tool Cleaning SSPC-SP3 NACE—none Visual Standards SSPC-Vis 1: BSt3, CSt3, DSt3	This method provides for the removal of loose rust and loose millscale (or any other loosely adherent contaminant) from the surface by power tools. Power tool cleaning does not require the removal of tight rust or millscale.
White Metal Blast Cleaning SSPC-SP 5-71° NACE 1 Visual Standards SSPC-Visa 1: ASa3, BSa3, CSa3, DSa3 NACE 1	White metal blast cleaning is used when a totally cleaned surface is required. This method of cleaning is defined as a blast cleaned surface with a gray-white uniform metallic color. It shall be free of *all* oil, grease, dirt, millscale, rust, corrosion products, oxides, paint, stains, streaks, or any other foreign matter across 100% of every square inch.
Commercial Blast Cleaning SSPC-SP 6-71° NACE 3 Visual Standards SSPC-Vis 1: CSa2, DSa2 NACE 3	A commercial blast cleaned surface is defined as one in which *all* oil, grease, dirt, millscale, rust, and old paint have been completely removed from the blast cleaned surface, *except* that slight streaks or discolorations caused by rust stain, millscale oxides, or slight tight residues of paint or coating may remain. If the surface is pitted, slight residues of rust or paint may remain in the bottom of the pits. The slight discolorations mentioned above are limited to one-third of every square inch.
Brush-Off Blast Cleaning SSPC-SP 7-71° NACE 4 Visual Standards SSPC-Vis 1: BSa1, CSal. DSal NACE 4	A brush-off blast is defined as one in which all oil, grease, dirt, rust scale, loose millscale, loose rust, and loose paint or coatings are completely removed. Tight millscale and tightly adherent rust and paint or coatings may remain as long as the entire surface has been exposed to the abrasive blasting.

TABLE VI (*Continued*)

Pickling SSPC-SP 8 NACE—none Visual Standards none	This method of surface preparation is defined as completely removing all millscale, all rust, and all rust scale by chemical reaction, by electrolysis, or by both.
Near-White Blast Cleaning SSPC-SP 10-71° NACE 2 Visual Standards SSPC-Vis 1: ASa2-1/2, BSa2-1/2, CSa2-1/2, DSa2-1/2 NACE 2	This degree of blast cleanliness is defined as one from which all oil, grease, dirt, millscale, rust, corrosion products, oxides, paint, or any other foreign matter have been completely removed from the surface. Permitted to remain are very light shadows, very slight streaks, or discolorations caused by rust stain, millscale oxides, or slight tight residues of paint or coating. Only 5% of every square inch may contain the slight discolorations mentioned above.

Use of inorganic zinc primers requires white metal grit blasting to remove all surface contamination, as inorganic zinc has both a chemical bond and a mechanical bond to the surface. Epoxy primers can be applied over commercial grade surfaces for land based exposures, but require near-white metal surfaces to maintain performance offshore.

B. Primers

The primer is by far the most important coat in the protection of steel. The topcoats really are only there to protect the primer and to give color and appearance. Inhibitive and protective primers are used for marine applications.

Inhibitive primers contain substances to resist the effect of contaminants on the steel, such as rust and salt, and to resist disbondment under corrosive conditions and cathodic protection. Metallic zinc primers are the most common protective primers used for marine service.

Inorganic zinc primers are based upon various silicate binders. Several types exist, both self curing and post curing. The binder provides substantial corrosion resistance by itself, as well as serving as a strong, adherent matrix for the zinc metal. The zinc is present as fine powder and must be present in sufficient amounts to provide metal-to-metal contact between both the zinc particles themselves and the steel surface. Zinc is anodic to steel and provides cathodic protection to the steel in the same manner as galvanizing. If a break develops in the coating, the zinc reacts before the iron, resulting in protection of the iron for long periods. Both laboratory tests and field experience indicate that inorganic zinc primers can at least double the life of a coatings system and often increase it ten times.

To be effective, however, inorganic zinc must be applied to a clean surface so that it can contact the steel. This requires that the steel be grit blasted to a white metal finish, removing all rust and contaminants. This makes inorganic zinc the choice of primers for new construction and when total repainting is performed, but makes it not suitable for touchup or maintenance painting. The requirement that white metal grit blasting be used also makes inorganic zinc an expensive coating. Its performance, however, justifies its use for many applications.

Zinc epoxy primer is an alternative to inorganic zinc for use as a primer when conditions are not appropriate for inorganic zinc. Zinc epoxy primer uses a polyamid epoxy binder with a slightly lower zinc loading to achieve many of the properties of inorganic zinc.

Zinc epoxy primer may be applied over old paint, provided that it does not lift it. This makes zinc epoxy a good primer for maintenance painting. Zinc epoxies require less stringent surface preparation than inorganic zinc primers —near-white metal grit blasting for offshore applications and commercial grade grit blasting in less severe environments.

The epoxy binder provides some protection to the zinc, allowing moderate exposure of the primer to the marine environment without corrosion of the zinc and the generation of zinc salts. Zinc salts cause intercoat adhesion problems and blistering if they are present. Thus the time between primer and topcoat application can be extended for zinc epoxies.

C. Topcoats

Topcoats for steel mainly to protect the primer and to add color and appearance. Many types of topcoats are in use. Several are discussed below.

Alkyds are the most common and versatile coatings in existence but are not much used in marine applications due to poor performance over steel. Alkyds are the product of reaction of a polybasic acid, a polyhydric alcohol, and a monobasic acid or oil. The number of combinations is very large, and thus a large range of performances is available. Alkyds are used in marine service in places not considered to be severe environments. Examples include interior coatings for cabins, quarters, engine rooms, kitchens, heads, and some superstructures.

Alkyds have poor performance over steel due to their oil bases. As corrosion proceeds on steel, hydroxyl ions are generated at cathode sites. Hydroxyl ions saponify oils, dissolving the coating and turning them into soaps.

Vinyls also have a broad range of desirable properties as a coating. Vinyls are solvent based coatings which form a tight homogeneous film over the substrate. They are easy to apply by brush, roller, and spray. Intercoat adhesion is excellent due to the solvent based nature of the coatings. Vinyls do not oxidize or otherwise age, and are inert to acids, alkalis, water, cement, and alcohols. They do soften slightly when covered with some crude oils. Vinyls dry quickly, and they can be recoated in a short time, often in minutes, depending upon the solvent used. Vinyls are also flexible and can accommodate motion of the steel beneath it, such as when a ship or platform is launched.

Vinyls were extensively used on ships and offshore platforms for many years, and they still are in many areas. Their use, however, has given way to epoxies in most marine applications, however, because the coating is thin and not very strong. Film thickness is typically only 2 mils per coat, and vinyls cannot withstand mechanical abuse. They also have difficulty in covering rough previously corroded surfaces.

Chlorinated rubber paints are based upon natural rubber which has been reacted with chlorine to give a hard high quality resin soluble in various solvents. Chlor-rubbers have been used for many years because of their low moisture permeability, strength, resistance to UV degradation, and ease of application. Chlor-rubbers were used to paint ships because of its ease of application and repairability, tolerance of poor surface preparation, fast drying characteristics, and relatively good wearability and abrasion resistance. They are still used on ships and are the standard paint system for containers. Modern fleet owners, however, have phased out chlor-rubbers in favor of higher quality epoxy systems.

Expoxies have developed into the workhorses of modern marine coatings because they are adherent, strong, can be applied in thick layers, are inexpensive, and have good chemical and solvent resistance.

These properties allow epoxies to last from 7–12 years on ships, offshore platforms, and coastal applications when applied over a suitable primer, either inhibited epoxy or inorganic zinc.

Since epoxies are chemically cured, a wide range of properties can be achieved by using various molecular weights for the resins to be cured, the choice of curing agent, and the type of pigment or fillers used.

IV. Cathodic Protection

Cathodic protection can be defined as a technique to reduce or eliminate the corrosion rate of a metal by making it the cathode of an electrochemical cell and passing sufficient current to it to reduce its corrosion rate. All cathodic protection systems require the following components:

(a) Direct current electrical power source
(b) Anode
(c) Cathode
(d) Return circuit
(e) Electrolyte

A. Types of systems

Two common types of systems for cathodic protection are used: impressed current or active systems, and sacrificial anode or passive systems. Sacrificial anodes are naturally anodic to steel and thus they require no external current source. Zinc anodes are used for cathodic protection of ships hulls (Figure 9) and buried pipelines (Figure 10). Aluminum anodes (actually aluminum plus

Fig. 9. Zinc anodes used to control corrosion on a small fishing boat.

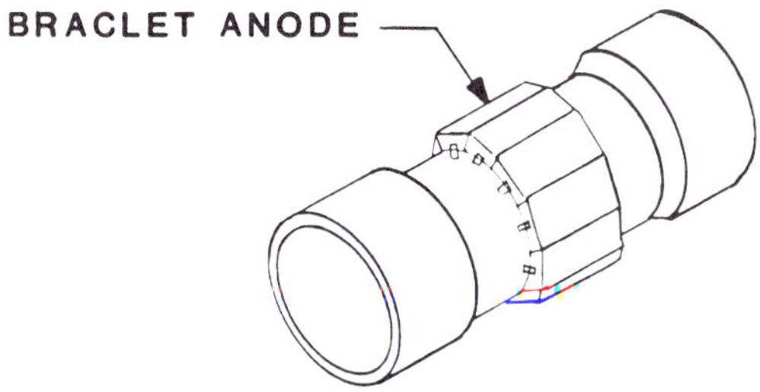

Fig. 10. Zinc bracelet anode used at a field joint on an offshore pipeline (*15*).

zinc alloys) are used on offshore platforms (Table VII). Impressed current anodes need not be naturally anodic to steel—their principal requirement is that they have low consumption rates at high current densities. Table VIII lists the metal consumption rates of various impressed current anodes.

Ceramic anodes and transition metal oxides on conductive metal substrates are gaining popularity for cathodic protection applications. These impressed current anodes have been used for years in the electrochemical processing industry, but their application to cathodic protection is new. These anodes have many of the same advantages (light weight, high current capacity) as the precious metal anodes. These materials are too new to appear in industry standards like the ones cited in Table VIII.

B. *Protection Criteria*

Cathodic protection surveys are necessary to determine if a vessel on a structure is protected from corrosion. Protection is assumed at a cathodic potential of −0.805 volts releative to a silver/silver chloride electrode is

TABLE VII

ENERGY CHARACTERISTICS OF SACRIFICIAL ANODE ALLOYS

Alloy	Energy Capability (amp-hr/lb)	Consumption Rate (lb/amp-yr)
Al-Zn-Hg	1250–1290	7.0–6.8
Al-Zn-In	760–1090	11.5–8.0
Al-Zn-Sn	420–1180	20.8–7.4
Zinc	370	23.7

TABLE VIII

CONSUMPTION RATES OF IMPRESSED CURRENT ANODE MATERIALS

Material	Typical Anode Current Density (amps/ft^2)[a]	Consumption Rate (per amp-yr)
Pb-6%Sb-1%Ag	15–20	0.1–0.2 lb
P_+(plated on substrate)	50–100	0.006 gm
P_+(wire or clad)	100–500	0.01 gm
Graphite	1–4	0.5–1.0 lb
Fe—14% Si-4% Cr	1–4	0.5–1.0 lb

[a] 1 A/ft^2 = 10.8A/m^2

achieved. Other criteria for determining if cathodic protection has been achieved are available (*11*), but the protection potential of −0.805 volts versus Ag/Ag Cl is the most widely-used criterion. Copper and zinc reference electrodes are sometimes used, but the silver/silver chloride reference electrode has become the industry standard (*11*).

C. *Environmental Variables*

The corrosivity of ocean waters varies throughout the world. The key variables that affect corrosivity and cathodic protection current requirements are temperature, pressure, and dissolved gases (primarily oxygen and carbon dioxide). Table IX shows industry standards for cathodic protection current densities on fixed offshore platforms. The current densities are the highest for Cook Inlet, Alaska, and the North Sea—two locations with low water temperatures, high dissolved gas concentrations, and rough seas.

Most offshore structures are unpainted below the water line. Cathodic protection causes pH shifts in the water. The resulting higher pH favors deposition of calcareous deposits on the structure surface. These deposits "polarize" the structure—in effect they lower the exposed metal surface area and reduce the current required for cathodic protection.

D. *Design Considerations*

Most cathodic protection design procedures are empirical in nature and are based on accepted industry practice.

TABLE IX

DESIGN CRITERIA FOR MARINE CATHODIC PROTECTION SYSTEMS[a]

Production Area	Water Resistivity (ohm-cm)	Water Temperature (°C)	Typical Design Current Density (mA/m^2)
Gulf of Mexico	20	22	54–65
U.S. West Coast	24	15	76–106
Cook Inlet	50	2	380–430
North Sea	26–33	0–12	86–216
Persian Gulf	15	30	54–86
Indonesia	19	24	54–65

[a] Adapted from NACE standard RP-01-76. The numbers in this table are used to illustrate the world wide variability of marine waters. Many operators have found that these numbers are inadequate for the initial current densities required for mineral deposition (calcareous deposit formation) on new structures.

Ships and pipelines normally have protctive coatings as their primary means of corrosion control. Cathodic protection systems are then sized so that an adequate electric current will be delivered to polarize the structure to the desired level. This is done for new structures by estimating the percentage of bare steel that defects of "holidays" in the protective coating will produce. Once the estimated "bare" steel is calculated, anodes are sized to provide adequate current densities for the design life of the system.

Marine pipelines are normally protected using sacrificial bracelet anodes which are fitted to the pipeline on the lay barge after welding and coating. Sacrificial anodes are preferred because these pipelines are normally remote from any power sources. Zinc is used because it is cheaper than aluminum. Marine pipelines require antibuoyancy coatings, so the added weight of zinc is also an advantage. The anodes are located at field joints where the pipe is welded together. Figure 10 shows a schematic of a zinc bracelet anode used on offshore pipelines.

Ships are returned to drydocks, so the size (and weight) of anodes can be reduced from what would be necessary for permanent anodes on offshore pipelines or platforms. Anodes are concentrated near the bow and stern where coating damage is most likely to occur. The stern is also the location where galvanic couples (propeller to hull) are possible. Relatively small anodes are placed on ships in these locations. Small anodes are desirable to minimize the drag effects caused by turbulence due to anode protrusions.

Both ships and pipelines are most often protected by sacrificial anodes. Pipelines rely on sacrificial anodes because they are frequently located too far from power supplies, but nearshore pipelines, e.g., those in Cook Inlet, can use shore-based electric power for cathodic protection. Some large super-tankers use impressed current cathodic protection.

Impressed current cathodic protection systems can cause problems by providing too much current causing "overprotection" in some instances. This can cause blistering in organic coating systems. Organic coatings can be debonded due to the formation of hydrogen gas bubbles underneath coatings. This coating debonding can produce increased surface areas requiring more cathodic protection. Coating debonding is controlled by placing dielectric shields between the impressed current anode and the hull (Figure 11). Large shields are sometimes fabricated from glass-reinforced epoxy, which is molded directly on the ship's hull. At one time coating debonding was a major concern for pipelines. Most modern pipeline coatings are resistant to debonding, and this is no longer a problem.

Hydrogen embrittlement of steel due to cathodic protection is sometimes a concern. This has been a problem on case-hardened shafts, bolts, and other high-strength attachments. Most structural steel and pipeline materials are

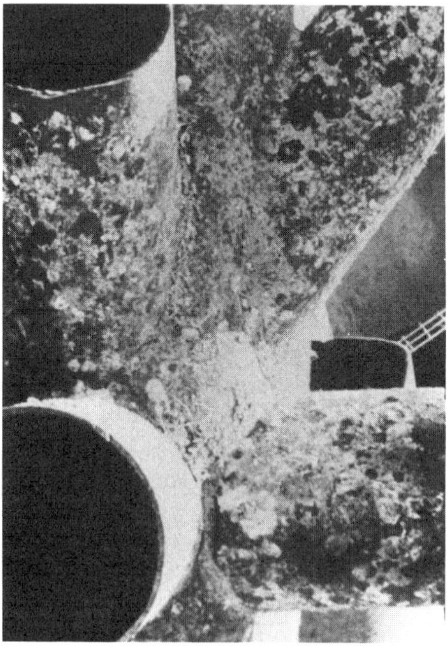

Fig. 11. Calcareous deposits on a node recovered from an offshore steel structure (*15*).

relatively low strength and have minimum susceptibility to hydrogen embrittlement.

The environment near a cathode becomes more alkaline (less acidic) due to the reduction reactions which occur at cathodes. This can cause cathodic corrosion of aluminum.

E. *Offshore Structures*

Corrosion protection on offshore production platforms is unusual because most platforms are not painted below the waterline. The cathodic protection system causes a pH shift in the water, which becomes more basic (higher pH). Most minerals are less soluble in bases than they are in near-neutral environments. Neutral water has a pH of 7, the pH of seawater is slightly basic and averages approximately 7.8. The higher pH near the cathode causes minerals to precipitate onto the steel surface and form a protective scale or calcareous deposit. Depending on water depth, temperature, velocity, etc., this protective scale may be calcium carbonate or magnesium hydroxide or a mixture of these and other minerals (*12*).

The technology of offshore cathodic protection is rapidly changing. Many of these changes are required because water depths are increasing. Thus offshore structures are now being built in deeper, colder water where mineral deposits are less likely to form.

The current densities recommended by NACE for various worldwide locations were shown in Table IX. Many operators are questioning the validity of the numbers shown in Table IX. Current trends are to increase initial current densities to several times those recommended in Table IX. This allows a mineral deposit to form on the structure (Figure 11), which can then be maintained using values similar to those shown in Table IX. Initial current densities are high, as 100–200 mA/ft^2 (approximately 1000–2000 mA/m^2) may be required to deposit a cathodic coating (calcareous deposit) on a bare steel platform in the Gulf of Mexico. This is over 20 times the long-term current density recommended in Table IX.

Many operators have preferred to use sacrificial anodes on offshore platforms, because sacrificial anode systems are simpler, rugged, and provide instant protection as soon as the platform is launched and do not depend on external electric power supplies. Surveys of the reliability of impressed current systems have led to the conclusion that they do not perform as well as sacrificial anode systems (*13*). Another reason may be the fact that impressed current systems can be turned off, and many operators are convinced that cathodic protection interferes with television reception, especially during the football (and hurricane) season.

Unfortunately, the weight of sacrificial anodes is becoming a serious consideration for deep-water platforms. Impressed current systems are gaining favor. Some operators have introduced hybrid designs, where the primary cathodic protection system uses impressed current, and a sacrificial anode system is used to protect the platform after launching and before the electrical system on the platform becomes oprational. Conoco's Murchison platform has the best-publicized hybrid cathodic protection system (*14*).

In past years, the inefficiencies associated with cathodic protection design were not serious. Water depths were shallow, and most designers overdesigned their cathodic protection systems to insure they would work. This was justified based on economics. A typical cathodic protection system is only 1%–2% of the total capital cost of a new platform, whereas a retrofit may cost as much as the platform itself. Early platforms in Cook Inlet and the North Sea were underdesigned, and the costs of retrofits led to the efforts which produced NACE RP-01-76 (*11*).

Typical designs are highly inefficient. The anode placement shown in Figure 12 places the anodes away from the nodes—welded or cast locations with the highest structural loading. These nodes are the locations where stresses are highest, corrosion is most likely to occur, and the consequences of corrosion are likely to be most serious. Figure 13 shows where corrosion fatigue cracking was actually found on one North Sea platform (*15*). Doughnut-shaped anodes, similar to the zinc bracelet anodes shown in Figure 10, could be located at nodes and would result in improved cathodic protection efficiency.

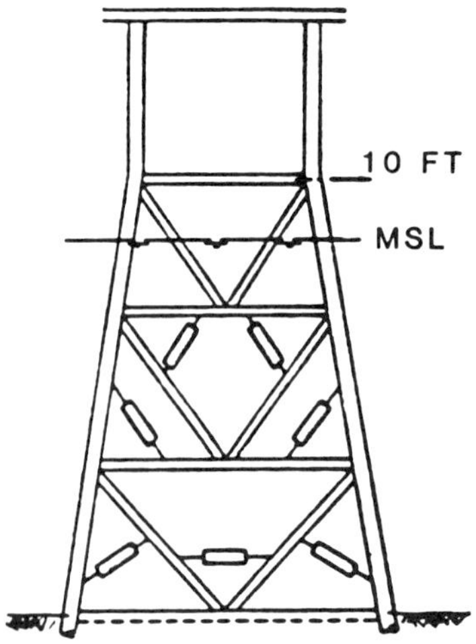

Fig. 12. Typical sacrificial anode system for cathodic protection of an uncoated offshore steel jacket platform (*13*).

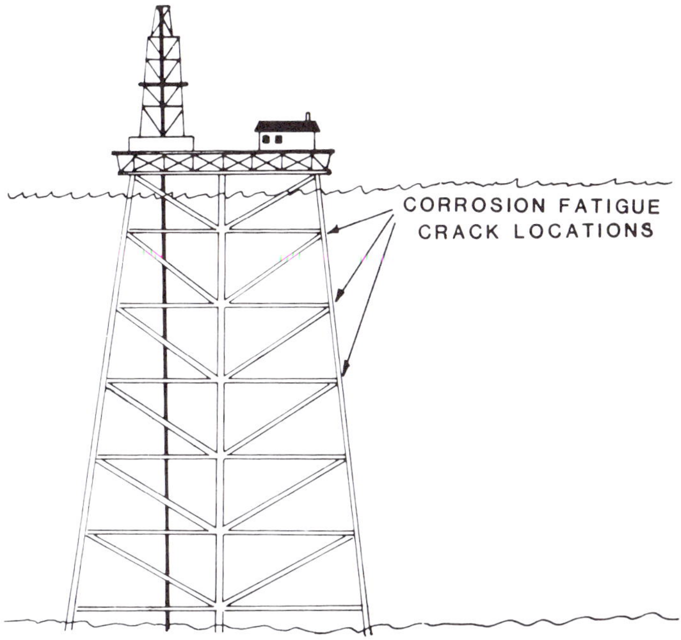

Fig. 13. Crack locations caused by corrosion fatigue on an offshore platform (*12*).

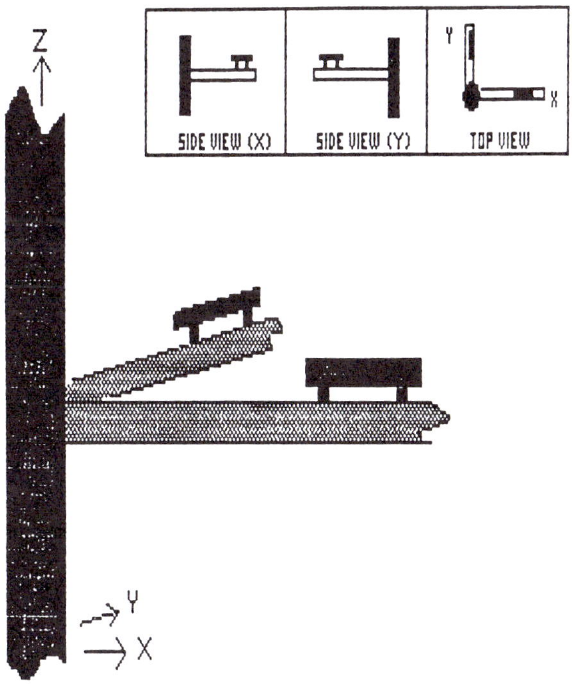

Fig. 14. Offshore node geometry generated by a small computer graphics package (*19*).

F. Computer-Aided Cathodic Protection Design Techniques

Computer-aided cathodic protection designs have been tried by several organizations in recent years. These computer-aided designs are of two types. Small computers have been used to make the types of calculations (wetted surface area versus anode consumption, etc.) which have commonly been used for cathodic protection design. For these systems the computer is a time saver which allows a greater number of alternatives to be considered but does not change the actual methodology of design. An alternative approach is to use numerical techniques (finite element, finite difference, boundary integral, etc.) to model the potential-current distribution field in the region on an offshore structure. Large computers can generate complex analyses of various alternative designs (*16–18*). These computerized designs have found limited acceptance because of the expense associated with the designs and the time delay caused by communications difficulties between the operator, the cathodic protection designer, and the computer expert.

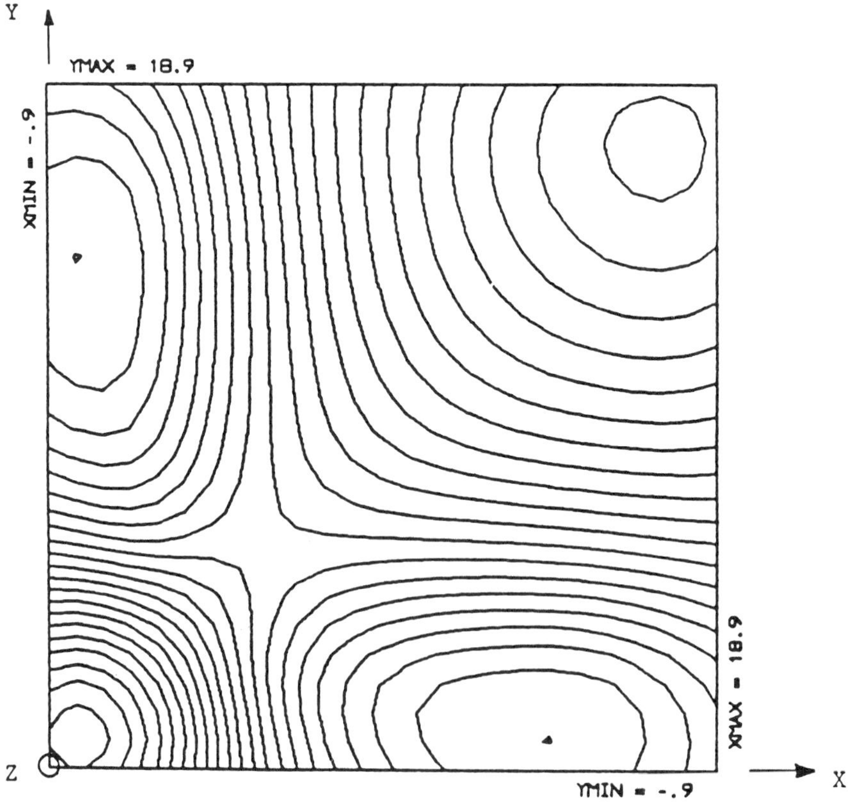

Fig. 15. Voltage contour map generated by a small computer graphics package (*19*).

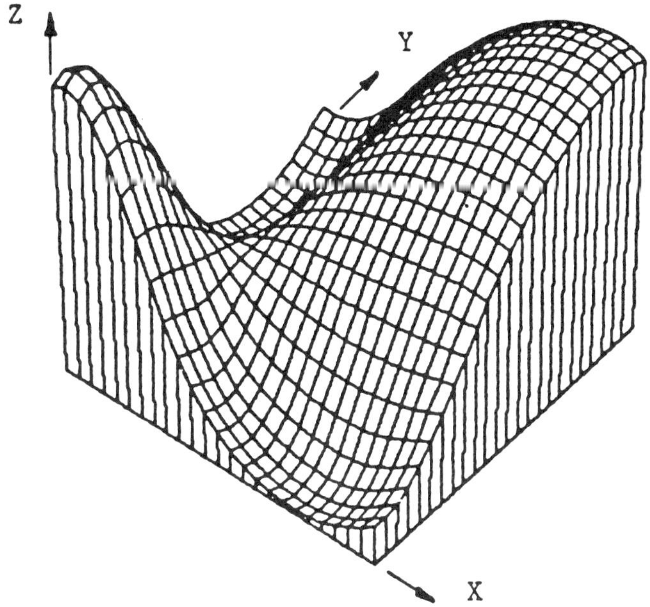

Fig. 16. Surface map showing the voltage distribution calculated by a small computer (*19*).

Efforts are underway to develop software so that small computers can be used with these numerical programs. Figures 14–16 were generated using software developed for use with desk top personal computers. This software is user-friendly and allows generation of design plot by the corrosion engineer in a matter of minutes (*19*). The plots generated by larger computers are more complex, but they take longer to generate and require communication between corrosion and computer experts. The plots shown in Figures 14–16 are adequate for many purposes and are being used by several companies for marine cathodic protection design.

V. Closure

Details of the three corrosion protection techniques used on marine structures have been presented. Details of materials selection, protective coatings, and cathodic protection systems are reviewed and suggestions to avoid common practical pitfalls are offered. In addition, state of the art computer-aided cathodic protection system design methods are introduced to illustrate the benefits which might be derived from more efficient corrosion protection systems.

References

1. Unified Numbering System for Metals and Alloys, Society of Automotive Engineers, Warrendale, PA, 1977.
2. Craig, B. D. (1984). *Practical Oil Field Metallurgy*. PennWell Books, Tulsa, OK.
3. Van Vlack, L. (1985). *Elements of Materials Science and Engineering*. Addison-Wesley Publishing, Reading, MA.
4. Heidersbach, R. (1986). "Corrosion." In: *An Attorney's Guide to Engineering* (Matthew Bender) New York, Chapter 3.
5. Allen, B., Heidersbach, R., and Mealy, S. (1980). "Cathodic Protection of Stainless Steels Against Crevice Corrosion in Seawater." *Proceedings, Offshore Technology Conference*, Paper 3856, Houston, TX.
6. Fontana, M. (1986). *Corrosion Engineering*. McGraw-Hill, New York.
7. Sedricks, J. (1979). *Corrosion of Stainless Steels*. Wiley, New York.
8. Crowe, C. R., and Hasson, D. F. (1981). "The Use of Titanium in Deep Sea Exploration." In: *Titanium for Energy and Industrial Applications*, (D. Eylon, ed). AIME, Warrendale, Pa., pp. 93–110.
9. Thomason, W. (1985). "Offshore Corrosion Protection with Thermal-Sprayed Aluminum." *Proceedings, Offshore Technology Conference*, Paper 4971, Houston, TX.
10. Joosten, M., Salama, M., and Myers, R. (1984). "Corrosion Fatigue of Aluminum Sprayed, High Strength Steel Immersed in Seawater." *Materials Performance* **July**, 22.
11. Anon (1983). *Recommended Practice: Corrosion Control on Steel, Fixed Offshore Platforms Associated with Petroleum Production*, NACE Standard RP-01-76. NACE, Houston, 1983 edition.
12. England, H., and Heidersbach, R. (1981). "The Effects of Water Depth on Cathodic Protection of Steel in Seawater." *Proceedings, Offshore Technology Conference*, Paper 4154, Houston, TX.
13. Boening, D. (1976). "Offshore Cathodic Protection Experience and Economic Reassessment." *Proceedings, Offshore Technology Conference*, Paper 2702, Houston, TX.
14. Levings, E., Finnegan, J., McKie, W., and Strommen, R. (1983). "The Murchison Platform Cathodic Protection System." *Proceedings Offshore Technology Conference*, Paper 4565, Houston, TX.
15. Smart, J. (1980). "Corrosion Failure of Offshore Steel Platforms." *Materials Performance* **May** 41.
16. Fu, J., Chow, J., and Chan, S. (1983). "A New Versatile Method for Designing Cathodic Protection Systems for Marine Structures." *Proceedings, Offshore Technology Conference*, Paper 4567, Houston, TX.
17. Heidersbach, R., Fu, J., and Erbar, R. editors (1986). *Computers in Corrosion Control*. NACE, Houston, TX.
18. Strommen, R. (1982). "Computer Modeling of Offshore Cathodic Protection Systems Utilized in CP Monitoring." *Proceedings, Offshore Technology Conference*, Paper 4367, Houston, TX.
19. Haroun, M. (1986). Cathodic Protection Modeling of Nodes in Offshore Structures, MS Thesis, Oklahoma University, July 1986.

6

Metallic Materials for Marine Structures

JAMES F. JENKINS

Materials Science Division
Naval Civil Engineering Laboratory
Port Hueneme, California

ISBN 0-12-341828-3

I. Introduction

Metals are one of man's most widely used materials. Their combination of high strength and ductility with ease of forming and fabrication makes the use of metallic materials a vital part of the modern world. However, just as strength and ductility are inherent properties of most metals, so is corrosion. Only by understanding the corrosion process, the behavior of materials in a wide variety of environments, and the means available to mitigate its effects can corrosion be controlled.

Seawater is perhaps the most universally corrosive environment known to man. Its high ionic content makes it an excellent conductor of electric current, a vital link in the electrochemical process responsible for corrosion in water. Seawater also contains almost every known chemical element, in addition to many organic compounds and the complex chemistry of life itself. Due to its inherent complexity and its interactions with the atmosphere, the earth's land masses, and the ocean floor, seawater is highly variable with respect to geographical location, depth, and time.

In this work, a limited review of the performance of metallic materials in seawater is presented. With over 15,000 commercially available alloys, many of which have not been tested in any marine environment, it is impossible to present a comprehensive description of the performance of all metallic materials in all possible marine environments. An attempt has been made to both describe the performance of commonly used, or potentially useful, alloys and to show how the performance of other alloys can be determined.

The corrosion behavior of metallic materials is, of course, only one factor in their successful application, but should be considered in all phases of design, construction, maintenance, and repair of marine structures in order to insure adequate reliability, long life, and overall cost effectiveness.

II. Methods for the Selection of Materials for Marine Structures

There are many methods which can be used for the selection of materials for marine structures. In the selection of materials for marine structures, many factors such as strength, toughness, availability, formability, weldability, cost, and corrosion performance must be considered. Corrosion performance is often the most difficult to evaluate. Unlike mechanical factors, corrosion factors are more difficult to incorporate into the design process. The evaluation of mechanical factors is usually quantitative. Corrosion factors are usually subjective judgments. The amount and type of corrosion which can be tolerated, the amount and type of corrosion which is likely to occur, and the impact of the corrosion damage on system performance can often only be roughly estimated.

It is impractical to give specific guidelines for the selection of materials for a specific application. The selection of materials depends upon a large number of criteria. As there are usually no "ideal" materials for a given application, materials selection is usually a compromise. The most important criteria for the selection of materials for a specific application may be reliability and function or, more commonly, economy of construction or life cycle costs. Thus, rather than attempt to give specific guidelines for the selection of materials for specific applications, several methods for the selection of materials based upon consideration of corrosion and the advantages and disadvantages of each method are presented.

It is vital that corrosion and materials selection be considered as early as possible in the design process. Successful designs are those that integrate all aspects of materials selection into the design process, rather than complete the design, then select the materials. There is a synergistic relationship between materials selection, corrosion, and design.

A. Determining Required Material Performance and Characteristics

The first steps in the selection of materials for marine applications are: (1) the determination of the environment to which the materials will be exposed and (2) the definition of required material performance characteristics. In the early stages of the design process, there is often considerable latitude in both of these factors. As the design process proceeds, it becomes less practical to vary these factors to fit available materials. After completion of the design, it is often found that there are no materials that truly fit all of the requirements. Then, either a compromise must be made in one or more of the selection criteria or the design must be changed. When a compromise in the selection criteria is made, it is often the corrosion performance criteria that are modified, as the initial function of the structure or equipment is seldom affected by corrosion. Corrosion then becomes a maintenance problem. The entire life cycle of the structure or equipment should be considered at the design stage. Proper design should have, as a target, the lowest life cycle cost commensurate with the required system performance and reliability.

Another important consideration in the selection of materials for a given application is the potential effectiveness of corrosion control measures. The entire system must be considered. In many cases, a material that may suffer unacceptable attack if unprotected will perform in an entirely satisfactory manner if protected. For example, stainless steels are subject to pitting and crevice corrosion in many marine environments and are unsatisfactory for many applications, particularly where resistance to crevice corrosion is an important factor. However, when cathodically protected, even by simple coupling to carbon steel, they have performed well in many applications. Likewise, carbon steels which corrode at high rates in the intertidal and splash zones of marine structures, such as offshore platforms, can be successfully used in conjunction with a protective coating system.

The determination of the environment to which the materials will be exposed may seem to be straightforward. However, it can be, and usually is, complex. A structure must often survive exposure to a wide variety of environments. For example, what environmental characteristics should be considered for a deep water temperature recorder which is recovered and reused? The recorder will be exposed to: (1) deep water environments of varying characteristics depending upon location and depth; (2) surface seawater during emplacement and retrieval; (3) the marine atmosphere during preparation for emplacement and retrieval; (4) cleaning solutions after retrieval (hopefully); and (5) varying atmospheric conditions during storage both at sea and on land.

The environmental considerations in Table I will influence the rate and type of corrosive attack. Such considerations must be taken into account when attempting to apply corrosion data, whether from experience or experiment. Corrosion behavior of many metals can vary considerably with

TABLE I

CORROSION-RELATED MARINE ENVIRONMENTAL CONSIDERATIONS

Type of Exposure	
Atmospheric	Immersion
Effects of splash or spray	Continuous or intermittent
Orientation—horizontal or vertical or inclined	Velocity—maximum, average
Can water be trapped & held?	Effects of depth—oxygen, pH, temperature
Temperature—mean & extremes	Is bacterial attack possible?
Sunlight	Biofouling effects
Pollution	Pollution
Stress—static & cyclic	Stress—static & cyclic
Effects of welding	Effects of welding
Rainfall—rinsing action	Crevices
Is there Galvanic coupling?	Is there Galvanic coupling?
Duration of exposure	Duration of exposure

only minor variations in environment. It is often only possible to estimate the likely range of attack.

After the environment to which the material will be exposed has been evaluated, other application requirements such as strength, weight, cost, and availability should be determined. All factors can then be used to estimate the amount and type of corrosion that is likely to occur and its likely impact.

B. *Selection Based Upon Experience with Similar Structures in Similar Environments*

Consideration of the information in Table I shows why experience with similar structures in similar environments is the best basis for the selection of materials. If it worked well before, it is likely to work well again under the same conditions. This method, however, is not foolproof. The performance upon which the selection is based must be carefully evaluated. Poor performance is seldom well advertised. All of the factors that lead to a successful experience must be understood before that experience can be applied to a new situation. For example, aluminum alloys, which show good performance in surface seawater, may show reduced resistance to attack in deep ocean waters where differences in environment adversely affect the corrosion of many aluminum alloys. Time of exposure is also quite important. If the duration of experience is short, long-term performance may be substantially different than that projected. Corrosion rates may either increase, decrease, or remain fairly constant with time. Materials that exhibit an incubation period where little or no measurable corrosion occurs followed by rapid propagation of attack, once initiated, should be carefully evaluated before they are selected based upon short-term experience.

C. *Selection Based Upon Experience with Similar Structures in Different Environments*

Lacking an experience base with similar structures in similar environments, the second most reliable method for the estimation of corrosion performance of materials is still the past performance of materials in service. In this case, the more similar the conditions the better. However, care must be used in establishing which conditions are indeed similar. For example, near surface immersion in the arctic and in the tropics would seem to be very different. However, in the case of steel, the effects of differences in temperature and dissolved oxygen content of the seawater counterbalance each other. In the arctic, the temperature is low but the oxygen content is high. In the tropics, the temperature is high but the dissolved oxygen content is low. The overall effect is such that the corrosion behavior of steel is similar in either of these seemingly dissimilar environments. Again, as in the assessment of performance of materials outlined above, it is important that all of the pertinent factors in the exposure and in the behavior of the material are understood before performance is used as a criteria for materials selection.

D. Selection Based Upon In Situ Corrosion Testing

If the differences between service experience and the projected use of the material being evaluated are great, the service experience may not be a reliable guide to the estimation of material performance. In this case, limited service experience should be replaced, or better, augmented by *in situ* corrosion testing. *In situ* tests should be performed in as similar an environment to the projected service exposure as possible. Again, time of exposure is an important factor. Projection of long-term corrosion behavior based upon short duration tests can be misleading. The configuration of the test specimens used should model accurately the geometric factors that will be present in service. It is also important that the techniques used to evaluate the specimens after exposure are appropriate for the type and amount of attack that occurred. Pertinent evaluation methods for each type of attack are described in the following sections on corrosion behavior of specific materials. It is also difficult to evaluate the impact of the amount and type of corrosion damage experienced in the testing of corrosion specimens on the serviceability of a material for a specific application. It is this difficulty in interpreting corrosion data that makes the use of service experience extremely valuable. Only through service experience can a reliable determination be made of the type and extent of corrosion that can be tolerated. Corrosion testing is extremely useful in establishing the relative corrosion behavior of materials, in studying the effect of environmental variables on corrosion, and in other corrosion studies. However, it should not be used as the sole means for estimating corrosion performance unless service performance is not available. If service experience is not available, it is advisable to perform a specific corrosion test designed to model actual service conditions as closely as possible.

E. Selection Based Upon Laboratory or Electrochemical Corrosion Testing

Laboratory corrosion tests should not be used as the sole basis for the selection of materials for marine environments. The ocean environment is very difficult to simulate under laboratory conditions (*1*). Even when actual seawater is used, the corrosion performance of many materials has been found to be substantially different under laboratory and field conditions. The more unnatural the test environments are, the less likely it is that the results will provide a reliable estimate of field performance. For example, attempts to study the effect of pH on corrosion behavior under laboratory conditions require that the pH of the test solution used be changed from its normal value through the addition of acid or base. pH shifts so obtained dramatically affect the buffering capacity of the resulting solution. The performance of many materials in seawater is often more dependent upon the buffering capacity of the seawater than upon pH itself. In natural environments, where the pH can vary between wide limits, the buffering capacity is fairly constant. Whether the laboratory corrosion tests are evaluated using conventional measurements of corrosion damage or are evaluated using electrochemical techniques, the results of such tests are of very limited value in predicting service performance. Attempts to accelerate corrosion under laboratory conditions are extremely difficult to apply to the estimation of service performance. In order to accelerate the corrosion, the conditions of exposure are intentionally changed. The results of such tests are therefore even further removed from service performance than under more natural laboratory conditions. While the results of laboratory tests are of very limited utility in predicting service performance of materials, they are extremely useful in studying the mechanisms of corrosion so that service behavior and the results of *in situ* corrosion tests can be better understood.

F. *Other Bases for Material Selection*

Other bases for the estimation of corrosion performance of materials in marine environments, such as predictions based upon corrosion theory or the performance in non-marine environments, should only be used when no other data are available or can be obtained. The selection of materials for marine environments is best made through the rational utilization of all available corrosion data, an understanding of the corrosion phenomena involved, and the incorporation of corrosion as a significant factor in the materials selection process. When incorporated into the design process at an early stage, the design and materials can work together to provide the required performance at the least cost.

III. Materials for Structures in Surface and Near-Surface Environments

A. *The Surface and Near-Surface Environment*

The corrosion of materials in the surface and near-surface environment varies significantly with location. This is due to the variation of chemical and physical characteristics of the marine environment from site to site. In some cases, the variation in performance between sites is small, in other cases it is large. In order to predict the performance of materials in any specific location, the environmental characteristics of the site must be established. Generalizations about the variation of corrosion performance with location (i.e., latitude) can be grossly misleading. Only by actual on-site testing can corrosion performance be accurately measured. Lacking on-site data, corrosion performance can best be estimated by referring to the performance of the materials at sites with similar environmental characteristics and estimating the effects of differences in the environments on corrosion of the materials being considered.

There is a wide variability of surface and near-surface marine environments. Dexter and Culbertson (2) give an excellent review of this variability. The primary variables of importance in the characterization of a surface or near-surface marine environment are:

(a) Salinity
(b) Temperature
(c) Dissolved oxygen
(d) pH
(e) Tidal velocity
(f) Fouling
(g) Pollution

1. Open Ocean Sites

The salinity of open ocean surface waters is usually in the range of 32 to 36 parts per thousand. In locations where there is unusual evaporation, the formation of sea ice, or surface runoff, salinity can be outside this normal range. The ionic content of the dissolved salts is remarkably constant (*3*). The average temperature of surface seawater ranges from $-1.5°C$ to $+28°C$ (*4*). The average and extremes of temperature at any site can fall outside these limits. Surface and near-surface waters in the open ocean are normally fully saturated with oxygen. The solubility of oxygen is higher in cold water than in warm water and ranges from 38 ml/L at 0°C to 23.6 ml/L at 24°C in seawater with 36 parts per thousand salinity (*4*). The pH of surface seawater usually falls in the range of 8.1 to 8.3 (*4*).

The velocity of seawater due to open ocean currents is usually too low to have a direct influence on corrosion, but the variations in environment associated with ocean currents can be significant. Since primary sources of nutrients for marine organisms are surface runoff from the land or upwelling associated with coastlines, the amount of fouling activity in the open ocean is normally less than in near-shore environments (5). In open ocean sites which are removed from the influence of the land, the surface seawater is relatively free from suspended sediments and pollution.

2. Near-shore Sites

Due to the influence of the land, the environment at near-shore sites is more varied than at open ocean sites. The salinity may be reduced considerably by surface runoff and may be seasonally variable. In some cases where enclosed basins are formed, such as the Mediterranean and Red Seas, the salinity of surface waters may be higher than in the open ocean due to high rates of evaporation. The temperature of surface seawater at near-shore sites is more variable than in the open ocean due to the direct influence of the land and due to upwelling of currents which can bring cold water from depth to the surface. The oxygen content of surface and near-surface seawater at near-shore sites is usually near saturation for the temperature and salinity at the site. However, due to local periodic "blooms" of certain species of algae or dinoflagellates or other organisms, the oxygen content of near-surface water may be reduced significantly below the saturation level if the organisms consume oxygen. If the organisms produce oxygen, the amount of dissolved oxygen present can actually exceed saturation. The pH of near-shore surface waters is normally in the same range as open ocean waters but can be significantly reduced by dilution from surface runoff and significantly increased when unusual amounts of dissolved carbon dioxide are consumed by plant photosynthesis (*6*). Tidal velocities may be significant in many near-shore sites and can significantly affect corrosion rates through increasing the availability of dissolved oxygen, by reducing the stability of protective films and, in some cases, by direct scouring action, particularly at the seawater-sediment interface. As the nutrient levels in near-shore environments are usually higher than in open ocean environments, the amount of fouling is also usually higher. The amount and type of fouling encountered at different near-shore sites is highly site specific. For fixed structures, the corrosiveness of the seafloor soil environment may be an important factor. The corrosiveness of the seafloor soil may either be higher or lower than the corrosiveness of the seawater at the site. The corrosiveness of the seafloor soil is extremely site specific. The corrosion-related characteristics of a seafloor soil environment include the classification of the soil (sand, gravel, silt, etc.) and the mineral type of the soil (quartz sand, coral, etc.). However, these characteristics are often more highly dependent upon the presence and activity of sulfate-reducing bacteria and the presence of hydrogen sulfide in the sediments. The differential environment at the soil-water interface can often result in accelerated attack in this zone. Local pollution can also affect the corrosiveness of near-shore sites either by direct chemical attack or by its influence on other environmental variables or on fouling.

3. Harbor and Estuarine Sites

The environment and the corrosion bhavior of materials at any specific harbor and estuarine sites is likely to be unique. The salinity, temperature, oxygen content, pH, fouling, and pollution may be highly variable. Tidal currents may be significant. The bottom sediments are often highly polluted and are often highly corrosive due to the presence of sulfate-reducing bacteria and hydrogen sulfide.

B. Performance and Selection of Irons and Steels

1. General Corrosion Performance

Corrosion of a wide variety of common irons and steels in surface and near surface seawater environments is usually uniform, usually occurs at a rate usually between 2 to 5 mils per year (mpy), and the rate of corrosion usually decreases with time. It is remarkable that this behavior is so similar at different sites, considering the wide range of environments encountered. This is due to the self-compensating effects of the variations of the environmental parameters encountered. For example, in cold waters the temperature related diffusion rates are low, this tends to reduce corrosion activity. However, in cold waters the amount of dissolved oxygen is high, which tends to increase the corrosion activity. The normal decrease in corrosion rates with time is due to the accumulation of both biofouling and corrosion products on the surface of the metal, which reduces the availability of oxygen and thus reduces the rate of corrosion. Typical reduction in corrosion rates with time are shown in Figure 1 (*7*). Pitting is usually associated with pollution or the presence of sulfate-reducing bacteria. When pitting occurs, the penetration due to pitting can be as great as ten times the average corrosion rate calculated by weight loss. Irons and steels are also susceptible to other forms of corrosion such as galvanic corrosion, stress-corrosion cracking, hydrogen embrittlement, corrosion fatigue, and fretting. Resistance to these forms of corrosion must be considered, as appropriate, for the specific application.

2. Significant Performance Parameters

In evaluating the performance of steel for use in surface environments, the corrosion rate, the variation of the corrosion rate with time, and the

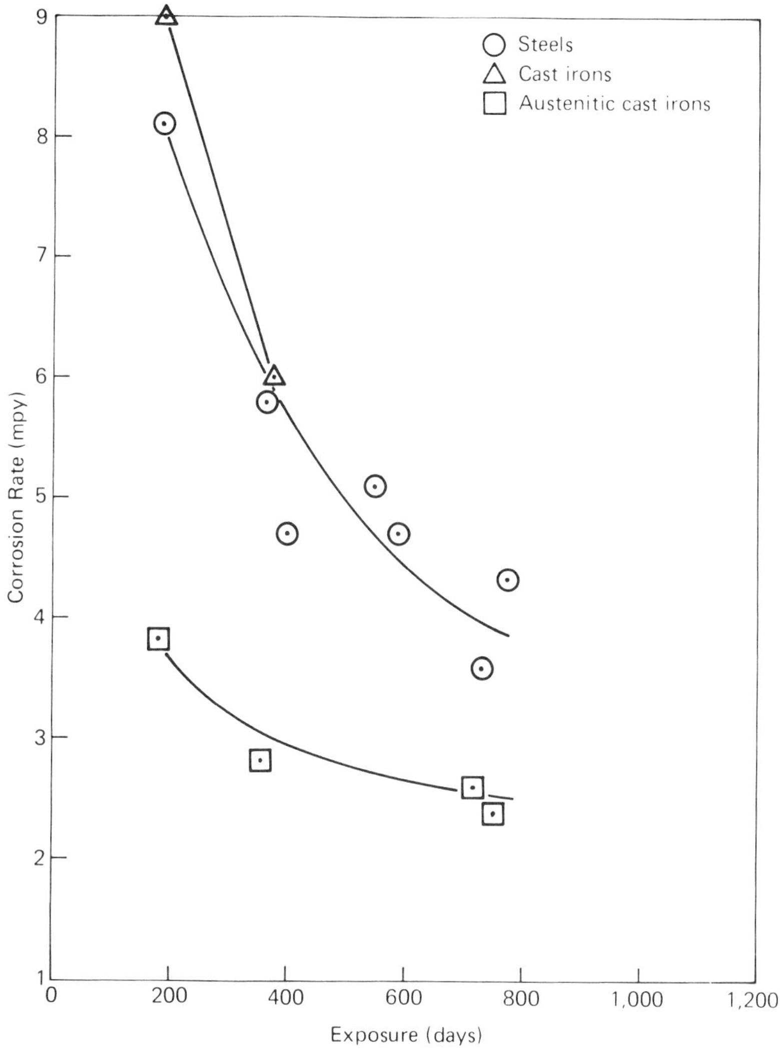

Fig. 1. Corrosion rates of irons and steels in surface seawater versus time of exposure.

occurrence and intensity of pitting are important parameters. However, steel is seldom used in marine environments unless it is protected by coatings, cathodic protection, or both. Thus, both the performance of the protective system and the consequences of failure of the protective system, either locally in the case of coatings or periodically in the case of cathodic protection systems, are often the most important considerations in the evaluation of the performance of irons and steels in marine environments. It is the ability of protective systems to mitigate the corrosion of steels, coupled with the relatively small impact of imperfect protection and the low cost of steel, that makes steel such an excellent candidate for many marine applications.

3. Performance of Commonly Used Ferrous Alloys

The performance of a wide variety of ferrous alloys falls within a narrow range in many near surface environments. Tables II and III show the composition and performance of selected irons and steels in surface environments. The data presented are taken from Reinhart (*7*).

4. Typical Applications

Steel and ferrous alloys are the most widely used structural materials for marine construction. When unprotected, the corrosion of these materials can be high, but corrosion can be reduced to acceptable levels through the use of protective coatings or cathodic protection. They are low in cost and easy to form and fabricate.

C. Performance and Selection of Aluminum Alloys

1. General Corrosion Performance

The corrosion of aluminum and aluminum alloys in surface and near surface waters is different in both degree and kind than the corrosion of ferrous materials. Pitting, crevice corrosion, and other nonuniform types of attack, such as intergranular corrosion and exfoliation, are the most common types of corrosion attack that occur on aluminum and aluminum alloys in surface seawater. Because of the nonuniform nature of this attack, corrosion rates based upon measurement of weight loss should not be used to evaluate the performance of these alloys. Aluminum and aluminum alloys, due to their active electrochemical potential, are also subject to accelerated attack through electrical coupling with a wide variety of materials. Despite their susceptibility to corrosion, aluminum alloys have several advantages, principally their light weight, which make them excellent candidates for marine applications. The corrosion of this group of materials can be minimized through design to reduce the incidence and degree of attack and allowance for the attack which is likely to occur. Corrosion of aluminum alloys can also be mitigated through the application of protective coatings and the use of cathodic protection.

2. Significant Performance Parameters

In order to evaluate the performance of aluminum and aluminum alloys, resistance to pitting, crevice corrosion, and other forms of nonuniform corrosion must be evaluated. The method of evaluation must match the types of corrosion which occur. Considerable judgment must be applied to the evaluation of corrosion test samples. Visual observations should be carefully recorded and reported. The depth and frequency of pitting that occurs must also be measured and reported. The impact of the type of corrosion that occurs on the intended application must also be considered. For example, the impact of pitting on an open framework structure is minimal, whereas the same amount and type of attack would have a much more significant impact on a tank or pressure vessel.

3. Performance of Commonly Used Alloys

There are significant differences between the performances of the different groups of aluminum alloys, as well as between different alloys in each group. Generally the alloys in the 1000, 3000, and 5000 series are the most corrosion resistant. Alloys in the 6000 series show somewhat more attack, and alloys in the 2000 and 7000 series are the least resistant to attack. Tables IV and V give the chemical composition and corrosion performance of several commonly used aluminum alloys. The data in Tables IV and V are taken from Reinhart (*7*), Van Horn (*8*), and Ailor (*9*).

4. Typical Applications

Aluminum alloys have gained increasing acceptance for many marine applications in surface waters. 5000 series alloys, particularly 5083, 5086, and 5456, have been used for hulls of ships ranging from small pleasure crafts to large vessels such as the Alcoa Seaprobe. Successful application of aluminum alloys involves careful attention to construction details to avoid crevices and galvanic coupling, and often in the application of cathodic protection.

D. Performance and Selection of Copper Alloys

1. General Corrosion Performance

In surface and near-surface marine environments, many copper alloys corrode uniformly and at low rates. Crevice attack is minimal on many

TABLE II

COMPOSITION OF SELECTED IRONS AND STEELS[a]

Alloy	C	Mn	P	S	Si	Ni	Cr	Mo	Other
Armco iron	—	0.02	—	—	—	—	—	—	—
Wrought iron	0.02	0.06	0.13	0.01	0.13	—	—	—	2.5 Slag
AISI 1010	0.12	0.50	0.004	0.023	0.060	—	—	—	—
Copper steel	—	0.40	0.01	—	0.02	0.01	0.03	—	0.28 Cu
ASTM A-36	0.24	0.70	0.011	0.027	0.055	—	—	—	—
HSLA No. 1	0.18	0.86	0.014	0.023	0.28	0.05	0.64	0.18	0.047 V 0.00028 B 0.020 Ti
HSLA No. 2	0.12	0.30	0.015	0.025	0.27	2.34	1.25	0.20	0.17 Co
HSLA No. 4	0.07	0.38	0.11	0.025	0.54	0.31	0.88	—	—
HSLA No. 5	0.14	0.78	0.020	0.025	0.23	0.74	0.56	0.42	0.22 Cu 0.36 V 0.0041B
HSLA No. 7	—	0.43	0.12	—	0.13	0.54	—	—	—
HSLA No. 11	—	0.69	0.08	—	—	0.50	0.26	—	0.30 Cu
HSLA No. 12	0.14	0.26	0.011	0.009	0.27	2.60	1.55	0.46	0.02 V
HS No. 1	0.11	0.78	0.008	0.006	0.29	5.03	0.56	0.42	0.05 V
HS No. 2	0.002	0.018	0.004	0.005	0.05	12.20	5.07	3.12	0.21 Ti 0.25 Al
HS No. 3	0.28	0.29	0.005	0.005	0.10	8.26	0.53	0.47	3.82 Co 0.15 V
18 % Ni maraging	0.02	0.10	0.005	0.007	0.14	17.92	—	4.78	8.75 Co 0.003 B 0.94 Ti 0.17 Al
AISI 4130	0.29	0.42	—	—	—	—	0.90	0.18	—
AISI 4340	0.40	0.73	0.013	0.014	0.27	1.77	0.62	0.24	—
AISI 502	0.06	0.48	0.020	0.010	0.33	—	4.75	0.55	—

Ni cast iron	—	0.68	—	—	2.47	1.56	—	—	—
Ni-Cr cast iron #2	—	0.86	—	—	1.99	3.22	0.98	—	—
Ductile cast iron #1	—	0.35	—	—	2.50	0.91	—	—	—
Si cast iron	—	—	—	—	14.5	—	—	—	—
Si-Mo cast iron	—	—	—	—	14.0	—	—	3.5	—
Austenitic cast iron, Type 1	—	1.4	—	—	2.05	15.8	1.79	—	6.71 Cu
Austenitic cast iron, Type 2	—	1.01	—	—	2.29	18.2	2.04	—	—
Austenitic cast iron, Type 3	—	0.6	—	—	1.15	28.4	2.87	—	—
Austenitic cast iron, Type 4	—	0.56	—	—	5.43	29.7	4.94	—	—
Austenitic cast iron, D-2	—	0.94	—	—	3.0	21.4	2.26	—	—
Austenitic cast iron, D-2b	—	0.96	—	—	2.0	20.8	3.19	—	—
Austenitic cast iron, D-2c	2.45	2.12	0.017	—	2.38	22.34	0.08	—	—
Austenitic cast iron, D-3	—	0.5	—	—	1.83	29.8	2.70	—	—

[a]Remainder—Fe

TABLE III

PERFORMANCE OF SELECTED IRONS AND STEELS IN SURFACE SEAWATER

Alloy	Exposure (Days)	Corrosion Rate (mpy)[a]	Type of Corrosion[b]
Armco iron	181	6.9	U, CR (12)
Armco iron	366	7.1	G
Wrought iron	181	5.3	U
Wrought iron	364	4.0	G
Wrought iron	723	4.0	G
Wrought iron	763	4.8	G
AISI 1010	181	9.1	U
AISI 1010	366	8.0	G
AISI 1010	398	8.2	U
AISI 1010	588	7.9	G
Copper steel	181	9.0	G
Copper steel	366	6.0	G
ASTM A-36	181	10.7	G, C(B)
ASTM A-36	398	6.2	G
ASTM A-36	540	6.3	G
ASTM A-36	588	5.8	G
HSLA No. 1	181	9.7	G, P
HSLA No. 1	398	5.2	G, P
HSLA No. 1	588	4.7	G, P
HSLA No. 2	181	6.8	G, P
HSLA No. 2	398	4.5	U, P
HSLA No. 2	540	4.4	G, P
HSLA No. 4	181	11.0	G
HSLA No. 4	366	8.0	G
HSLA No. 5	181	8.9	U
HSLA No. 5	366	8.0	G
HSLA No. 5	398	6.0	G, P
HSLA No. 5	540	5.4	G, P
HSLA No. 7	181	11.0	G
HSLA No. 7	366	8.0	G
HSLA No. 10	181	11.0	G
HSLA No. 10	366	8.0	G
HSLA No. 12	181	8.5	U, P
HSLA No. 12	398	4.2	G, P
HSLA No. 12	540	4.9	G, P
HSLA No. 12	588	4.3	G, P
HS No. 1	181	9.9	U
HS No. 1	398	4.7	G, P
HS No. 1	540	4.5	G, P
HS No. 1	588	4.2	G, P
HS No. 2	181	8.2	U
HS No. 2	398	3.5	G, P
HS No. 2	588	3.3	G, P

TABLE III (*continued*)

Alloy	Exposure (Days)	Corrosion Rate (mpy)[a]	Type of Corrosion[b]
HS No. 3	398	5.0	U, P
HS No. 3	540	3.8	G, P
HS No. 3	588	4.0	G, P
18% Ni maraging	181	5.4	U, P
18% Ni maraging	398	3.0	U, P
18% Ni maraging	588	3.1	G, C, P
1.5% Ni steel	181	11.0	G
1.5% Ni steel	366	8.0	G
3% Ni steel	181	11.0	G
3% Ni steel	366	8.0	G
5% Ni steel	181	8.0	G
5% Ni steel	366	7.0	G
9% Ni steel	181	10.0	I-P
9% Ni steel	366	8.0	G
AISI 502	181	7.0	P (18)
AISI 502	398	4.4	G, P (30)
AISI 502	540	4.1	G, P (45)
Gray cast iron	366	2.6	G
Ni cast iron	181	8.5	U
Ni cast iron	366	7.6	G
Ni-Cr cast iron	181	6.7	U
Ni-Cr cast iron	366	5.2	U
Ni-Cr cast iron #2	181	8.5	U
Ni-Cr cast iron #2	366	4.9	G
Ductile cast iron #1	181	10.0	U
Ductile cast iron #1	366	6.2	CR (24)
Silicon cast iron	181	< 0.1	E
Silicon cast iron	366	< 0.1	ET
Si-Mo cast iron	181	< 0.1	NC
Si-Mo cast iron	366	< 0.1	ET
Austenitic cast iron, Type 1	181	4.1	U
Austenitic cast iron, Type 1	366	2.7	U
Austenitic cast iron, Type 2	181	5.8	U
Austenitic cast iron, Type 2	366	2.9	U
Austenitic cast iron, Type 3	181	5.0	U
Austenitic cast iron, Type 3	366	2.8	U
Austenitic cast iron, Type 4	181	3.8	U

(*continued*)

TABLE III (*continued*)

Alloy	Exposure (Days)	Corrosion Rate (mpy)[a]	Type of Corrosion[b]
Austenitic cast iron, Type 4	364	2.4	G
Austenitic cast iron, Type 4	723	2.0	G
Austenitic cast iron, D-2	181	4.3	G
Austenitic cast iron, D-2	366	2.4	G
Austenitic cast iron, D-2b	181	4.1	G
Austenitic cast iron, D-2b	366	2.7	G
Austenitic cast iron, D-2c	181	3.9	U
Austenitic cast iron, D-2c	364	3.2	G
Austenitic cast iron, D-2c	723	3.1	U
Austenitic cast iron, D-2c	763	2.8	U
Austenitic cast iron, D-3	181	4.3	G
Austenitic cast iron, D-3	366	3.2	G

[a] mpy = mils penetration per year calculated from weight loss.
[b] Symbols signify the following types of corrosion:

C = Crevice
P = Pitting
U = Uniform
G = General
E = Edge
CR = Cratering
NC = No visible corrosion
ET = Etched

Numbers after symbols refer to maximum depth of attack in mils.

copper alloys. Where crevice attack occurs on copper alloys, the attack is usually due to the metal ion concentration cell type of attack. The corrosion in metal ion concentration cells is localized just outside the crevice area and often has less serious consequences than oxygen concentration cell attack, where the attack is localized within the crevice area. Figure 2 illustrates the difference between these two forms of attack. Copper alloys in general are not resistant to attack by seawater containing large amounts of sulfide, and, therefore, their performance is better in open ocean conditions than in harbor or estuarine sites or where seawater can be trapped and stagnate. One unique property of several copper alloys is their fouling resistance. Toxic copper compounds are produced by the corrosion of many of the copper alloys, and when the rate of release of copper exceeds a certain critical value, commonly in the order of 5 mg/dm^2/day, the surfaces remain free from macrofouling (*3*).

2. Significant Performance Parameters

For those copper alloys that corrode uniformly, corrosion rates based upon weight loss can be used to measure their performance. For those alloys that corrode nonuniformly, the measurement of corrosion performance must be carefully matched to both corrosion behavior and intended use. This is particularly true when dealloying or parting corrosion is experienced. This form of corrosion can only be adequately measured by sectioning the sample and measuring the depth of attack visually or under a microscope. Dealloying is often self limiting in depth and may have only minimal effect upon thick wall castings. Where corrosion by high velocities are expected, test results should include consideration of both flow velocity and the effects of turbulence and entrained gases or solids.

3. Performance of Commonly Used Alloys

Copper alloys can be conveniently grouped into several types as listed in Table VI. The composition and performance of several commonly used copper alloys is given in Tables VI and VII. The data are taken from Leidheiser (*10*) and Reinhart (*11*).

4. Typical Applications

Due to the cost of copper alloys, they are not widely used for structural applications but have wide applications in piping, valves, fittings, heat exchangers, propellers, and other mechanical equipment. They are frequently

TABLE IV

COMPOSITION OF SELECTED ALUMINUM ALLOYS[a]

Alloy	Si	Fe	Cu	Mn	Mg	Cr	Ni	Zn	Ti	Other
1100-H14	0.14	0.55	0.14	—	—	—	—	0.06	—	—
2014-T6	0.91	0.53	4.23	0.80	0.32	0.03	0.01	0.08	0.02	—
2219-T81	0.20	0.30	6.3	0.30	< 0.02	—	—	0.10	0.06	0.10V 0.17Zr
2219-T87	0.08	0.12	6.54	0.26	< 0.02	< 0.02	<0.02	0.03	0.02	0.10V 0.15Zr
3003-H14	0.20	0.58	0.13	1.05	0.03	0.01	0.01	0.02	0.01	—
Alclad 3003-H12	0.06	0.70	0.20	1.25	—	—	—	0.10	—	—
Core										
Cladding	*b*	*b*	0.10	0.10	0.10	—	—	1.0	—	—
5050-H34	0.18	0.64	0.04	—	1.19	—	—	—	—	—
5052-O	0.10	0.17	0.01	0.03	2.20	0.22	—	—	0.01	—
5052-H34	0.13	0.30	0.02	—	2.31	0.22	—	—	—	—
5083-O	0.17	0.31	0.06	0.71	4.78	0.14	—	0.08	0.03	—
5083-H113	0.40	0.40	0.10	0.65	4.5	0.15	—	0.25	0.15	—
5086-O	0.13	0.28	0.06	0.33	3.94	0.15	< 0.02	0.10	< 0.02	—
5086-H112	—	—	—	0.45	4.0	0.15	—	—	—	—
5086-H32	0.15	0.25	0.05	0.32	3.75	0.12	—	0.12	0.01	—
5086-H34	0.40	0.50	0.10	0.45	4.0	0.15	—	0.25	0.15	—
5454-H32	*c*	*c*	0.10	0.75	2.7	0.13	—	0.25	0.20	—
5454-H34	0.13	0.22	0.03	0.71	2.75	0.12	< 0.02	0.02	< 0.02	—

5456-H321	*d*	*d*	0.10	0.75	5.0	0.13	—	0.25	0.20	—
5456-H343	0.10	0.16	0.10	0.80	5.0	0.07	—	0.01	0.05	—
6061-T6	0.60	0.70	0.27	0.15	1.0	0.25	—	0.25	0.15	—
7002-T6	0.20	0.40	0.75	0.15	2.50	−.15	—	3.5	0.03	—
Alclad 7002-T6										
Core										
Cladding	0.10	0.15	0.75	0.18	2.5	0.20	—	3.5	—	—
(7072)	—	—	—	—	—	—	—	1.0	—	—
7039-T6	0.11	0.17	—	0.22	2.8	0.21	—	4.0	0.02	—
7075-T6	0.10	0.14	1.40	0.08	2.10	0.30	—	5.1	0.07	—
Alclad 7075-O										
Core										
Cladding	0.17	0.25	1.53	0.06	2.45	0.22	—	5.65	—	—
(7072)	*d*	*d*	0.05	< 0.01	0.10	0.03	—	1.34	—	—
7079-T6	0.14	0.12	0.64	0.20	3.50	0.16	—	4.30	0.03	—
Alclad 7079-T6										
Core										
Cladding	—	—	0.60	0.20	3.30	0.17	—	4.30	—	—
(7072)	—	—	—	—	—	—	—	1.0	—	—
7178-T6	0.50	0.70	2.0	0.03	2.75	0.30	—	6.8	0.10	—

[a] Composition in % by weight-Remainder-Aluminum

b – Si + Fe = 0.70
c – Si + Fe = 0.40
d – Si + Fe = 0.30

TABLE V

PERFORMANCE OF SELECTED ALUMINUM ALLOYS IN SURFACE SEAWATER

Alloy	Exposure (Days)	Corrosion			
		Rate (mpy)	Maximum Pit Depth (mils)	Crevice Depth (mils)[a]	Type[a]
1100-H14	181	1.4	0	S	S-C
1100-H14	366	0.6	13	13	C; P
2024-O	181	3.8	32	32	C; P
2024-O	366	4.1	34	34	C; P
2219-T81	181	3.5	24	0	IG; G-P
2219-T81	398	2.5	26	0	P
2219-T81	540	1.4	48	32	C; P
2219-T81	588	4.4	62	43	C; E; P
3003-H14	181	1.1	33	0	E; P
3003-H14	398	1.0	21	0	P
3003-H14	540	0.3	34	75	C; P
3003-H14	588	2.0	65	0	P
Alclad 3003-H12	181	1.0	I	0	I-P
Alclad 3003-H12	398	1.1	16	0	P
Alclad 3003-H12	540	0.3	16	0	P
Alclad 3003-H12	588	1.8	17	0	P
5052-O	181	1.2	I	0	I-P
5052-O	366	0.6	5	5	C; P
5083-O	365	0.173	20.9	—	P
5083-O	730	0.120	18.9	—	P
5083-O	1825	0.066	7.0	—	P
5083-O	3650	0.059	24.0	—	P
5086-O	365	0.155	18.1	—	P
5086-O	730	0.122	7.5	—	P
5086-O	1825	0.087	2.0	—	P
5086-O	3650	0.057	1.2	—	P
5083-H113	181	1.2	I	3	C; IG(E); I-P
5083-H113	398	0.4	0	0	ET
5083-H113	540	0.7	36	114	S-C; P
5083-H113	588	0.3	11	I	I-C; P
5086-H32	181	1.0	I	0	IG(E); I-P
5086-H32	398	0.4	20	0	P
5086-H32	588	0.2	22	I	I-C; P
5086-H34	181	1.2	6	0	IG; P

TABLE V (*continued*)

Alloy	Exposure (Days)	Corrosion: Rate (mpy)	Maximum Pit Depth (mils)	Crevice Depth (mils)[a]	Type[a]
5086-H34	398	0.8	27	I	I-C; P
5086-H34	540	0.3	0	0	I-C; ET
5086-H34	588	1.6	47	43	S-C; S-P
5086-H112	181	1.1	I	0	I-P
5454-H32	398	0.5	8	0	P
5454-H32	540	0.3	I	0	I-P
5454-H32	588	0.7	39	0	P
5454-H34	365	0.171	17.6	—	P
5454-H34	730	0.141	19.9	—	P
5454-H34	1825	0.091	7.1	—	P
5454-H34	3650	0.059	20.0	—	P
5456-H321	181	1.2	12	0	P
5456-321	398	0.6	16	0	P
5456-H321	540	0.9	14	8	C; P
5456-H321	1825	0.126	37	—	P; IG(E)
5456-H321	3650	0.064	44	—	P; IG(E)
6061-T6	181	1.0	I	8	C; I-P
6061-T6	398	0.7	16	0	E; P
6061-T6	540	0.3	23	I	I-C; P
7039-T64	181	1.1	0	0	ET
7039-T64	398	1.1	22	I	I-C; P
7039-T64	540	0.3	6	3	C; P

[a] Symbols for types of corrosion:

S = Severe
C = Attack at crevice
P = Pitting
E = Attack at edge
ET = Etched
IG = Intergranular attack
S-C = Severe crevice attack
G-P = General pitting
I-P = Incipient pitting
I-C = Incipient crevice attack
IG(E) = Intergranular attack at edge

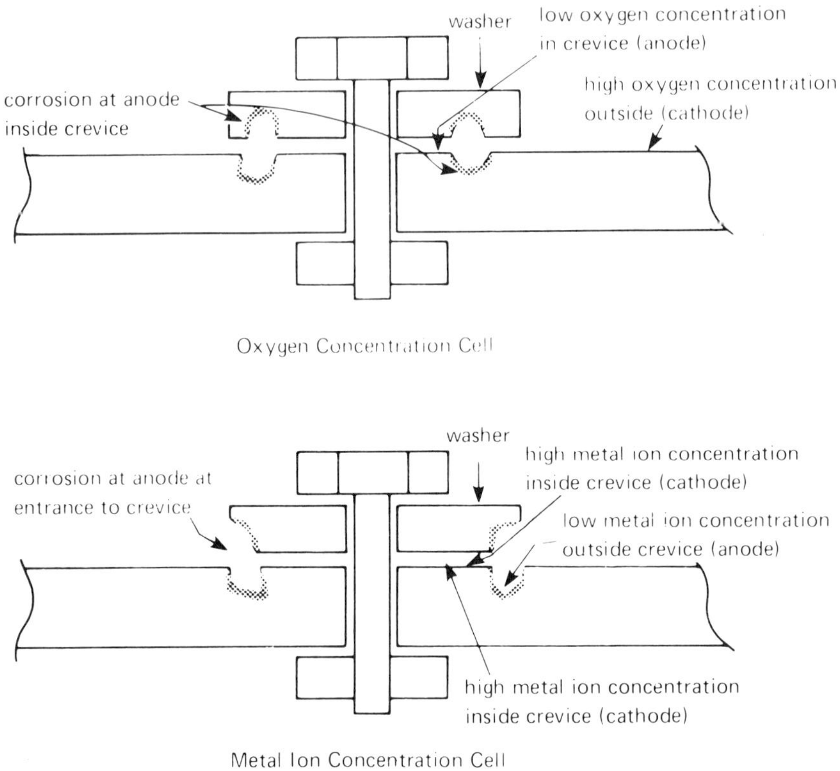

Fig. 2. Forms of crevice corrosion.

used as castings. The combined resistance of some copper alloys, particularly the copper-nickel alloys, to both fouling and to attack by high velocity seawater has led to their wide use as heat exchanger tubes and in seawater piping systems. The resistance to fouling has also led to limited, but highly successful, use of copper-nickel as hull sheathing for small fishing vessels. The reduction in fuel costs and the decreased requirements for hull cleaning has been shown to outweigh the increased cost for the copper-nickel (*12*).

E. Performance and Selection of Stainless Steels

1. General Corrosion Performance

The term "stainless steel" refers to such a wide range of alloys that it is impossible to generalize their corrosion performance as a group. Stainless

TABLE VI

COMPOSITION OF SELECTED COPPER ALLOYS

Alloy	CDA No.	Cu	Zn	Sn	Ni	Al	Mn	Fe	Other
Copper, O free	102	99.96	—	—	—	—	—	—	—
Be-Cu	170	97.80	—	—	0.05	—	—	—	1.90 Be 0.25 Co
Commercial bronze	220	90	10	—	—	—	—	—	—
Yellow brass	270	68.48	31.50	< 0.05	—	—	—	—	—
Muntz metal	280	60.69	39.29	—	—	—	—	< 0.02	—
Arsenical Admiralty	443	71.79	27.77	1.0	—	—	—	0.01	0.027 As
Ni-Mn bronze	868	54.58	34.48	0.7	3.77	1.73	3.06	1.66	0.02 Pb
G bronze	—	88.0	10.0	2.0	—	—	—	—	—
P bronze A	510	94.64	< 0.10	4.94	—	—	—	< 0.05	0.26 P
Al bronze 5%	606	95.0	—	—	—	5.0	—	—	—
Al bronze 7%	614	90.11	0.15	—	—	6.59	—	3.15	< 0.02 Pb
Al bronze 13%	—	83.0	—	—	—	—	13.0	4.0	—
Ni-Al bronze #2	—	80.0	—	—	5.0	10.0	0.5	4.0	—
Si bronze A	655	95.49	—	—	—	—	1.18	< 0.02	3.28 Si
Cu-Ni 90-10	706	89.04	—	—	9.42	—	0.53	1.16	—
Cu-Ni 70-30	715	68.61	—	—	30.53	—	0.33	0.53	—
Cu-Ni 70-30	716	64.02	—	—	29.95	—	0.75	5.27	—

TABLE VII

PERFORMANCE OF SELECTED COPPER ALLOYS IN SURFACE SEAWATER

Alloy	CDA No.	Exposure (Days)	Corrosion Rate (mpy)	Type[a]
Copper, O free	102	398	1.1	G; P(37)
Copper, O free	102	540	0.9	G; P(22)
Copper, O free	102	588	0.9	G; P(20)
Be-Cu	172	181	0.1	ET
Be-Cu	172	364	1.1	U
Be-Cu	172	723	0.7	U
Be-Cu	172	763	0.7	U
Commercial bronze	220	181	1.1	U
Commercial bronze	220	366	1.1	P(4)
Yellow brass	270	181	2.1	U
Yellow brass	270	366	1.3	U
Muntz metal	280	181	2.4	DZ
Muntz metal	280	398	3.1	DZ; P(6)
Muntz metal	280	540	3.4	DZ; I-P
Muntz metal	280	588	3.3	DZ; I-P
Arsenical Admiralty	443	181	1.3	U
Arsenical Admiralty	443	366	1.3	U
Arsenical Admiralty	443	608	1.1	U; I-P
Ni-Mn bronze	868	364	0.7	DZ
Ni-Mn bronze	868	723	2.9	DZ
Ni-Mn bronze	868	763	3.0	DZ
G bronze	—	181	1.3	G
G bronze	—	366	1.2	CR(9)
P bronze A	510	181	1.1	U
P bronze A	510	366	1.3	CR(5)
P bronze A	510	588	1.3	P(15); CR(3)
Al bronze 5%	606	181	1.1	G

Al bronze 5%	606	366	0.7	G
Al bronze 7%	614	181	2.9	NU; DA
Al bronze 7%	614	366	0.6	G
Al bronze 7%	614	588	0.9	SL-DA; CR(44); P(20)
Al bronze 13%	—	181	2.1	S-DA
Al bronze 13%	—	366	1.9	S-DA
Ni Al bronze #2	—	181	1.0	U
Ni Al bronze #2	—	366	0.4	U
Si bronze A	655	181	1.8	U
Si bronze A	655	398	1.1	G
Si bronze A	655	540	2.5	P(30); C(15)
Si bronze A	655	588	0.9	P(9)
Cu-Ni 90-10	706	181	1.1	NU
Cu-Ni 90-10	706	366	0.6	U
Cu-Ni 90-10	706	608	0.5	U
Cu-Ni 70-30	715	181	0.5	U
Cu-Ni 70-30	715	366	0.4	G
Cu-Ni 70-30	715	398	0.2	P(7)
Cu-Ni 70-30	715	608	0.4	I-P
Cu-Ni 70-30	716	181	0.8	I-P; C(5)
Cu-Ni 70-30	716	398	0.7	U; P(17)
Cu-Ni 70-30	716	608	0.6	P(18); CR(13)

[a]Symbols for corrosion type:

U = Uniform attack
G = General attack
P = Pitting (numbers in parenthesis indicate maximum pit depth in mils)
DZ = Dezincification
NU = Nonuniform
DA = Dealloying
CR = Crevice (numbers in parenthesis indicate maximum crevice attack in mils)
I-P = Incipient pitting
S-DA = Severe dealloying
SL-DA = Slight dealloying

steels depend upon a thin transparent oxide film for their corrosion resistance. The film formed on many alloys commonly referred to as "stainless steels" is unstable in chloride containing environments, and the corrosion resistance of these alloys in seawater is poor. Other alloys have more stable films and, if properly used, can give satisfactory resistance to corrosion in surface seawater. Many new alloys that fall into the general classification of "stainless steels" have been developed specifically for resistance to corrosion in seawater and, although long-term field experience is limited, have shown good performance in many applications.

Stainless steels are subject to many forms of localized corrosion such as crevice corrosion, pitting, and stress-corrosion cracking. Measurement of corrosion rates through weight loss is normally not applicable to evaluation of the corrosion performance of stainless steels. Crevice corrosion is a problem for even the most corrosion-resistant alloys. Crevices formed by metal-to-metal contact, metal-to-nonmetal contact, through fabricating operations such as shearing and welding, and by the attachment of fouling organisms can all result in the initiation of crevice attack. Metal-to-nonmetal crevices are generally more detrimental than metal-to-metal crevices, as they are generally tighter (*13*). The increased availability of oxygen and limitation of fouling in high velocity seawater improves the performance of many alloys in applications where velocity conditions are continuously present. However, when stagnant conditions occur, the reduction on available oxygen can result in rapid attack.

2. SIGNIFICANT PERFORMANCE PARAMETERS

As for other alloys that are susceptible to nonuniform attack, characterization of the corrosion performance of the stainless steels is difficult. Corrosion rates based upon weight loss have little value. There have been several crevice corrosion resistance tests proposed but none have found universal acceptance for characterizing the corrosion resistance of stainless steels. It is important to determine the temperature of the seawater in both the proposed application and the test environment, as seawater temperature has been found to be a critical parameter in the crevice corrosion resistance of many alloys (*13*). In evaluation of resistance to crevice corrosion it is often more reasonable to consider the probability of initiation of attack and the consequences of the attack as a primary consideration than it is to rely on depth versus time predictions. One factor that makes it extremely difficult to evaluate the crevice corrosion resistance of the more resistant alloys is that they exhibit long incubation periods where there is little or no evidence of corrosion followed by rapid attack once the protective films within the crevices break down. Also, the relative areas of material exposed within and without the crevice is an important factor in the initiation and propagation of crevice attack (*3*). Resistance to pitting is also difficult to evaluate, primarily as it is difficult to expose test specimens without crevices. The activity associated with crevice attack is often sufficient to inhibit other forms of attack such as pitting through cathodic protection. Stress corrosion is difficult to evaluate particularly in alloys with little resistance to crevice attack. It is difficult to design a stress corrosion test specimen without crevices. Crevices may either directly affect the potential of the specimen or may result in corrosion, which changes the stress intensity within the specimen. Intergranular corrosion is associated with improper welding or heat treatment and must be evaluated essentially on a case by case basis. Stress corrosion cracking seldom, if ever, occurs in austenitic grades of stainless steel at ambient temperatures (*14*). However, stress-corrosion cracking and hydrogen embrittlement can occur on ferritic and martensitic stainless steels, particularly the high strength alloys (*13*). Duplex stainless steels appear to be less susceptible to stress-corrosion cracking than fully austenitic alloys, but welding can cause stress corrosion cracking problems if the duplex structure of the alloy is destroyed in the vicinity of the weld (*13*).

3. Performance of Commonly Used Alloys

As the resistance to crevice attack is a significant factor that is difficult to quantify, a simple listing of alloys in order of their relative probability of initiation of crevice attack is useful for comparing alloys. Such a listing is given in Table VIII (*3*). Data for the composition and performance of common alloys are given in Tables IX, X, and XI. The data for this table are taken from Sedricks (*13*), LaQue (*3*), and Reinhart (*15*).

4. Typical Applications

There have been many successful applications of stainless steel in surface and near-surface seawater (*16*). In many cases, these successful applications are due as much or more to the careful selection and use of the alloys than to their inherent corrosion resistance. Cathodic protection can mitigate crevice corrosion of austenitic stainless steel in many instances, but the potentials required to inhibit crevice attack on ferritic and martensitic stainless steels will likely result in hydrogen embrittlement or blistering (*3*). As many materials, particularly carbon steel and aluminum alloys, are sufficiently anodic to many stainless steels to prevent crevice attack, combinations of stainless steels with these materials or the use of other forms of cathodic

TABLE VIII

RATING OF PROBABILITY OF INITIATION OF CREVICE CORROSION OF STAINLESS STEELS AND SELECTED ALLOYS IN SURFACE SEAWATER

Most Resistant	Inconel 625
Most Resistant	Hastelloy C-276
Most Resistant	Titanium alloys
Most Resistant	Al-6X stainless steel
Most Resistant	IN-748 nickel alloy
	MP-35N
	70-30 copper-nickel
	90-10 copper nickel
	Tin bronze
	Aluminum bronze
	Yellow brass
	Aluminum brass
	Silicon bronze
	Copper
	Monel alloys
	Austenitic cast irons
	Cast irons
	Carbon & low alloy steels
	Incoloy 825
	Carpenter 20 stainless steel
	Type 317 stainless steel
	Type 316 stainless steel
	26-Cr-Mo stainless steel
	Incoloy 800
	Type 310 stainless steel
	Inconel 600
	Type 304 stainless steel
	Type 347 stainless steel
	Type 321 stainless steel
	Type 301 stainless steel
	Precipitation Hardening Stainless Steels
	Type 303 stainless steel
	Type 430 stainless steel
	Type 440 stainless steel
	Type 440F stainless steel
	Type 410 stainless steel
	Type 416 stainless steel

TABLE IX

COMPOSITION OF SELECTED STAINLESS STEELS

Alloy	C	Mn	P	S	Si	Ni	Cr	Mo	Other
AISI 201	0.14	7.0	—	0.009	—	4.5	16.5	—	—
AISI 304	0.06	1.36	0.024	0.013	0.43	10.0	18.8	—	—
AISI 304L	0.03	1.24	0.028	0.023	0.68	10.20	18.7	—	—
AISI 310	0.04	1.78	—	—	—	20.9	25.3	—	—
AISI 316	0.06	1.61	0.021	0.016	0.40	13.60	18.3	2.41	—
AISI 316L	0.03	1.29	0.015	0.019	0.51	13.10	17.5	2.32	—
AISI 321	0.06	2.0	—	—	—	10.5	18.5	—	0.47 Ti
AISI 347	0.04	1.19	—	—	—	11.3	18.1	—	0.83 (Cb + Ta)
20-Cb	0.04	0.79	0.018	0.004	0.67	28.38	19.8	2.06	3.11 Cu 0.77 (Cb + Ta)
AISI 410	0.13	0.43	0.019	0.05	0.45	0.010	12.3	—	—
AISI 430	0.07	0.47	0.029	0.011	0.36	—	16.4	—	—
AISI 630	0.03	0.24	0.017	0.011	0.59	4.17	15.92	—	3.23 Cu 0.24 Cb
AISI 631	0.07	0.48	0.017	0.018	0.42	7.42	17.12	—	1.19 Al
AISI 635	0.05	0.54	0.011	0.006	0.57	6.89	16.8	—	0.14 Al
18Cr-14Mn-0.5N	0.07	14.3	0.021	0.03	0.67	0.27	18.4	—	0.48

[a] Remainder—Fe

TABLE X

PERFORMANCE OF SELECTED AUSTENITIC STAINLESS STEELS IN SURFACE SEAWATER

		Corrosion			
Alloy	Exposure (Days)	Rate (mpy)	Maximum Pit Depth (Mils)	Crevice Depth (Mils)	Type[a]
AISI 201	182	< 0.1	0	I	I-C
AISI 201	366	0.6	0	0	S-E
AISI 304	181	1.8	210	0	E; P(PR); T(500)
AISI 304	366	0.4	34	33	C; P
AISI 304	540	0.7	42	103	C; P; T(180)
AISI 304	588	0.5	0	138	C; T(110)
AISI 304L	181	0.7	97	0	E; P; T(400)
AISI 304L	366	0.5	50	0	P(PR)
AISI 304L	398	1.0	115	0	E; P(PR); T(1100)
AISI 304L	540	0.7	115	115	C(PR); E; P(PR); T(1500)
AISI 310	181	< 0.1	0	I	I-C
AISI 310	366	< 0.1	0	50	C(PR)
AISI 316	181	1.3	230	0	E; P(PR); T(1250)
AISI 316	398	0.4	154	20	C; E; P; T(1350)
AISI 316	540	0.3	0	63	C; T(70)
AISI 316	588	0.2	0	130	C; T(1500)
AISI 316L	181	0.3	125	0	E; P(PR); T(1400)
AISI 316L	366	< 0.1	0	I	I-C
AISI 316L	398	< 0.1	0	0	SL-E
AISI 321	181	< 0.1	0	2	C
AISI 321	366	< 0.1	22	0	P
AISI 347	181	0.8	50	50	C(PR); P(PR)
AISI 347	366	0.7	50	50	C(PR); P(PR)
20-Cb	181	< 0.1	0	5	C; SL-E
20-Cb	398	< 0.1	14	0	P; SL-E
20-Cb	540	< 0.1	24	0	P
20-Cb	588	< 0.1	0	21	C

[a]Symbols for type of attack

E = Edge attack

P = Pitting (numbers in parentheses refer to maximum pit depth in mils, or PR for perforation)

T = Tunneling attack (numbers in parentheses refer to maximum length of tunneling in mils)

I-C = Incipient crevice attack

S-E = Severe edge attack

SL-E = Slight edge attack

TABLE XI

PERFORMANCE OF SELECTED FERRITIC AND MARTENSITIC STAINLESS STEELS IN SURFACE SEAWATER

		Corrosion			
Alloy	Exposure (Days)	Rate (mpy)	Maximum Pit Depth (Mils)	Crevice Depth (Mils)	Type[a]
AISI 410	181	2.1	40	40	C(PR); E P(PR); T(1000)
AISI 410	366	3.0	50	50	C(PR); P(PR)
AISI 430	181	1.7	50	50	C(PR); P(PR)
AISI 430	366	1.1	50	50	C(PR); P(PR)
AISI 430	540	0.7	50	50	C(PR); P(PR); T(4450)
AISI 430	588	0.9	50	50	C(PR); P(PR); T(3900)
AISI 630[b]	181	3.0	112	112	C(PR); P(PR); E; T(1800)
AISI 630[b]	398	1.1	112	112	C(PR); P(PR); E; T(1000)
AISI 631[c]	181	2.9	125	125	C(PR); P(PR); S-E; T(2500)
AISI 635	181	0.9	40	275	C(PR); E; P; T(1000)
AISI 635	398	0.6	40	40	C; E; P; T(1200)
18Cr-14Mn-0.5N	181	4.9	50	50	C(PR); P(PR)
18Cr-14Mn-0.5N	366	2.6	50	50	C(PR); P(PR)

[a] Symbols for type of attack

E = Edge attack
P = Pitting (numbers in parentheses refer to maximum pit depth in mils, or PR for perforation)
T = Tunneling attack (numbers in parentheses refer to maximum length of tunneling in mils)
I-C = Incipient crevice attack
S-E = Severe edge attack
SL-E = Slight edge attack

[b] H925 condition

[c] RH1050 condition

protection can result in successful, reliable designs. The use of free machining grades of austenitic stainless steels should be avoided as the corrosion resistance of these alloys is very poor in marine environments (*6*). Many of the newly developed marine grades of stainless steel have shown excellent performance in many tests and applications (*17*), but care must still be exercised in their use, as crevice attack and attack under stagnant conditions has been experienced. One high strength fastener material, A-286, has also shown excellent corrosion resistance in many surface and near-surface marine environments (*18*).

F. Performance and Selection of Titanium Alloys

1. General Corrosion Performance

Due to the extreme stability of the oxide film on titanium alloys, these materials are highly resistant to most forms of corrosion in surface and near-surface seawater, with the notable exception of stress corrosion cracking. While crevice corrosion of titanium alloys in seawater at elevated temperatures has been encountered, the alloys appear to be fully resistant at ambient seawater temperatures (*19*).

2. Significant Performance Parameters

Evaluation of the resistance of titanium alloys to stress corrosion cracking involves a critical assessment of the magnitude and distribution of the stresses that will be encountered in service and, perhaps more importantly, upon the size, distribution, and orientation of flaws that can serve as points of initiation of stress corrosion cracking. Fracture toughness considerations can be used to establish a critical stress intensity for a particular situation.

3. Performance of Commonly Used Alloys

Critical stress intensities for propagation of cracks in several commercial and experimental titanium alloys are presented in Table XII (*20*).

4. Typical Applications

Titanium alloys have been successfully used in many applications where their extreme corrosion resistance has offset their high cost. Due to their resistance to both high velocity and stagnant seawater, they have been used

TABLE XII

CRITICAL STRESS INTENSITIES FOR STRESS CORROSION CRACKING OF TITANIUM ALLOYS IN SURFACE SEAWATER

Alloy	Yield Strength	K_{ISCC}
6Al-4V	120 KSI	70 KSI in^{-2}
5Al-2.5Sn	125 KSI	50 KSI in^{-2}
7Al-2Cb-1Ta	113 KSI	40 KSI in^{-2}
8Al-2Cb-1Ta	112 KSI	30 KSI in^{-2}
13V-11Cr-3Al	127 KSI	22 KSI in^{-2}

for heat exchanger tubes in power plants and critical marine power heat exchangers (*21*). Except for problems with the tube sheet and shell materials caused by galvanic coupling to the titanium, most of these applications have been highly successful. A notable exception to this success has been problems with hydriding of the titanium tubes where there was overzealous application of cathodic protection to mitigate tubesheet corrosion (*22*). Titanium has also been successfully applied in sheet type heat exchangers. Titanium has also been used for a wide variety of critical applications such as pumps, high speed propellers, wire ropes, and pressure hulls for submarines and submersibles, where the reliability inherent in corrosion-resistant materials is required.

G. *Performance and Selection of Other Metallic Materials*

1. NICKEL ALLOYS

Several high nickel alloys have been shown to have excellent resistance to corrosion in seawater. Inconel 625, Hastelloy C, and Hastelloy C-276 have been shown by both test and application to be essentially immune to corrosion in surface and near-surface seawater. Recently, however, crevice corrosion of Inconel 625 has been detected in ambient temperature seawater under extreme crevice conditions (*23*). The use of these alloys in critical applications where their cost is outweighed by the consequences of corrosion is often justified. Other high nickel alloys such as Inconel 718 and Incoloy 825 have shown resistance to seawater corrosion superior to that of many stainless steels. Monel 400 and Monel K-500 nickel-copper alloys have been successfully used in many marine applications. Although both alloys will suffer from corrosion, they are commonly used in contact with other materials sufficiently anodic to them that corrosion is mitigated. In addition, crevice corrosion of both alloys is of the metal ion concentration cell type,

TABLE XIII

COMPOSITION OF SELECTED NICKEL ALLOYS

Alloy	Ni	C	Mn	Fe	S	Si	Cu	Other
Nickel 200	99.5	0.05	0.29	0.04	0.006	0.07	0.02	—
Monel 400	65.17	0.11	1.06	0.90	0.007	0.10	32.62	—
Monel K-500	65.00	0.15	0.60	1.00	0.005	0.15	29.50	2.80 Al
								0.50 Ti
Inconel 600	75.26	0.06	0.18	7.25	0.008	0.27	0.38	16.0 Cr
Inconel 718	52.50	0.04	0.20	18.0	0.007	0.20	0.10	19.0 Cr
								5.2 Cb
								3.0 Mo
								0.8 Ti
								0.6 Al
Incoloy 825	41.12	0.05	0.82	30.9	0.010	0.31	1.61	21.1 Cr
								2.94 Mo
								1.00 Ti
								0.14 Al
Inconel 625	61.0	0.05	0.15	3.00	0.007	0.30	0.10	22.0 Cr
								9.0 Mo
								4.0 Cb
Hastelloy C	55.86	0.05	0.52	6.32	0.009	0.62	—	16.7 Mo
								15.3 Cr
								3.53 W
								0.96 Co
								0.26 V
								0.01 P
Hastelloy C-276	54.03	0.01	0.80	6.34	0.02	0.50	—	16.2 Mo
								15.3 Cr
								3.65 W
								2.3 Co
								0.31 V
								0.02 P

which is usually less damaging than the oxygen concentration cell attack experience on stainless steels.

Tables XIII and XIV present data on the composition and corrosion performance of selected nickel alloys taken from Reinhart (*24*) and Cabot Corporation (*25*).

2. COBALT ALLOYS

Cobalt based superalloys such as Stellite Alloy 6B and MP-35N have shown excellent resistance to attack in seawater coupled with high strength

TABLE XIV

PERFORMANCE OF SELECTED NICKEL ALLOYS IN SURFACE SEAWATER

Alloy	Exposure (Days)	Corrosion Rate (mpy)	Corrosion Maximum Pit Depth (Mils)	Corrosion Crevice Depth (Mils)	Corrosion Type[a]
Nickel 200	181	1.9	45	50	C; P
Nickel 200	398	1.9	125	—	P(PR); T
Nickel 200	540	1.5	125	—	P(PR); T
Nickel 200	588	1.5	125	—	P(PR); T
Monel 400	181	0.6	28	15	C; P
Monel 400	398	0.8	39	—	P
Monel 400	540	0.9	17	—	P; E
Monel 400	588	0.8	29	—	P
Monel K-500	181	1.0	8	12	C; P
Monel K-500	366	3.6	30	30	C(PR); P(PR)
Inconel 600	181	0.2	105	—	E; P
Inconel 600	366	0.6	50	50	C(PR); P(PR)
Inconel 600	540	0.5	67	—	P
Inconel 718	181	< 0.1	—	—	NC
Inconel 718	540	< 0.1	—	—	NC
Incoloy 825	181	< 0.1	44	33	C; P
Incoloy 825	398	< 0.1			NC
Incoloy 825	540	< 0.1	43	24	C; P
Incoloy 825	588	< 0.1	18	—	P
Incoloy 825	608	< 0.1	54	—	P
Inconel 625	181	< 0.1	—	—	NC
Inconel 625	398	< 0.1	—	—	NC
Inconel 625	540	< 0.1	—	—	NC
Inconel 625	588	< 0.1		—	NC
Hastelloy C	181	< 0.1	—	—	NC
Hastelloy C	398	< 0.1	—	—	NC
Hastelloy C	608	< 0.1	—	—	NC
Hastelloy C276	181	< 0.1	—	—	NC
Hastelloy C276	398	< 0.1	—	—	NC
Hastelloy C286	608	< 0.1	—	—	NC

[a] Symbols for type of attack

C = Crevice attack (numbers in parenthesis refer to maximum pit depth in mils, or PR for perforation)

E = Edge attack

P = Pitting (numbers in parenthesis refer to maximum pit depth in mils, or PR for perforation)

T = Tunneling attack

NC = No visible corrosion

TABLE XV

COMPOSITION OF SELECTED COBALT BASED SUPERALLOYS[a]

Alloy	Ni	Si	Fe	Mn	Cr	Mo	W	C
Stellite 6B	3.00	2.00	3.00	2.00	28.00–32.00	1.50	3.50–5.50	0.90–1.40
MP-35N	35	—	—	—	20	10	—	

[a] Balance—Cobalt

TABLE XVI

PROPERTIES OF SELECTED COBALT BASED SUPERALLOYS

	Property			
Alloy	UTS (KSI)	YS (KSI)	Elastic Modulus psi × 10^6	Hardness Rockwell C
Stellite 6B[a]	140	90	30.4	37
MP-35N[b]	295	285	34.05	51

[a]Solution treated and aged 3 hours @ 1500°F.
[b]Work strengthened and aged 4 hours @ 1100°F.

and abrasion resistance. When their unique combination of corrosion resistance and mechanical properties are required, their use may be appropriate despite their high cost. Tables XV and XVI give the composition and properties of these alloys taken from Cabot Corporation (*26*) and Latrobe Steel Corporation (*27*).

IV. Materials for Structures in Deep Ocean Environments

A. The Deep Ocean Environment

1. DEPTH

As will be shown in the corrosion performance data to follow, there can be significant differences between the performance of materials in surface and

TABLE XVII

MAJOR ENVIRONMENTAL PARAMETERS WHICH VARY WITH DEPTH

Dissolved oxygen content
pH
Salinity
Conductivity
Temperature
Carbonate equilibrium
Pressure
Type and extent of biofouling
Bottom sediment type
Flow velocity

near-surface waters and their performance at depth. However, the differences are not due directly to depth but to the variability of other environmental parameters with depth. Table XVII presents a list of the major environmental parameters that are known to affect corrosion and that can vary with depth.

2. Dissolved Oxygen and pH

There has been too little *in situ* testing in the deep ocean to specifically identify the contribution of changes in any single environmental parameter to changes in the corrosion behavior of materials with depth. In some cases environmental parameters do not vary independently. For example, as shown in Figure 3, the oxygen content and pH can, and often do, vary in a parallel fashion, with zones with the lowest oxygen content having the lowest pH. Thus, discrimination between the effects of these two parameters on a material such as aluminum, where the reduction in either pH or in oxygen content are suspected to have similar effects, is difficult. As described previously in this report, testing which artificially attempts to alter these parameters can give misleading results. Thus, in this report, an engineering approach is used. It has been found that the performance of a wide variety of alloys can be correlated with the oxygen content of the environment at depth. Whether or not the oxygen content of the water is the primary variable controlling the corrosion rate remains to be answered. However, measurements of the oxygen content of the water at a given site can be used to predict the performance of many alloys in that environment.

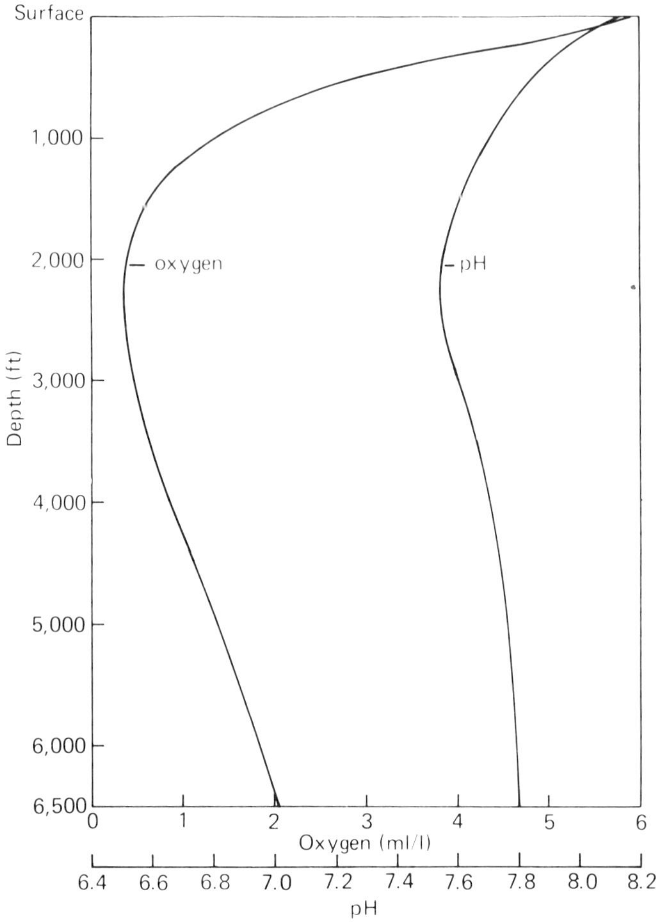

Fig. 3. Variation of oxygen content and pH of seawater versus depth.

3. Location

The variability of environment with depth differs greatly with location. For example, in the Pacific Ocean off the coast of Southern California the oxygen level initially decreases with depth and reaches an extremely low minimum value at approximately 2500 feet of depth. At depths below 2500 feet, the oxygen level increases slowly with increasing depth and reach levels approximately midway between their surface values and the minimum values at 2500 feet. The variation of pH with depth at this site follows a similar pattern. In contrast, at a typical site in the Atlantic Ocean there is a reduction in oxygen content from the surface to a depth of approximately 2500 feet. However, the minimum oxygen level is approximately two-thirds that of the level at the surface. At depths below 2500 feet, the oxygen level and pH increase, and at depths below 4000 feet they actually exceed their levels at the surface. Thus, if oxygen levels are used to predict performance of materials at depth, it would be predicted that materials exposed at a depth of 4000 feet in the Atlantic would be similar to their performance in near-surface waters. This is indeed the case. The corrosion rate of steel after one year at the surface was found to be 5.5 mpy, whereas the corrosion rate after the same period at depth was found to be 5.6 mpy (*28*). In contrast, the corrosion rate of steel after one year at the level of minimum oxygen in the Pacific is only 1 mpy (*29*).

4. Salinity, Temperature, and Pressure

Salinity of seawater can also vary with depth, generally decreasing with depth. The range of the variability of salinity with depth is less than its variability in surface waters. Temperature usually decreases with depth, but as there is no atmospheric exchange, the effects of temperature on the dissolved gas content of seawater that occur at the surface are absent at depth. The equilibrium between carbon dioxide and carbonate ion is also shifted with depth, with deep water generally being less saturated with respect to carbonate than surface waters. This effect may not only directly affect the corrosion behavior of materials but may also affect the requirements for cathodic protection of structures at depth.

5. Deep Ocean Fouling and Bottom Sediments

The amount and variety of marine biofouling–fouling also decreases with depth. At depths below 600 feet, the amount of biofouling is minimal unless the exposure is within a few feet of the bottom sediments. However, some fouling organisms have been found on structures retrieved from the ocean at all depths if the exposure periods were more than one year. The bottom sediments at depth are more likely to be very fine than near the surface. Anaerobic bacteria are almost universally present in deep ocean sediments. At extreme depths, due to the variation in carbonate equilibrium with depth, the sediments tend to be depleted in calcium carbonate and rich in silicates.

6. Other Considerations

There are many other possible environmental variables in the deep ocean that may affect the behavior of materials at depth. One consideration is the lack of appreciable currents or other disturbing influences at depth. Deterioration which would render a structure unable to withstand the rigors of exposure near the surface would have much less effect on a structure at depth. Attempts to salvage structures from depths which, upon initial inspection, appear to be relatively intact, have shown that the amount of corrosion has been substantial. Often sufficient corrosion has occurred to cause the structure to fail during retrieval. Another possible effect is the presence of a layer of water directly above the bottom sediments that is substantially different in composition from the bulk composition of the water above. This is particularly true when alkaline sediments are present and the currents are low.

B. Performance and Selection of Irons and Steels

1. General Corrosion Performance

As shown in Figure 4, the corrosion rate of steel is a direct function of the oxygen content of seawater. In exposures at sites with low oxygen content, the corrosion rate of steel can be expected to be low, with a minimum in the range of 1 mpy. At sites with high oxygen contents, the corrosion rate of steel can be expected to be high, independent of direct effects of depth. Nearly all steels, low alloy steels, alloy steels, and cast irons follow this general trend. As in the case of the corrosion rates of steels in surface waters, the corrosion rates of steel in deep ocean environments decrease with time as shown in Figure 5. Due to the dearth of fouling organisms at depth, the steels are generally less susceptible to pitting in deep ocean waters than when exposed near the surface.

2. Significant Performance Parameters

As the corrosion of steel remains uniform at depth, corrosion rates based upon weight loss are appropriate for measurement of their performance.

3. Performance of Commonly Used Alloys

The composition of several commonly used irons and steels was given in Table II. The performance of these alloys at depth, taken from Reinhart (*7*), Boyd and Fink (*19*), Reinhart (*29*), and Waldron et al. (*30*) are given in Table XVIII.

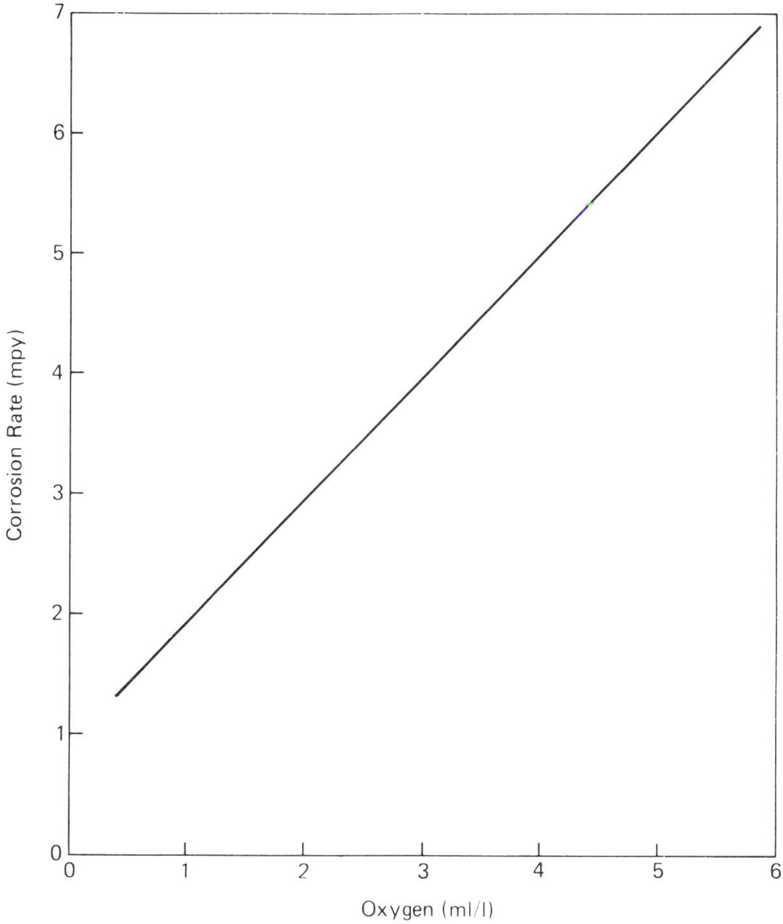

Fig. 4. Corrosion rate of irons and steels in seawater after one year of exposure versus oxygen content.

4. Typical Applications

Steels have been widely used as a structural material at depth. They have been successfully used for large underwater instrument platforms (*31*), submersible hulls (*32*), instrument canisters, wire rope, and cable (*33*). As in surface applications, they are usually coated and often cathodically protected as well. Although the reduction in carbonate scale formation may interfere with cathodic protection at depth, the dearth of fouling organisms results in

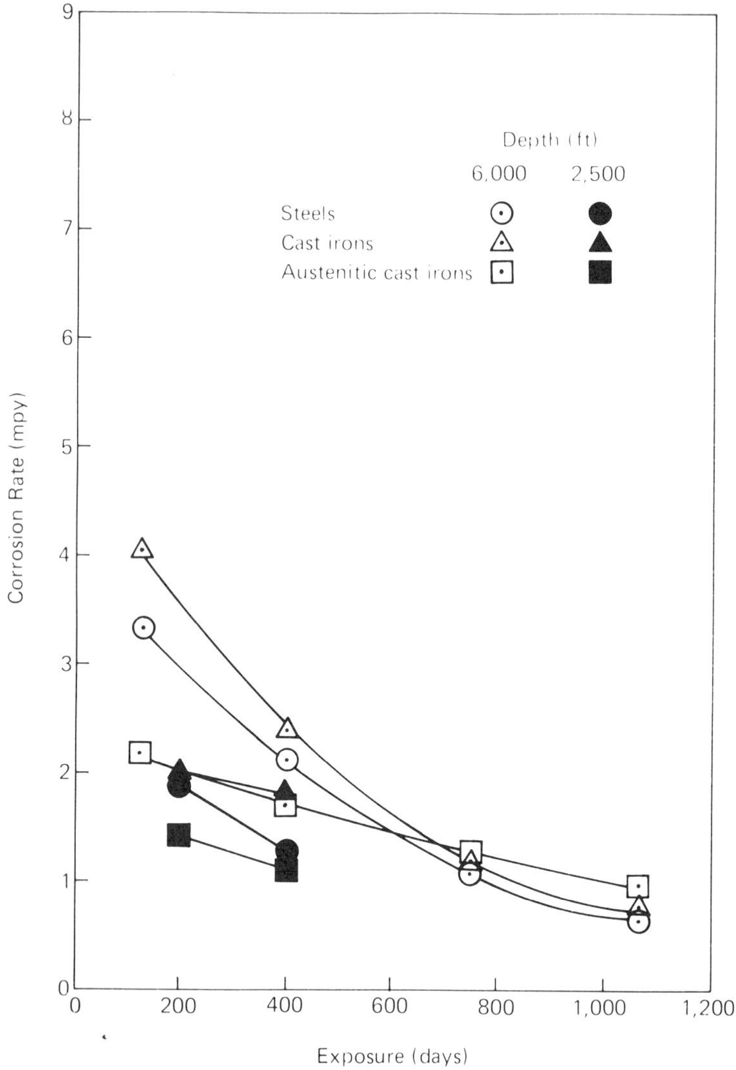

Fig. 5. Corrosion rates of irons and steels in the deep ocean versus time of exposure.

TABLE XVIII

PERFORMANCE OF SELECTED IRONS AND STEELS AT DEPTH

Alloy	Exposure Days	Exposure Depth (ft)	Corrosion Rate (mpy)[a]	Type of Corrosion[b]
Armco iron	123	5,640	3.1	U
Armco iron	304	6,780	1.5	U
Armco iron	751	5,640	0.8	G
Armco iron	1064	5,300	0.7	U
Armco iron	197	2,340	1.9	G
Armco iron	402	2,370	1.4	G
Wrought iron	123	5,640	2.6	U
Wrought iron	403	6,780	1.4	U
Wrought iron	751	5,640	0.9	U
Wrought iron	1064	5,300	0.6	U
Wrought iron	197	2,340	2.0	U
Wrought iron	402	2,370	1.5	G
AISI 1010	123	5,640	3.0	U
AISI 1010	403	6,780	1.5	U
AISI 1010	751	5,640	0.9	U
AISI 1010	1064	5,300	0.8	U
AISI 1010	197	2,340	1.5	U
AISI 1010	402	2,370	1.2	U
Copper steel	123	5,640	1.9	U
Copper steel	403	6,780	2.1	G
Copper steel	751	5,640	1.4	G
Copper steel	1064	5,300	0.5	U
Copper steel	197	2,340	2.0	G
Copper steel	402	2,370	1.1	U
ASTM A-36	123	5,640	3.1	U
ASTM A-36	403	6,780	1.5	U
ASTM A-36	751	5,640	0.9	U
ASTM A-36	1064	5,300	0.6	U
ASTM A-36	197	2,340	1.7	U
ASTM A-36	402	2,370	1.3	U
HSLA No. 1	123	5,640	2.9	U
HSLA No. 1	403	6,780	2.0	U
HSLA No. 1	751	5,640	0.9	U
HSLA No. 1	1064	5,300	0.6	U
HSLA No. 1	197	2,340	1.4	U
HSLA No. 1	402	2,370	1.0	U
HSLA No. 2	123	5,640	4.7	U
HSLA No. 2	403	6,780	2.1	U
HSLA No. 2	751	5,640	0.9	U
HSLA No. 2	1064	5,300	0.5	U

(continued)

TABLE XVIII (*Continued*)

Alloy	Exposure Days	Exposure Depth (ft)	Corrosion Rate (mpy)[a]	Type of Corrosion[b]
HSLA No. 2	197	2,340	1.4	U
HSLA No. 2	402	2,370	1.3	U
HSLA No. 4	123	5,640	3.6	U
HSLA No. 4	403	6,780	3.3	U
HSLA No. 4	751	5,640	1.2	U
HSLA No. 4	1064	5,300	0.3	U
HSLA No. 4	197	2,340	1.4	U
HSLA No. 4	402	2,370	1.1	G
HSLA No. 5	123	5,640	3.1	U
HSLA No. 5	403	6,780	2.7	U
HSLA No. 5	751	5,640	1.4	U
HSLA No. 5	1064	5,300	0.9	U
HSLA No. 5	197	2,340	1.4	U
HSLA No. 5	402	2,370	1.1	U
HSLA No. 7	123	5,640	3.5	C(4); U
HSLA No. 7	403	6,780	1.5	G
HSLA No. 7	751	5,640	0.8	G
HSLA No. 7	1064	5,300	0.8	U
HSLA No. 7	197	2,340	2.3	G
HSLA No. 7	402	2,370	1.4	G
HSLA No. 10	123	5,640	4.1	C(9); U
HSLA No. 10	403	6,780	1.8	G
HSLA No. 10	751	5,640	0.9	G
HSLA No. 10	1064	5,300	0.9	U
HSLA No. 10	197	2,340	2.1	G
HSLA No. 10	402	2,370	1.5	G
18% Ni maraging	123	5,640	3.6	U
18% Ni maraging	403	6,780	1.6	U
18% Ni maraging	197	2,340	1.3	G
18% Ni maraging	402	2,370	3.5	G
1.5% Ni steel	123	5,640	3.5	U
1.5% Ni steel	403	6,780	1.7	G
1.5% Ni steel	751	5,640	1.0	G
1.5% Ni steel	1064	5,300	0.7	U
1.5% Ni steel	197	2,340	1.9	U
1.5% Ni steel	402	2,370	1.5	U
3% Ni steel	123	5,640	3.4	U
3% Ni steel	403	6,780	1.9	C(2); G
3% Ni steel	751	5,640	0.9	G
3% Ni steel	1064	5,300	0.9	U
3% Ni steel	197	2,340	1.7	G
3% Ni steel	402	2,370	1.3	G
5% Ni steel	123	5,640	2.8	U

TABLE XVIII (*Continued*)

Alloy	Exposure		Corrosion Rate (mpy)[a]	Type of Corrosion[b]
	Days	Depth (ft)		
5% Ni steel	403	6,780	2.8	C(6); G
5% Ni steel	751	5,640	1.1	G
5% Ni steel	1064	5,300	0.7	U
5% Ni steel	197	2,340	1.7	G
5% Ni steel	402	2,370	1.3	U
9% Ni steel	123	5,640	5.6	U
9% Ni steel	403	6,780	2.9	C(9); G
9% Ni steel	751	5,640	1.1	G
9% Ni steel	1064	5,300	4.6	U
9% Ni steel	197	2,340	1.9	G
9% Ni steel	402	2,370	1.6	G
AISI 4130	123	5,640	2.3	U
AISI 4130	403	6,780	2.7	U
AISI 4130	751	5,640	2.2	U
AISI 4130	1064	5,300	1.3	U
AISI 4130	197	2,340	2.3	U
AISI 4130	402	2,370	1.0	U
AISI 4340	123	5,640	2.7	U
AISI 4340	403	6,780	2.2	U
AISI 4340	751	5,640	0.8	U
AISI 4340	1064	5,300	1.3	U
AISI 4340	197	2,340	1.9	U
AISI 4340	402	2,370	1.0	U
AISI 502	123	5,640	5.9	P; C(21)
AISI 502	403	6,780	2.3	C(22)
AISI 502	751	5,640	2.8	E; C(50); P(36)
AISI 502	1064	5,300	2.6	C
AISI 502	197	2,340	1.4	P; C(23)
AISI 502	402	2,370	0.8	P; C(16)
Gray cast iron	123	5,640	4.2	U
Gray cast iron	403	6,780	1.8	U
Gray cast iron	751	5,640	1.2	G
Gray cast iron	1064	5,300	0.8	U
Gray cast iron	197	2,340	2.0	G
Gray cast iron	402	2,370	1.7	U
Ni cast iron	123	5,640	4.4	U
Ni cast iron	403	6,780	2.9	U
Ni cast iron	751	5,640	1.4	G
Ni cast iron	1064	5,300	0.9	G
Ni cast iron	197	2,340	2.2	G
Ni cast iron	402	2,370	1.5	U
Ni-Cr cast iron #2	123	5,640	4.3	U

(*continued*)

TABLE XVIII (*Continued*)

Alloy	Exposure Days	Exposure Depth (ft)	Corrosion Rate (mpy)[a]	Type of Corrosion[b]
Ni-Cr cast iron #2	403	6,780	1.8	U
Ni-Cr cast iron #2	751	5,640	1.3	G
Ni-Cr cast iron #2	1064	5,300	0.7	U
Ni-Cr cast iron #2	197	2,340	1.9	G
Ni-Cr cast iron #2	402	2,370	1.8	U
Ductile cast iron #1	123	5,640	3.1	U
Ductile cast iron #1	403	6,780	3.4	G
Ductile cast iron #1	751	5,640	1.0	G
Ductile cast iron #1	1064	5,300	0.6	U
Ductile cast iron #1	197	2,340	1.9	G
Ductile cast iron #1	402	2,370	2.9	U
Si cast iron	123	5,640	< 0.1	NC
Si cast iron	403	6,780	< 0.1	NC
Si cast iron	751	5,640	< 0.1	NC
Si cast iron	1064	5,300	< 0.1	NC
Si cast iron	197	2,340	< 0.1	NC
Si cast iron	402	2,370	< 0.1	NC
Si-Mo cast iron	123	5,640	< 0.1	NC
Si-Mo cast iron	403	6,780	< 0.1	NC
Si-Mo cast iron	751	5,640	< 0.1	NC
Si-Mo cast iron	1064	5,300	< 0.1	NC
Si-Mo cast iron	197	2,340	< 0.1	NC
Si-Mo cast iron	402	2,370	< 0.1	NC
Austenitic cast iron, Type 1	123	5,640	2.4	G
Austenitic cast iron, Type 1	403	6,780	1.0	U
Austenitic cast iron, Type 1	751	5,640	0.5	G
Austenitic cast iron, Type 1	1064	5,300	0.5	U
Austenitic cast iron, Type 1	197	2,340	1.8	G
Austenitic cast iron, Type 1	402	2,370	1.5	U
Austenitic cast iron, Type 2	123	5,640	2.4	G
Austenitic cast iron, Type 2	403	6,780	2.2	U
Austenitic cast iron, Type 2	751	5,640	1.6	G
Austenitic cast iron, Type 2	1064	5,300	1.0	G

TABLE XVIII (*Continued*)

Alloy	Exposure Days	Exposure Depth (ft)	Corrosion Rate (mpy)[a]	Type of Corrosion[b]
Austenitic cast iron, Type 2	197	2,340	1.1	G
Austenitic cast iron, Type 2	402	2,370	0.7	U
Austenitic cast iron, Type 3	123	5,640	1.9	G
Austenitic cast iron, Type 3	403	6,780	1.8	U
Austenitic cast iron, Type 3	751	5,640	1.9	G
Austenitic cast iron, Type 3	1064	5,300	1.2	U
Austenitic cast iron, Type 3	197	2,340	0.8	G
Austenitic cast iron, Type 3	402	2,370	0.6	U
Austenitic cast iron, Type 4	123	5,640	1.8	G
Austenitic cast iron, Type 4	403	6,780	2.0	U
Austenitic cast iron, Type 4	751	5,640	2.0	U
Austenitic cast iron, Type 4	1064	5,300	1.2	G
Austenitic cast iron, Type 4	197	2,340	0.9	U
Austenitic cast iron, Type 4	402	2,370	0.9	G
Austenitic cast iron, D-2	123	5,640	2.6	G
Austenitic cast iron, D-2	403	6,780	1.2	U
Austenitic cast iron, D-2	751	5,640	1.3	G
Austenitic cast iron, D-2	1064	5,300	1.1	U
Austenitic cast iron, D-2	197	2,340	1.2	G
Austenitic cast iron, D-2	402	2,370	1.1	U
Austenitic cast iron, D-2b	123	5,640	2.1	G

(continued)

TABLE XVIII (*Continued*)

Alloy	Exposure Days	Exposure Depth (ft)	Corrosion Rate (mpy)[a]	Type of Corrosion[b]
Austenitic cast iron, D-2b	403	6,780	1.6	U
Austenitic cast iron, D-2b	751	5,640	1.2	G
Austenitic cast iron, D-2b	1064	5,300	1.0	G
Austenitic cast iron, D-2b	197	2,340	1.4	G
Austenitic cast iron, D-2b	402	2,370	0.9	U
Austenitic cast iron, D-2c	189	5,900	3.3	U
Austenitic cast iron, D-2c	402	2,370	1.8	U
Austenitic cast iron, D-3	123	5,640	1.9	G
Austenitic cast iron, D-3	403	6,780	2.7	G
Austenitic cast iron, D-3	751	5,640	2.1	G
Austenitic cast iron, D-3	1064	5,300	1.2	U
Austenitic cast iron, D-3	197	2,340	0.9	G
Austenitic cast iron, D-3	402	2,370	0.7	U

[a] MPY = mils penetration per year calculated from weight loss
[b] Symbols signify the following types of corrosion

C = Crevice (numbers in parentheses refer to maximum depth of crevice attack in mils)
P = Pitting (numbers in parentheses refer to maximum pit depth in mils)
U = Uniform
G = General
E = Edge
NC = No visible corrosion

improved coatings performance. In deep water sites with low oxygen content, the reduction in corrosion rates has resulted in improved service life of steel structures.

C. Performance and Selection of Aluminum Alloys

1. General Corrosion Performance

In the deep ocean, the types of corrosion found on aluminum alloys are similar to those found at near surface sites. Pitting, crevice corrosion, and other nonuniform types of attack such as intergranular corrosion and exfoliation, are, as in surface water, the most common types of corrosion attack that occur on aluminum and aluminum alloys in the deep ocean. As in the case of steels, there appears to be no direct effect of depth on the corrosion behavior of aluminum alloys. Where correlations were possible, it was found that the performance of these alloys could be correlated with oxygen content of the seawater. Figure 6 shows such a correlation between the corrosion behavior of aluminum alloy 6061 and the oxygen content of the seawater. Note that while the corrosion rate based upon weight loss is little affected, the depth of pitting increases by a factor of seven at the depth with lowest oxygen content.

2. Significant Performance Parameters

As in the evaluation of the performance of these alloys in surface seawater, corrosion rates based upon measurement of weight loss should not be used to evaluate the performance of these alloys. Since the forms of corrosion experienced on aluminum alloys are similar to the forms of corrosion in surface waters, the same parameters are applicable for measuring their performance.

3. Performance of Commonly Used Alloys

As in the corrosion performance of aluminum alloys at the surface, the general corrosion performance of aluminum alloys can be ranked as follows: the alloys in the 1000, 3000, and 5000 series are the most corrosion resistant; alloys in the 6000 series show somewhat more attack; and alloys in the 2000 and 7000 series are the least resistant to attack. Table IV gave the chemical composition of several commonly used aluminum alloys. The performance of

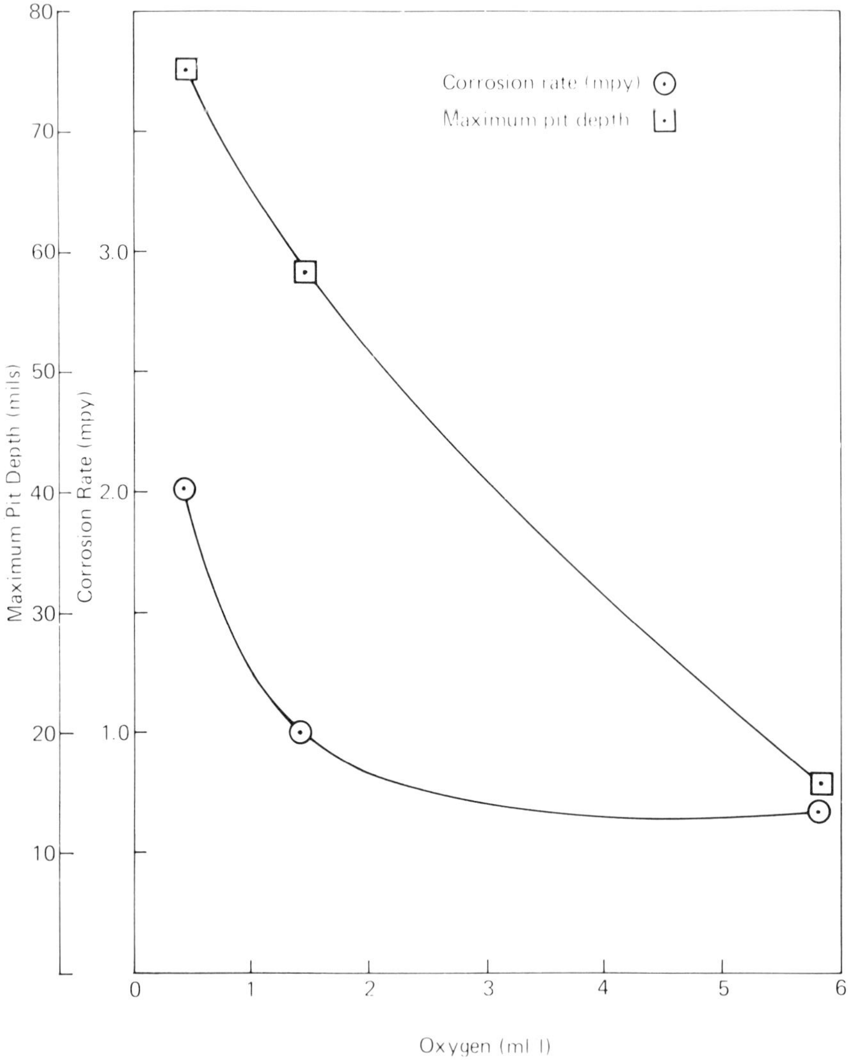

Fig. 6. Corrosion behavior of aluminum alloy 6061 after one year of exposure versus oxygen content.

these alloys in the deep ocean, taken from Boyd and Fink (*19*), Lennox (*28*), Waldron et al. (*30*), and Reinhart (*34*) is given in Table XIX.

4. Typical Applications

As in surface seawater, the light weight, high strength-to-weight ratio of some alloys, and their ease of fabrication makes aluminum alloys excellent candidates for some deep ocean applications. Where the effects of pitting and other nonuniform types of attack are minimal, such as in open framework structures, aluminum alloys have been successfully used for long-term exposures in the deep ocean (*35*). In applications where such effects are significant, the long-term use of aluminum alloys in deep ocean waters with low oxygen content is limited. However, even when nonuniform attack is significant, such as in the case of submersible pressure hulls (*36*) and instrument canisters, aluminum alloys have been successfully applied for short-term exposures. For such applications, the aluminum alloys are commonly protected through a combination of coatings and cathodic protection. In some applications where pitting attack is particularly significant and the use of protective coatings is precluded, the performance of aluminum alloys in waters with low oxygen content has been shown to be inferior to their performance in similar applications when exposed to near-surface waters (*37*).

D. Performance and Selection of Copper Alloys

1. General Corrosion Performance

While there were some copper alloys that corroded more slowly at deep ocean sites with low oxygen (Figure 7), the differences in corrosion rates were considered to be insignificant for most applications. As for exposure in surface waters, the corrosion rates of many copper alloys decreased with increasing time of exposure, as shown in Figure 8. There appeared to be little difference between the deep ocean and near-surface behavior of those copper alloys that were susceptible to nonuniform attack.

2. Significant Performance Parameters

As in surface exposures, corrosion rates based upon weight loss can be used to measure the performance of those alloys that corrode uniformly. For those alloys that corrode nonuniformly, such as the alloys the suffer from parting

TABLE XIX

PERFORMANCE OF SELECTED ALUMINUM ALLOYS AT DEPTH

Alloy	Exposure		Corrosion			
	Days	Depth	Rate (mpy)	Maximum Pit Depth (mils)	Crevice Depth (mils)	Type[a]
1100-H14	123	5,640	2.0	39	0	P
1100-H14	403	6,780	4.2	0	62	C(PR)
1100-H14	751	5,640	4.5	0	62	C(PR)
1100-H14	1064	5,300	3.0	0	62	C(PR)
1100-H14	197	2,340	5.6	0	62	C(PR)
1100-H14	402	2,370	1.6	0	62	C(PR)
2024-O	123	5,640	5.2	28	0	P
2024-O	403	6,780	6.2	62	0	P(PR)
2024-O	751	5,640	2.8	65	62	C(PR); P(PR)
2024-O	1064	5,300	1.9	62	62	(PR)
2024-O	197	2,340	3.1	19	62	C(PR)
2024-O	402	2,370	3.0	62	62	C(PR)
2219-T81	123	5,640	2.0	16	15	C; IG; G-P
2219-T81	403	6,780	3.6	35	38	C; E; IG; G-P
2219-T81	751	5,640	2.4	30	24	C; E; IG; G-P
2219-T81	1064	5,300	1.6	47	47	C; SL-E; IG; G-P
2219-T81	197	2,340	2.6	25	0	SL-E; IG; G-P
2219-T81	402	2,370	4.5	78	69	S-C; S-E; IG; S-P
3003-H14	123	5,640	0.5	27	32	C; P
3003-H14	403	6,780	3.9	125	66	S-C; P(PR)
3003-H14	751	5,640	2.3	125	125	C(PR); P(PR)
3003-H14	1064	5,300	2.0	125	125	C(PR); S-E; P(PR)
3003-H14	197	2,340	2.4	48	28	C; S-E; P
3003-H14	402	2,370	1.4	91	93	S-C; S-E; P
Alclad 3003-H12	123	5,640	0.2	18	15	C; SL-E; P
Alclad 3003-H12	403	6,780	0.4	13	14	C; SL-E; P
Alclad 3003-H12	751	5,640	0.3	13	13	C; E; P
Alclad 3003-H12	1064	5,300	0.5	20	13	C; P
Alclad 3003-H12	197	2,340	2.2	15	13	C; P
Alclad 3003-H12	402	2,370	2.2	14	15	C; P
5052-O	123	5,640	3.7	0	65	C(PR)
5052-O	403	6,780	4.5	0	62	C(PR)
5052-O	751	5,640	2.2	65	65	C(PR); P(PR)
5052-O	1064	5,300	3.1	0	65	C(PR)
5052-O	197	2,340	1.8	0	65	C(PR)
5052-O	402	2,370	0.4	0	20	C
5083-H113	123	5,640	0.9	65	0	P
5083-H113	403	6,780	4.0	59	0	S-E; P
5083-H113	751	5,640	1.6	83	0	E; P
5083-H113	197	2,340	0.5	23	0	E; P
5083-H113	402	2,370	0.6	I	52	S-C; I-P

TABLE XIX (*Continued*)

	Exposure		Corrosion			
Alloy	Days	Depth	Rate (mpy)	Maximum Pit Depth (mils)	Crevice Depth (mils)	Type[a]
5086-H32	189	5,900	0.1	33	15	C; SC-P
5086-H32	360	4,200	0.1	47	0	E; P
5086-H32	402	2,370	0.4	I	18	C; I-P
5086-H34	123	5,640	0.1	0	0	ET
5086-H34	403	6,780	0.6	I	53	S-C; I-P
5086-H34	751	5,640	2.0	72	60	S-C; E; S-P
5086-H34	1064	5,300	0.9	73	9	S-C; S-E; S-P
5086-H34	197	2,340	0.7	29	I	I-C; SL-E; P
5086-H34	402	2,370	0.6	I	5	C; I-P
5454-H32	123	5,640	0.1	7	0	P
5454-H32	403	6,780	0.9	38	0	E; P
5454-H32	751	5,640	0.9	65	0	E; P
5454-H32	197	2,340	0.7	24	0	SL-E; P
5454-H32	402	2,370	0.3	I	39	C; I-P
5456-H321	123	5,640	< 0.1	0	SL	SL-C; SL-ET
5456-H321	403	6,780	1.0	I	50	S-C; E; I-P
5456-H321	751	5,640	2.0	53	25	C; E; P
5456-H321	1064	5,300	1.0	43	37	C; E; P
5456-H321	197	2,340	2.7	17	36	C; E; P
5456-H321	402	2,370	1.1	41	44	S-C; E; P
6061-T6	123	5,640	< 0.1	5	10	C; P
6061-T6	403	6,780	1.0	58	55	S-C; S-P
6061-T6	751	5,640	1.1	65	65	S-C; E; S-P
6061-T6	1064	5,300	0.8	77	66	S-C; E; S-P
6061-T6	197	2,340	1.2	47	42	C; P
6061-T6	402	2,370	2.0	75	66	S-C; S-P
7039-T6	123	5,640	0.7	3	0	P; XF
7039-T6	403	6,780	—	—	—	EX-XF
7039-T6	751	5,640	—	—	—	EX-XF
7039-T6	197	2,340	1.9	27	0	P; XF
7039-T6	402	2,370	—	—	—	EX-XF

[a] Symbols for types of corrosion:

P = Pitting
E = Attack at edge
C = Attack at crevice
ET = Etched
IG = Intergranular attack
XF = Exfoliation
G-P = General pitting
S-C = Severe crevice attack
S-P = Severe pitting
S-E = Severe attack at edge
I-P = Incipient pitting
I-C = Incipient crevice attack
SL-E = Slight attack at edge
EX-XF = Extreme exfoliation
C(PR) = Crevice attack to perforation
C(PR) = Pitting to perforation
SL-ET = Slight etching

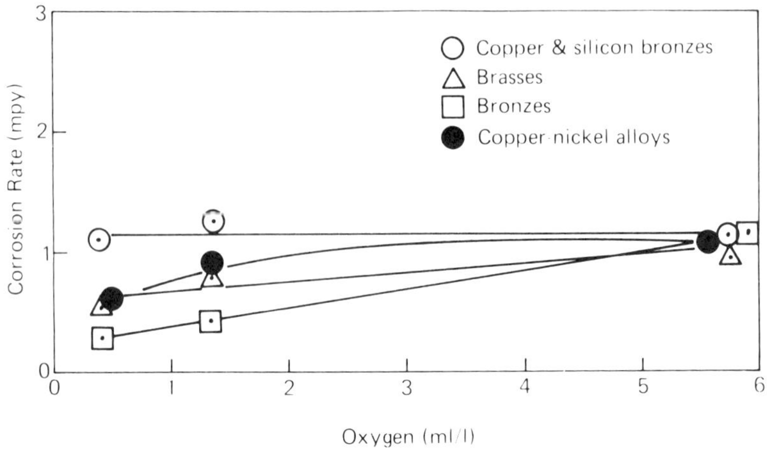

Fig. 7. Corrosion rate of copper alloys after one year of exposure versus oxygen content.

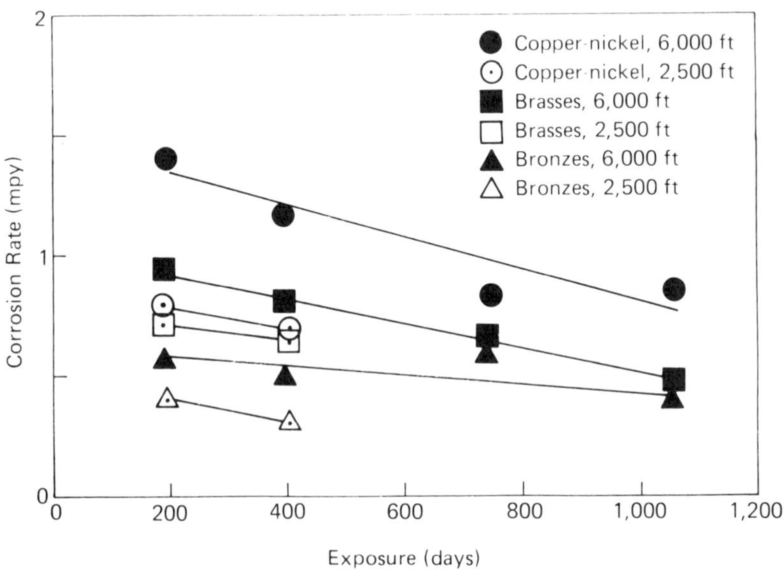

Fig. 8. Corrosion rate of copper alloys at depth versus time of exposure.

corrosion, the measurement of corrosion performance must be appropriate for the form of attack.

3. Performance of Commonly Used Alloys

The composition of several commonly used copper alloys was given in Table VI. The deep ocean corrosion behavior of these materials is given in Table XX. Data in Table XX were taken from Reinhart (*11*), Boyd and Fink (*19*), and Waldron et al. (*30*).

4. Typical Applications

Due to their high strength and good corrosion resistance, beryllium copper alloys have been widely used for long-term deep ocean applications. Trans-oceanic communication cable repeater housings have been fabricated from beryllium copper for over 30 years and have given excellent service. Recently developed spinoidal copper nickel chromium alloys, which also have high strength and good corrosion resistance, appear to also be excellent candidates for such applications. Due to the lack of applications requiring high heat transfer in the deep ocean and lack of a requirement for fouling resistance, the use of copper alloys for other deep ocean applications has been limited.

E. Performance and Selection of Stainless Steels

1. General Corrosion Performance

It is difficult to establish correlations between environment and performance of stainless steels because the quantitative measurements of performance show such a large scatter and because of the limited number of deep ocean corrosion experiments. In general, the stainless steels as a group are more susceptible to the initiation of crevice and tunneling corrosion at depth; the weight loss due to these forms of attack is usually less at depth. This reduction in weight loss is probably due to the reduction in the difference between the environment within and without a crevice when the oxygen content of the seawater is low. This phenomenon was studied in detail in the deep ocean by Wheatfall, who exposed Type 304 stainless steel specimens with varying ratios of crevice and boldly exposed areas. Figure 9 shows the results of this experiment and clearly shows the effect of variations in this area ratio on the severity of crevice attack and the effect of oxygen. The weight loss

TABLE XX

PERFORMANCE OF SELECTED COPPER ALLOYS AT DEPTH

Alloy	CDA No.	Exposure Days	Exposure Depth	Corrosion Rate (mpy)	Type[a]
Copper, O free	102	123	5,640	1.6	U
Copper, O free	102	403	6,780	1.2	U
Copper, O free	102	751	5,640	0.7	U
Copper, O free	102	1064	5,300	0.5	U
Copper, O free	102	197	2,340	0.8	U
Copper, O free	102	402	2,370	0.9	U
Be-Cu	172	123	5,640	0.5	U
Be-Cu	172	751	5,640	0.5	U
Be-Cu	172	197	2,340	0.2	U
Be-Cu	172	402	2,370	0.5	U
Commercial bronze	220	123	5,640	0.6	U
Commercial bronze	220	403	6,780	0.6	U
Commercial bronze	220	751	5,640	0.6	C(9)
Commercial bronze	220	1064	5,300	0.4	C(20)
Commercial bronze	220	197	2,340	0.3	U
Commercial bronze	220	402	2,370	0.2	SL-DZ
Yellow brass	270	123	5,640	1.4	U
Yellow brass	270	403	6,780	1.0	U
Yellow brass	270	751	5,640	2.5	SL-DZ
Yellow brass	270	1064	5,300	0.6	U
Yellow brass	270	197	2,340	0.9	U
Yellow brass	270	402	2,370	0.9	U
Muntz metal	280	123	5,640	1.6	SL-DZ
Muntz metal	280	403	6,780	2.6	SL-DZ
Muntz metal	280	751	5,640	3.2	G-DZ
Muntz metal	280	1064	5,300	2.3	S-DZ
Muntz metal	280	197	2,340	0.7	SL-DZ; P(10)
Muntz metal	280	402	2,370	0.7	SL-DZ
Arsenical Admiralty	443	123	5,640	1.0	U
Arsenical Admiralty	443	403	6,780	0.7	U
Arsenical Admiralty	443	751	5,640	0.6	U
Arsenical Admiralty	443	1064	5,300	0.5	U
Arsenical Admiralty	443	197	2,340	0.6	U
Arsenical Admiralty	443	402	2,370	0.6	U
Ni-Mn bronze	868	123	5,640	0.5	SL-DZ
Ni-Mn bronze	868	403	6,780	0.4	MD-DZ
Ni-Mn bronze	868	751	5,640	2.3	VS-DZ
Ni-Mn bronze	868	197	2,340	0.4	SL-DZ
Ni-Mn bronze	868	402	2,370	1.6	SL-DZ
G bronze	—	123	5,640	0.5	U
G bronze	—	403	6,780	2.6	U
G bronze	—	751	5,640	3.2	U

TABLE XX (*Continued*)

Alloy	CDA No.	Exposure		Corrosion Rate (mpy)	Type[a]
		Days	Depth		
G bronze		1064	5,300	2.3	U
G bronze	—	197	2,340	0.7	U
G bronze	—	402	2,370	0.7	U
P bronze A	510	123	5,640	0.6	U
P bronze A	510	403	6,780	0.2	ET
P bronze A	510	751	5,640	0.2	ET
P bronze A	510	1064	5,300	0.4	U
P bronze A	510	197	2,340	0.3	U
P bronze A	510	402	2,370	0.1	ET
Al bronze 5%	606	123	5,640	0.6	U
Al bronze 5%	606	403	6,780	0.2	SL-DA
Al bronze 5%	606	751	5,640	0.3	SL-DA
Al bronze 5%	606	1064	5,300	0.2	NU
Al bronze 5%	606	197	2,340	0.4	U
Al bronze 5%	606	402	2,370	0.2	U
Al bronze 7%	614	123	5,640	0.5	SL-DA
Al bronze 7%	614	403	6,780	0.7	SL-DA; C(12); P(12)
Al bronze 7%	614	751	5,640	0.5	MD-DA; C(7); P(12)
Al bronze 7%	614	1064	5,300	0.2	P(7)
Al bronze 7%	614	197	2,340	0.3	U
Al bronze 7%	614	402	2,370	0.2	U
Al bronze 13%	—	123	5,640	0.5	SL-DA
Al bronze 13%	—	403	6,780	0.6	S-DA
Al bronze 13%	—	751	5,640	1.9	S-DA
Al bronze 13%		1064	5,300	0.6	S-DA
Al bronze 13%	—	197	2,340	0.4	MD-DA
Al bronze 13%	—	402	2,370	0.3	MD-DA
Ni Al bronze #2	—	123	5,640	0.5	U
Ni Al bronze #2	—	403	6,780	0.2	I-P
Ni Al bronze #2	—	751	5,640	0.5	SL-DA
Ni Al bronze #2	—	1064	5,300	0.2	C(5)
Ni Al bronze #2	—	197	2,340	0.3	U
Ni Al bronze #2	—	402	2,370	0.2	U
Si bronze A	655	123	5,640	1.6	U
Si bronze A	655	403	6,780	1.2	DA
Si bronze A	655	751	5,640	0.8	S-DA
Si bronze A	655	1064	5,300	0.6	SL-DA
Si bronze A	655	197	2,340	0.8	U
Si bronze A	655	402	2,370	1.0	ET
Cu-Ni 90-10	706	123	5,640	1.6	U

(*continued*)

TABLE XX (*Contined*)

Alloy	CDA No.	Exposure Days	Exposure Depth	Corrosion Rate (mpy)	Type[a]
Cu-Ni 90-10	706	403	6,780	0.8	U
Cu-Ni 90-10	706	751	5,640	0.7	U
Cu-Ni 90-10	706	1064	5,300	0.7	U
Cu-Ni 90-10	706	197	2,340	0.8	U
Cu-Ni 90-10	706	402	2,370	0.6	U
Cu-Ni 70-30	715	123	5,640	1.2	U
Cu-Ni 70-30	715	403	6,780	1.2	U
Cu-Ni 70-30	715	751	5,640	0.7	U
Cu-Ni 70-30	715	1064	5,300	0.6	U
Cu-Ni 70-30	715	197	2,340	0.7	U
Cu-Ni 70-30	715	402	2,370	0.5	U
Cu-Ni 70-30	716	123	5,640	0.2	U
Cu-Ni 70-30	716	403	6,780	0.1	ET
Cu-Ni 70-30	716	751	5,640	0.5	C(16); P(16); DA
Cu-Ni 70-30	716	197	2,340	0.1	U
Cu-Ni 70-30	716	402	2,370	0.1	U

[a]Symbols for types of corrosion:

U = Uniform attack
P = Pitting (numbers in parentheses indicate maximum pit depth in mils)
C = Attack at crevice (numbers in parentheses indicate maximum penetration in mils)
ET = Etched
DA = Dealloying
NU = Nonuniform attack
I-P = Incipient pitting
G-DZ = General dezincification
S-DZ = Severe dezincification
S-DA = Severe dealloying
MD-DZ = Medium dezincification
MD-DA = Medium dealloying
VS-DZ = Very severe dezincification
SL-DZ = Slight dezincification
SL-DA = Slight dealloying

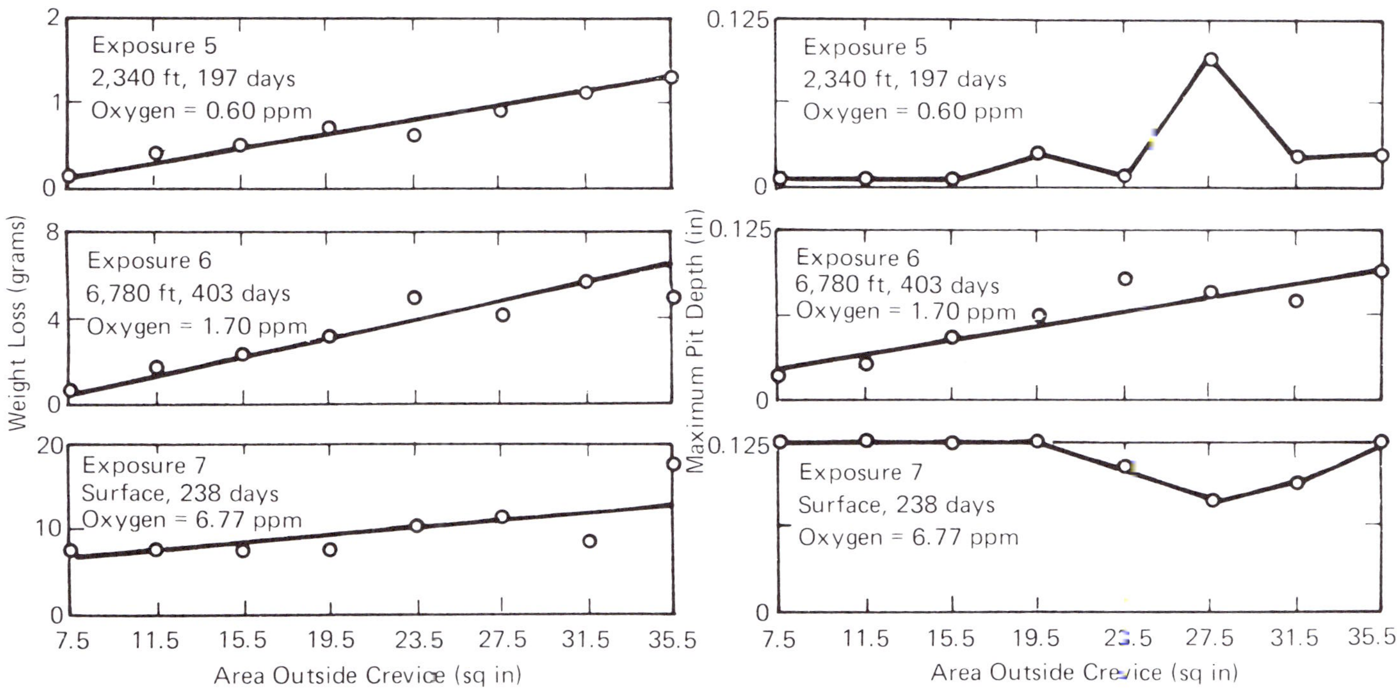

Fig. 9. Crevice corrosion of Type 304 stainless steel versus crevice-to-boldly exposed area ratio.

of the specimens increased as the area outside the crevice increased. The weight loss was an order of magnitude greater in highly oxygenated surface waters than in the low oxygen content waters at depth. The depth of attack, while greater in surface waters than in the low oxygen content waters at depth, was more erratic (*38*).

The absence of crevice forming hard shell fouling organisms at depth significantly reduces the incidence of surface pitting on the stainless steels at depth. The stress corrosion cracking of the precipitation hardening stainless steels appears to be somewhat more severe at the surface than at depth but is difficult to assess from the experiments performed due to the high rates of crevice corrosion of these alloys which made them difficult to hold in a stressed condition (*15*).

2. Significant Performance Parameters

As in the evaluation of the performance of stainless steels in surface seawater, the quantitative measurement of the nonuniform attack typical of stainless steels in seawater at any depth is difficult. For deep ocean environments, where the number of test exposures is limited, it is even more difficult to quantitatively assess the effect of any single parameter on the performance of stainless steels. It is even difficult to compare the performance of different alloys. Corrosion rates based upon weight loss have little value for most evaluations. Depth of attack and frequency of initiation are often the best means of quantitative measurement of corrosion behavior. Due to the nature of deep ocean tests, it is also difficult to determine the time to initiation of attack.

3. Performance of Commonly Used Alloys

The general listing of stainless steels given in Table VIII appears to be equally applicable to deep ocean exposures. Data for the composition were given in Table IX. Data on the performance of the stainless steels at depth are given in Table XXI. The data in Table XXI were taken from Reinhart (*15*), Lennox (*28*), and Wheatfall (*38*).

4. Typical Applications

There have been few successful long-term applications of stainless steel in the deep ocean. The use of stainless steels has been successful in short-term exposures, intermittent exposures, or where the stainless steels received

TABLE XXI

PERFORMANCE OF SELECTED STAINLESS STEELS AT DEPTH

	Exposure		Corrosion			
Alloy	Days	Depth	Rate (mpy)	Maximum Pit Depth (Mils)	Crevice Depth (Mils)	Type[a]
AISI 201	123	5,640	< 0.1	0	I	I-C
AISI 201	403	6,780	< 0.1	0	I	NC
AISI 201	751	5,640	< 0.1	0	2	C
AISI 201	1064	5,300	< 0.1	0	I	I-C
AISI 201	197	2,340	< 0.1	0	4	C
AISI 201	402	2,370	< 0.1	0	1	C
AISI 304	123	5,640	< 0.1	0	0	T(450)
AISI 304	403	6,780	0.5	210	0	E; P(PR); T(2000)
AISI 304	751	5,640	0.5	210	0	P(PR); T(1600)
AISI 304	1064	5,300	< 0.1	13	74	C; E; P; T(1750)
AISI 304	197	2,340	0.3	0	11	C; E; T(950)
AISI 304	402	2,370	0.4	210	0	E; P(PR); T(2000)
AISI 304L	123	5,640	0.2	0	SL	SL-C; T(200)
AISI 304L	403	6,780	< 0.1	115	12	C; E; P(PR) T(4850)
AISI 304L	751	5,640	0.3	115	0	P(PR); T(4000)
AISI 304L	1064	5,300	0.2	115	115	C(PR); E; P(PR); T(4000)
AISI 304L	197	2,340	0.2	0	8	C; T(800)
AISI 304L	402	2,370	0.4	115	0	P(PR); T(3000)
AISI 310	123	5,640	< 0.1	0	I	NC
AISI 310	403	6,780	< 0.1	0	I	I-C
AISI 310	751	5,640	< 0.1	0	0	NC
AISI 310	1064	5,300	< 0.1	0	I	I-C
AISI 310	197	2,340	< 0.1	0	I	I-C
AISI 310	402	2,370	< 0.1	0	14	C
AISI 316	123	5,640	< 0.1	0	10	SL-E; SL-T
AISI 316	403	6,780	< 0.1	0	I	NC
AISI 316	751	5,640	< 0.1	0	0	NC
AISI 316	1064	5,300	0.2	21	0	P
AISI 316	197	2,340	< 0.1	0	0	NC
AISI 316	402	2,370	0.1	230	0	E; P(PR); T(500)

(*continued*)

TABLE XXI (*Continued*)

	Exposure		Corrosion			
Alloy	Days	Depth	Rate (mpy)	Maximum Pit Depth (Mils)	Crevice Depth (Mils)	Type[a]
AISI 316L	123	5,640	< 0.1	0	SL	SL-C
AISI 316L	403	6,780	< 0.1	0	0	NC
AISI 316L	751	5,640	< 0.1	0	13	C
AISI 316L	1064	5,300	< 0.1	0	25	C
AISI 316L	197	2,340	< 0.1	0	18	C
AISI 316L	402	2,370	< 0.1	0	0	NC
AISI 321	123	5,640	< 0.1	0	0	NC
AISI 321	403	6,780	< 0.1	0	I	I-C
AISI 321	751	5,640	< 0.1	0	0	NC
AISI 321	1064	5,300	< 0.1	0	I	I-C
AISI 321	197	2,340	< 0.1	0	I	I-C
AISI 321	402	2,370	0.2	0	30	C(PR)
AISI 347	123	5,640	< 0.1	0	10	C
AISI 347	403	6,780	< 0.1	0	I	I-C
AISI 347	751	5,640	< 0.1	0	0	NC
AISI 347	1064	5,300	< 0.1	I	I	I-C; I-P
AISI 347	197	2,340	< 0.1	0	0	NC
AISI 347	402	2,370	< 0.1	0	I	I-C
20-Cb	123	5,640	0	0	0	NC
20-Cb	403	6,780	0	0	0	NC
20-Cb	751	5,640	< 0.1	0	26	C
20-Cb	1064	5,300	0	0	0	NC
20-Cb	197	2,340	< 0.1	0	I	I-C
20-Cb	402	2,370	< 0.1	0	0	NC
AISI 410	123	5,640	0.2	40	SL	SL-C; P(PR); T(2950)
AISI 410	403	6,780	0.2	40	40	C(PR); P(PR); T(6000)
AISI 410	751	5,640	0.5	40	40	C(PR); P(PR); T(10000)
AISI 410	1064	5,300	1.7	50	50	C(PR); P(PR)
AISI 410	197	2,340	0.3	0	10	C; T(7000)
AISI 410	402	2,370	0.5	40	40	C(PR); P(PR); T(6400)
AISI 430	123	5,640	0.5	0	10	C; T(1000)
AISI 430	403	6,780	0.2	137	30	C; P(PR); T(3730)
AISI 430	751	5,640	0.6	137	30	C; P(PR); T(5000)
AISI 430	1064	5,300	0.8	137	97	C; E; P(PR); T(7200)
AISI 430	197	2,340	0.5	0	17	C; T(2375)

TABLE XXI (*Continued*)

	Exposure		Corrosion			
Alloy	Days	Depth	Rate (mpy)	Maximum Pit Depth (Mils)	Crevice Depth (Mils)	Type[a]
AISI 430	402	2,370	0.6	137	20	C; E; P(PR); T(6000)
AISI 630[b]	403	6,780	< 0.1	0	0	NC
AISI 630[b]	197	2,340	0.1	0	0	NC
AISI 630[b]	402	2,370	< 0.1	0	1	C
AISI 631[c]	403	6,780	0.4	125	125	C(PR); P(PR); T(3000)
AISI 631[c]	197	2,340	0.3	I	0	I-P; T(1375)
AISI 631[c]	402	2,370	< 0.1	I	0	I-P; T(1250)
AISI 635	123	5,640	< 0.1	0	SL	SL-C
AISI 635	403	6,780	0.2	0	20	C
AISI 635	751	5,640	< 0.1	0	0	NC
AISI 635	1064	5,300	< 0.1	0	50	C
AISI 635	197	2,340	< 0.1	0	0	NC
AISI 635	402	2,370	0.3	0	275	C(PR)
18Cr-14Mn-0.5N	123	5,640	< 0.1	0	0	NC
18Cr-14Mn-0.5N	403	6,780	0.5	115	0	P(PR); T(2750)
18Cr-14Mn-0.5N	751	5,640	0.2	115	24	C; P(PR); T(2800)
18Cr-14Mn-0.5N	1064	5,300	0.2	115	0	P(PR); T(2250)
18Cr-14Mn-0.5N	197	2,340	0.8	115	109	C; P(PR); T(1250)
18Cr-14Mn-0.5N	402	2,370	0.8	115	0	P(PR); T(2000)

[a] Symbols for type of attack

E = Edge attack
P = Pitting (numbers in parenthesis refer to maximum pit depth in mils, or PR for perforation)
T = Tunneling attack (numbers in parentheses refer to maximum length of tunneling in mils)
C = Crevice attack (numbers in parentheses refer to maximum depth of crevice attack in mils, or PR for perforation)
NC = No corrosion
I-C = Incipient crevice attack
I-P = Incipient pitting
SL-E = Slight edge attack
SL-C = Slight crevice attack
SL-T = Slight tunneling

[b] H925 condition
[c] RH1050 condition

sufficient cathodic protection to inhibit crevice corrosion. Few of the newly developed marine grades of stainless steels have been either tested or used in deep ocean environments. Their performance in the deep ocean is likely to be similar to their performance in near-surface waters.

F. Performance and Selection of Titanium Alloys

1. General Corrosion Performance

With the exception of stress corrosion cracking, there appears to be no effect of corrosion on titanium or its alloys either in surface seawater or at depth. Due to the difficulty of performing fracture mechanics experiments in the deep ocean, there has been no determination of stress intensity factors for stress corrosion cracking of titanium alloys in the deep ocean. Results from tests of welded specimens of an alloy known to be extremely sensitive to stress corrosion cracking (13V–11Cr–3 Al) appear to indicate that the environment at the surface is more detrimental than at depth (*39*).

2. Typical Applications

Titanium alloys have been successfully used in several deep ocean applications where their extreme corrosion resistance and high strength to weight ratios are vital. Submersible pressure hulls fabricated from titanium alloys such as the 6Al-4V hull of the research submersible ALVIN have performed well for periods of over 10 years.

G. Performance and Selection of Other Metallic Materials

1. Nickel Alloys

The high nickel alloys Inconel 625, Hastelloy C, and Hastelloy C-276, which are essentially immune to corrosion in surface and near-surface seawater, are also immune to corrosion at depth. Due to the unavailability of maintenance or repair and the consequence of failure in the deep ocean, the use of these alloys is often well justified for many critical deep ocean

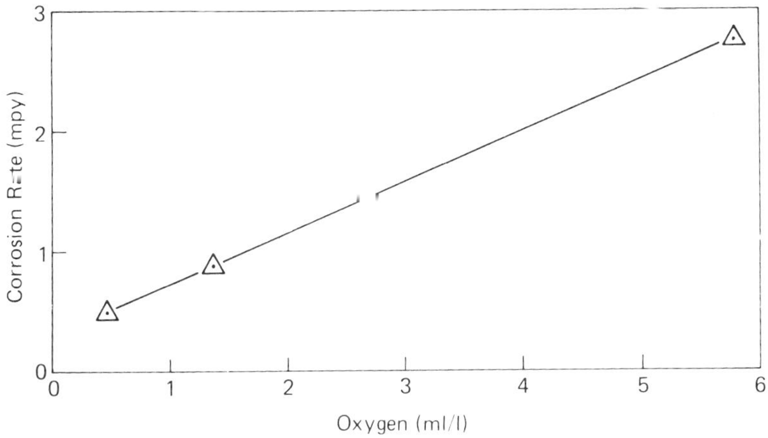

Fig. 10. Corrosion rate of Monel alloys versus oxygen content.

applications. Other high nickel alloys such as Inconel 718 and Incoloy 825, which have shown resistance to seawater corrosion superior to that of many stainless steels, appear to perform at least as well at depth, primarily due to the lack of crevice forming fouling organisms. The corrosion rate of Monel 400 and Monel K-500 nickel-copper alloys at depth is less than at the surface and appears to be related to the oxygen content of the seawater as shown in Figure 10 (*24*).

2. Cobalt Alloys

Cobalt based superalloys such as Stellite Alloy 6B and MP-35N, which show excellent resistance to attack in surface seawater, show similar performance at depth.

V. Validation of Material Performance

As described in Section II of this chapter, it is highly desirable to select materials for marine applications based upon their previous performance in similar applications. The use of such a technique, of course, presupposes that the information on the previous performance is valid. As corrosion is such a complex phenomenon, evaluation of its effects on materials is often difficult to quantify or even to assess reliably in a qualitative manner. Assessment of corrosion behavior of materials in marine environments is also often made more difficult by the often hostile environment. Professor John Issacs, formerly of Scripps Institution of Oceanography, was fond of saying that "everything that goes to sea becomes either lost, corroded, or seasick." This includes equipment, samples, and personnel involved in the assessment of corrosion.

A. In Situ Evaluation

By evaluating structures and equipment in place, requirements for bringing the structure or portions of it to the surface for evaluation are eliminated and the potential for disturbing the evidence of corrosion on the structure is minimized.

1. Diver Accessible Structures

As for the evaluation of all corrosion processes, a careful visual examination by a trained observer is the first, and possibly the most important, step. For shallow structures that are accessible by SCUBA divers using air, it is often more practical to teach a trained observer of corrosion phenomena to dive than it is to teach a trained diver to make valid observation of corrosion, but both methods have been successful.

The first step in a visual inspection is the location and documentation of the appearance of corrosion products on the surface. Underwater photography and video have both been successfully used in underwater inspections. Stereo photography has also proven useful. The location of corrosion products is often precluded by the presence of thick accumulations of fouling organisms where preparatory cleaning of the surfaces is necessary. When cleaning of the surface to remove fouling is necessary, or after the location and nature of the corrosion products have been documented, then selected areas of the surface should be cleaned. Cleaning of small areas can usually be accomplished using hand tools. Cleaning of larger areas usually requires the use of either power tools or water blasting. If the surface to be inspected is coated, care must be taken to minimize the impact of cleaning on coating integrity. Underwater coatings may be used to "touch up" small areas of coating damage if necessary.

After the surfaces have been cleaned, they should again be inspected visually and photographed as appropriate. Particular attention should be paid to nonuniform attack such as pitting. Crevice corrosion is often difficult to identify unless it has led to excessive attack. One possible visual clue to crevice attack is the accumulation of corrosion products adjacent to the crevice areas. When nonuniform corrosion is detected, its extent can usually be determined through the use of mechanical measuring gages. The measurement of pit depths is, of course, made from the existing surface, and allowance for uniform attack must be made. In order to evaluate uniform attack without sophisticated instrumentation, the diver must have access to both surfaces of the material and use calipers or similar measuring devices to determine the remaining metal thickness. Cracking, whether due to strictly mechanical causes or due to corrosion, may also be visible.

Several nondestructive techniques used for inspection of structures on land have been adapted for use by divers. Ultrasonic thickness gages (which eliminate the need for access to both surfaces of a material for determination of thickness loss due to corrosion), as well as techniques for the detection of cracks such as magnetic particle, ultrasonics, and X-rays, have been adapted for underwater use by divers. For evaluation of cathodic protection systems, or where galvanic or stray current corrosion is suspected, underwater voltmeters or portable reference electrodes with surface readouts can be used to establish corrosion related potentials.

It is also possible to install instrumentation on structures to monitor corrosion related phenomena. This is particularly true for monitoring of the performance of cathodic protection systems.

2. Structures Inaccessible to Divers

Primarily due to excessive depth, inspection of marine structures using divers can be impractical or impossible. In these cases, visual observations and measurement of corrosion can be made through the use of either manned submersibles or remotely operated vehicles (ROVs). These inspections are usually more limited in extent than diver inspection due to time limitations and cost. Surface cleaning is also more difficult, but in deep environments the dearth of fouling reduces the need for surface cleaning in many cases. Mechanical measurement of corrosion damage is also possible using submersibles or ROVs but is more difficult and time consuming. Both manned submersibles and ROVs can be equipped for photography and video recording of the appearance of the structure being inspected. They can also be equipped with nondestructive testing instrumentation for corrosion evaluations. The use of instrumented ROVs for the evaluation of cathodic protection systems on deep water pipelines has been particularly successful.

B. Evaluation of Retrieved Structures

In some cases, inspection of structures upon or after retrieval from the sea is more appropriate than *in situ* inspection. The structure may be retrieved for repair, maintenance, or salvage. In many cases, corrosion information obtained from "objects of opportunity" retrieved from the ocean is the only available information on corrosion of materials at a given site. Jenkins (*33*) gives specific procedures for the inspection of structures recovered from the sea.

During retrieval, corrosion products are often disturbed but, if the structure can be inspected soon after retrieval, the corrosion products can contribute significantly to the evaluation of the corrosion of the structure. Corrosion product samples should be taken in accordance with National Association of Corrosion Engineers Recommended Practice RP-01-73 (*40*), if possible. The appearance of many corrosion products will change rapidly and significantly after exposure of the structure to the atmosphere. This is particularly true for steel and aluminum retrieved from low oxygen environments where the corrosion products are compact and often in a low oxidation and highly hydrated state. The rapid expansion and color changes of these corrosion products upon exposure to the atmosphere is responsible for the folklore that materials retrieved from the deep ocean corrode more in a few days of exposure to the atmosphere after retrieval than they did in long duration exposures to the deep sea.

Photographs of the structure should be taken as soon as possible after retrieval in order to document the initial appearance of the corrosion products. Later, photography should be used to document changes in the appearance of the corrosion products as well as any pitting or other visually evident forms of attack.

As in underwater inspections, there is a wide variety of tools that can be used to evaluate the corrosion that has occurred on structures retrieved from the sea. Mechanical or nondestructive thickness measurements, measurements of pit depths, location of cracks, and assessment of crevice attack can all be performed on structures retrieved form the sea. When samples are removed from the structure, they should be removed in a manner that preserves the information contained in the sample. Flame-cutting is often the most expedient means of sample removal but is the most damaging. Disassembly of bolted section is the most desirable means of sample removal.

Sawing, chiseling, or other mechanical methods are also acceptable means of sample removal. If flame cutting must be used, a 2 inch to 3 inch border of "waste" material around the area of interest should also be removed. Cooling of the area of interest with a stream of water or wet rags will also reduce the adverse effects of flame-cutting.

Samples or structural elements removed from structures retrieved from the sea can be further evaluated through the use of standard metallurgical and analytical laboratory techniques appropriate for the type of corrosion attack present on the sample.

C. Analysis of Failed Structures

A multidisciplinary effort is almost always required in order to determine the cause of failure of a structure. In structural failures there is seldom a single cause of failure. In some cases, however, corrosion is the primary cause of failure. Flooding of an instrument package resulting from failure of an O-ring seal due to crevice corrosion is an example of a common failure where corrosion is the primary cause. The situation is usually less straightforward. A cooperative effort, including evaluation of mechanical, structural, electronic, oceanographic, meteorological, metallurgical, and corrosion factors and their interactions, is required to determine the causes of failures in such a manner that remedial action may be identified.

VI. Conclusions, Recommendations, and Acknowledgments

A. Conclusions

The corrosion of metallic materials in the sea is a complex and often poorly understood phenomenon. Considerable engineering judgment must be exercised in the selection and utilization of materials for marine use in order to minimize the impact of corrosion. For deep sea application the requirements for the performance of materials are often severe, the knowledge of their performance at depth is often limited, and the consequences of failure are often catastrophic. Even though man has brought his technology to the sea for over 10,000 years, he is just beginning to understand how to make his machines survive in this hostile environment.

B. Recommendations

Although there have been several attempts to catalog information on the performance of materials in the marine environment, they have met with only limited success. This is not due to lack of effort but due to lack of continuity of effort. A continuing effort to both collect existing data from the wide variety of worldwide sources and to establish a "clearinghouse" for the collection and dissemination of marine corrosion data is required and should be pursued.

The last major effort to evaluate the behavior of materials in deep ocean environments was completed in 1970. Due to both the need for additional data on the materials previously tested in the deep and the need to evaluate the deep ocean performance of newly developed materials, particularly the "marine grade" stainless steels, deep ocean testing of materials should be resumed.

Corrosion engineers should endeavor to integrate the use of corrosion technology throughout the design of marine structures and equipment at an early stage so that the design and materials can interact synergistically and result in efficient, reliable, and cost effective designs.

Acknowledgments

The author wishes to acknowledge the efforts of all of his colleagues who have "dipped bits of metal into the sea" and who have written down and published the results. Without their efforts, each new marine design would indeed be an adventure.

The author wishes to acknowledge the efforts of two persons in particular: Mr Fred Reinhart who taught me what I know about corrosion and how to find out what I don't know and Frank LaQue who taught me how to ask good questions.

References

1. Compton, K. G. (1970). *Corrosion* **26**, 448–449.
2. Dexter, S. C., and Culbertson, C. (1980). *Materials Performance* **19** (9), 16.
3. LaQue, F. L. (1975). *Marine Corrosion, Causes and Prevention.* Wiley, New York.

4. Sverdrup, H. U., Johnson, M. W., and Fleming, R. H. (1942). *The Oceans.* Prentice-Hall, New York.
5. Gerchakov, S. M., and Sallman, B. (1977). *Biofouling and Effects of Organic Compounds and Microorganisms on Corrosion Processes.* Dept. of Microbiology, University of Miami, School of Medicine.
6. Chandler, K. A. (1985). *Marine and Offshore Corrosion.* Butterworths, London.
7. Reinhart, F. M. (1976). *Technical Report R-834.* U.S. Naval Civil Engineering Laboratory, Port Hueneme, CA.
8. Van Horn, K. R., editor (1967). *Aluminum.* Vol. I. American Society for Metals, Metals Park, OH, pp. 209–276.
9. Ailor, W. H. (1974). In: *ASTM STP-558.* American Society for Testing and Materials, Philadelphia, PA, pp. 117–134.
10. Leidheiser, H. Jr. (1971). *The Corrosion of Copper, Tin and their Alloys.* Wiley, New York.
11. Reinhart, F. M. (1968). *Technical Note N-961.* U.S. Naval Civil Engineering Laboratory, Port Hueneme, CA.
12. International Nickel Company (1972). *Nickel Topics* **25** (4), 3–4.
13. Sedricks, A. J. (1979). *Corrosion of Stainless Steels.* Wiley, New York.
14. Money, K. L., and Kirk, W. W. (1978). *Materials Performance* **17** (7), 28.
15. Reinhart, F. M. (1971). *Technical Note 1172.* U.S. Naval Civil Engineering Laboratory, Port Hueneme, CA.
16. Moeller, G. E. (1977). *J. Soc. Petroleum Engr.* **17** (2), 101–110.
17. Morrow, H. editor (1980). *Proc. Symposium on Advanced Stainless Steels for Seawater Applications, Piacenza, Italy.* pp. 11–93.
18. Armco Steel Corp. (1971). *Product Data, Armco A-286*, Baltimore, MD.
19. Boyd, W. K., and Fink, F. W. (1978). *MCIC Report 78–37.* Metals and Ceramics Information Center, Colombus, OH.
20. Brown, B. F. (1977). *NBS Monograph 156.* National Bureau of Standards, Washington, DC.
21. Apblett, W. R., editor (1982). *Second Symposium on Shell and Tube Heat Exchangers.* American Society for Metals, Metals Park, OH.
22. Covington, L. C. (1979). *Corrosion* **35**, 378–382.
23. Morton, A. G. S. (1986). U.S. Naval Ship Research and Development Center, Annapolis, MD, personal conversation.
24. Reinhart, F. M. (1967a). *Technical Note N915.* U.S. Naval Civil Engineering Laboratory, Port Hueneme, CA.
25. Cabot Corporation (1981). *Hastelloy Alloy C-276.* Kokomo, IN.
26. Cabot Corporation (1983). *Haynes Wrought Wear Resistant Alloys.* Kokomo, IN.
27. Latrobe Steel Corporation (1981). *MP-35n Alloy.* Latrobe, PA.
28. Lennox, T. J., Jr. (1967). *NRL Report 1725.* Naval Research Laboratory, Washington, DC.
29. Reinhart, F. M. (1967b). *Technical Note N-900.* U.S. Naval Civil Engineering Laboratory, Port Hueneme, CA.
30. Waldron, L. J., Forgeson, B. W., Peterson, M. H., Brown, B. F. (1962). *NRL Memorandum Report 1383.* Naval Research Laboratory, Washington, DC.
31. Jenkins, J. F. (1973a). *Technical Note N-1284.* U.S. Naval Civil Engineering Laboratory, Port Hueneme, CA.
32. Gross, J. H. (1967). *Proc. Materials—Key to Effective Use of the Sea, New York 1967*, pp. 236–264.
33. Jenkins, J. F. (1976). *Technical Note N-1424.* U.S. Naval Civil Engineering Laboratory, Port Hueneme, CA.
34. Reinhart, F. M. (1969). *Technical Note N-1008.* U.S. Naval Civil Engineering Laboratory, Port Hueneme, CA.

35. Jenkins, J. F. (1973b). *Technical Note N-1267*. U.S. Naval Civil Engineering Laboratory, Port Hueneme, CA.
36. Lindberg, R. I. (1969). In: *ASTM STP-455*. American Society for Testing and Materials, Philadelphia, PA, pp. 88–92.
37. Larsen-Basse, J. (1984). *Materials Performance* **23**, 16–21.
38. Wheatfall, W. L. (1967). *MEL R&D Phase Report 429/66*. U.S. Navy Marine Engineering Laboratory, Annapolis, MD.
39. Reinhart, F. M. (1967c). *Technical Note N-921*. U.S. Naval Civil Engineering Laboratory, Port Hueneme, CA.
40. National Association of Corrosion Engineers (1973). *RP-01-73*. Houston TX.

7

Concrete Marine Structures

BEN C. GERWICK, JR.

Professor of Civil Engineering
University of California
Berkeley, California

ISBN 0-12-341828-3

I. Introduction

Pozzolanic concretes were used by the Romans in their construction of bridge piers across the mouths of rivers. Portland cement was invented by Joseph Aspin for use in lighthouse construction on the English coasts. The first use of reinforced concrete was in France by Lambot: It was a boat designed in concrete for durability and is still in service a century later. Thus, we see that the history of concrete is intimately bound up with marine structures. This trend has continued.

Today, prestressed concrete piles are utilized throughout the world for harbor structures. Reinforced concrete piers support our largest overwater bridges. Seventeen massive concrete platforms support offshore oil production in the North Sea, braving storm waves up to 30 meters in height (Figure 1), while in the Arctic Ocean prestressed lightweight concrete platforms support exploratory drilling, resisting the high pressures and impact of moving ice. Prestressed concrete sheet piles protect marinas whose floats are made of lightweight concrete. Floating concrete structures serve as offshore terminals for cryogenic storage, as floating container piers, and as the hull of floating process plants. Concrete seawalls protect our coasts. Concrete outfalls discharge effluent from waste water treatment plants into the ocean (Figure 2), and concrete intakes bring in cooling water to cool our power and industrial plants. While concrete must share with steel its role as the preeminent structural material for the marine environment, it has special properties that are especially well-suited to this service. It can also be effectively utilized in many cases in conjunction with structural steel, in hybrid designs, making best use of the properties of each material.

The positive properties of properly designed and constructed reinforced and prestressed concrete are:

Durability
High compressive strength

High fatigue endurance
Rigidity
Readily molded to any desired form
Economy

The negative properties, which can often be successfully overcome in practical use, but which require proper technology in design and construction are:

Early cracking under tensile forces
Time-dependent dimensional changes
Relatively heavy weight
Lack of durability if improperly designed and constructed

Recent years have seen revolutionary advances in the research and development of advanced concrete technology. Through such means as electron microscopy, we now are learning about the role of microcracks and the interface between the cement paste matrix and the aggregate. Chemical admixtures enable us to control the behavior of concrete so that we can now design special recipes for mixes to meet our varied needs. It is now practicable to produce concrete twice as strong as that available five years ago.

Fig. 1. The Ninian Central offshore drilling platform incorporates precast concrete shell elements (photo by author).

Fig. 2. Concrete outfall sewer under construction in the Pacific Ocean (courtesy of Morrison-Knudsen Inc. and Ocean Outfall Constructors).

Concrete marine structures are a composite material in which the cement paste, the aggregate, the reinforcing steel, and the prestressing all work as one. In other constructions, steel, polymer, or glass fibers are being utilized to enhance the tensile strength.

In the past two decades we have seen a remarkable illustration of the "pull-push," "carrot-stick" phenomenon, in which the exciting new needs for concrete structures in the marine environment have interacted with advanced technological and research developments. This process is continuing at an accelerated pace.

II. Structural Elements

Within the overall structural configurations, certain structural types of reinforced and prestressed concrete stand out as of special importance to marine structures.

(a) Shell elements: walls and slabs, singly or doubly-curved, designed to provide resistance to external loads in membrane compression as well as in

Fig. 3. Precast concrete bells for San Mateo-Hayward, California, bridge piers.

flexure (Figure 3). While technically still behaving as "thin shells," they often may have thicknesses of 500 to even 1000 mm or more.

Such shells or walls are especially well-suited to resist external hydrostatic loads. If configured properly, these loads will primarily be resisted in membrane compression, thus utilizing the full concrete section in its most effective mode. Shell walls can also be designed to resist internal pressures by the use of prestressing.

(b) Prestressed concrete beams, girders, and slabs configured in their traditional land-based forms but designed to act in monolithic action with their supporting structure.

(c) Prestressed plates, able to resist both positive and negative moments. They may also be designed to resist cyclic membrane tension-compression. A notable example is that of a large barge, whose hull bottom and deck are experiencing cyclic reversing forces under wave action.

(d) Prestressed and/or reinforced concrete walls designed to resist membrane shear, typical of that experienced by the bulkheads and sidewalls of floating structures or of the vertical diaphragms (frame) in a fixed offshore structure. Prestressing not only resists the principle tensile stresses but changes their direction of action so as to enable efficient participation by the passive reinforcing steel.

(e) Axially loaded members such as piles, columns, shafts, and struts. Care must be taken to provide proper confining reinforcement to resist the lateral tension forces developed as a result of Poisson's effect.

(f) Intersections and joints which are subject to rotation, especially cyclic rotation, as well as direct forces.

All of these above may be combined in a typical marine structure such as a gravity-base platform.

(g) Pipes, such as intakes and outfalls, where weight is an asset in providing stability, where circumferential rigidity is desirable, and where durability is required.

(h) Blocks and shapes for use as armor for revetments and breakwaters, where weight and durability are needed.

(i) Mass underwater concrete, to serve as footing blocks for bridge piers, and as the base slab for graving docks and pumping plants.

III. Materials

A. Concrete

While, as with structural uses on land, the concrete is basically made up of cement, water, sand, and either gravel or crushed rock, the mix proportions and the additives employed for concrete marine structures are rather specialized in order to meet the twin requirements of structural performance and durability.

1. Cement

To resist sulfate attack, the tricalcium aluminate (C3A) is usually limited to 8% to 10% (unless pozzolan is added). Conversely, to resist chloride penetration, with its potential attack on the reinforcement, a certain minimum of C3A is desirable, usually 5% to 6%. ASTM Type II and many Type I cements meet these conflicting requirements quite well. Sulfate-resisting cement, Type V, is not desirable for reinforced concrete in the marine environment because of its higher permeability to chloride penetration. However, Type V cement can give enhanced durability for unreinforced marine concrete in waters having high sulfate content. Alkalis in the cement (sodium and potassium oxides) must also be limited in order to prevent reaction with the aggregates.

Blast furnace slag can be used in conjunction with Portland cement to provide the desired cementitious properties. When the blast furnace slag is coarse-ground, e.g., 3700 cm^2/gm maximum, then it can be effectively used in proportions of slag to cement of 70-30 or even 85-15. For most offshore uses, excessively finely ground blast furnace slag to which gypsum has been added is undesirable because it develops excessive heat of hydration.

2. Pozzolans

These are available either as natural powders from volcanoes or, more commonly, as processed fly ash from coal-burning power plants. Consisting of finely-divided silica, they chemically combine with the calcium hydroxide from the cement hydration to increase long-term strength and durability. More significantly, they decrease the permeability of the concrete by a factor of 10^3. This improves the durability substantially.

Pozzolan additions for reinforced concrete should not exceed 20% to 30% by weight of cement, in order to leave enough uncombined gypsum from the cement to passivate the reinforcing steel. For unreinforced concrete, such as tremie concrete footings or breakwater armor units, up to 50% replacement can be employed.

Pozzolan is less costly than cement and generally increases workability.

Because it reacts more slowly than cement, it reduces the peak temperature. For the same reasons, it requires that adequate moisture be available during an extended period of hydration, thus often requiring special curing procedures. Of course, such moisture is often provided automatically in marine structures.

3. A Special Pozzolan is Condensed Silica Fume

This material is 100 times smaller than a cement grain. It not only combines chemically but also mechanically strengthens the matrix. Its water demand eliminates bleed. It increases cohesiveness of the mix and prevents segregation, but is somewhat stiffer to place.

The use of 6%-10% silica fume, by weight of cement, leads to dramatic increases in concrete strength, from 50%-100%. Structural lightweight concrete strengths of 7000 to 9000 psi are obtainable in standard practice; for normal sand-gravel or sand-crushed rock aggregates, 12,000 to 14,000 psi compressive strengths are obtainable. These refer to standard cylinder tests at 28 days age. Tensile strengths are also increased proportionally to the square root of the compressive strength. Impermeability and hence durability are increased by the use of silica fume. Tensile and shear strengths also increase, but only as the square root of the compressive strength, thus following normal concrete laws.

Silica fume, because of its larger surface area, requires the use of a high range water reducer (superplasticizer) to adequately reduce the water-cementitious ratio. Alternatively, double dosages of conventional water-reducing agents can be used.

4. Both Normal Weight and Lightweight Aggregates Are Employed in Marine Structures

Normal weight coarse aggregates may be gravel or crushed rock, with the latter producing slightly higher strength in rich mixes, due to its better surface characteristics for bonding.

The proportion of sand to coarse aggregates in concrete for marine structures usually differs from that for similar practice on land. For reasons of durability and workability, a high content of cementitious materials is usually employed. Hence the design strength can usually be achieved with a sand-coarse aggregate ratio of 40% to 50%. Such a proportion produces a very workable concrete, especially when relatively high slumps are generated by the use of superplaticizers.

Natural beach and desert aggregates may require washing with fresh or brackish water to remove chlorides and sulphates. These may have been deposited by direct action, e.g., waves, by salt fogs, or by evaporation from ground water.

Aggregates must not be reactive with the cement. Tests should be run on both the coarse aggregate and the sand, the latter being especially important because of the high specific surface. Opaline and flint inclusions in limestone are a common case of reactive aggregates. Flint and chert aggregates are also suspect. Otherwise destructive expansion and pop-outs may result.

Aggregates for use in seawater must be sound and resistant to sulfate attack. Desert aggregates, which have not been subjected to weathering in nature, may undergo excessive expansion when immersed in seawater.

Structural lightweight coarse aggregate is usually combined with natural sand. Lightweight coarse aggregate qualities vary widely: Water absorptions may vary from 5% to 20% and attainable strengths by a factor of two. Hence it is is important to specify the quality of the coarse aggregate. The highest qualities are obtained by sealed surface aggregates subjected to longer burning. For applications subject to freezing, an absorption less than 6% in 24 hours is desirable: This may be attained either in the burning process or by application of special coatings.

5. Chemical Admixtures are Added to Give Special Properties to the Concrete Mix

Water-reducing agents reduce the required water content of the mix, thus reducing the ratio of the water to the cementitious materials (W/C ratio) while maintaining or increasing slump and workability.

High range water reducing agents ("superplasticizers") are extremely effective in reducing the W/C ratio while maintaining workability, achieving otherwise unattainable values such as 0.35, 0.30, and even 0.28. However, concrete containing superplasticizers can suddenly suffer a severe reduction in slump about one and one-half hours after being introduced into the mix, resulting in a drastic loss of workability. This can be delayed by adding retarding admixtures, so many proprietary compounds incorporate these already intermixed.

Retarding admixtures retard the time of set. Their use is essential in hot weather and with mass concrete. Many water reducing admixtures are also retarders.

Air entrainment admixtures create tiny bubbles, the size of a cement grain, throughout the mix. They improve workability. Most importantly, they give resistance to freeze-thaw attack in service. Their efficacy for this latter depends on the spacing of the individual bubbles and their specific surface. These can only be measured by microscopic examination of hardened concrete. Although a somewhat inconvenient test, its use is growing because of the increasing use of concrete in marine structures in northern environments.

Typical air entrainment requirements are 5%–7% by volume, as measured in the fresh concrete, with a spacing factor of 0.20 mm and a surface area of air voids of 20 mm^2/mm^3. Plasticizing admixtures reduce the internal shear strength, thus promoting internal flow.

6. Water

Fresh water is required for reinforced and prestressed concrete in order to ensure the formation of a thin protective skin of ferric hydroxide on the surface of the reinforcing steel, passivating it against electrochemical corrosion. Chloride content of water is therefore usually limited to 500–600 ppm.

Seawater may be used for unreinforced concrete such as breakwater armor units and for special cases where corrosion is not a potential problem. Its use will give slightly higher strengths. To prevent flash set, a heavy dose of retarder should be added to the mix.

B. *Reinforcement*

1. Reinforcing Steel Bars

The majority of structural concrete in concrete marine structures is conventionally reinforced with deformed bars. Therefore, conventional practice is largely followed, with special considerations addressed below:

(*a*) *Bundled Bars*

The demands for concentrated resisting members leads to the need for large steel ratios in both compression and tension. Alternating loads, reversing from compression to tension on the same face, also lead to much higher steel ratios than normally encountered in practice. Therefore bundled bars are commonly employed. Bar sizes up to 32 mm diameter (No. 11) are bundled in groups of two, three, or four. Splices must be staggered.

(*b*) *Low Temperature Steel*

Reinforcing bars that are subjected to extreme low temperatures may be subject to brittle fracture under suddenly applied impact loads. While being embedded in concrete serves to minimize this problem, critical bars subjected to Arctic environments should be selected for ductile low temperature behavior.

(*c*) *Weldable and Ductile Steel Bars*

Some zones of typical marine structures are subject to damage, as from accidental impact or abrasion. To facilitate repairs, weldable reinforcing steel should be used in those zones.

Where field bends of bars are required, the steel should be selected for ductility. Non-ductile steel may develop microscopic fractures on the outer surfaces, which may then be highly susceptible to corrosion.

(*d*) *Corrosion Resistance*

The major problem with reinforcing steel in marine structures is its potential corrosion. Chlorides from the seawater penetrate the concrete cover to the surface of the steel. There they react with the ferric hydroxide passivation to form a soluble compound, which leaves the carbon steel bar unprotected. Oxygen, primarily in the above water zone, completes the circuit and ferrous ions go into solution at the anode. Many steps are available to prevent or minimize this process. The ones to be described here are coatings on the steel bars.

Epoxy coatings can be electrostatically deposited over the steel surface. The bars must thenceforth be handled with care to prevent abrading the surface. Fiber slings should be used in lifting. Abraded areas can be touched up in the field. The epoxy coating not only protects the steel from the electrolyte (seawater) but is impervious to the oxygen that is necessary to support the cathodic reaction.

Galvanizing (zinc coating) is another means to prevent corrosion. Even if the zinc coating is scratched or abraded, it will act as a sacrificial anode and protect the bare area. However, pure zinc will react with the cement to give off hydrogen gas bubbles. This can be prevented by washing the zinc after galvanizing with a chromate solution.

Corrorion-resisting steel bars containing nickel are under development in Japan. Stainless steel bars have also been used on occasion in very critical areas. Extreme care must be taken that they are not electrically connected with ordinary steel as, depending on the alloy, they may cause accelerated corrosion of the carbon steel bars or embedments. Dissimilar metals and stray currents are other causes of corrosion.

It is now practicable to consider use of cathodic protection to prevent corrosion of reinforcing steel in concrete structures. For concrete which is permanently underwater, sacrificial anodes can be used. The reinforcing bars must, of course, all be electrically connected so that the anode can be effective. Sacrificial anodes have the advantage that they cannot cause too great a change in polarity. Impressed current improperly controlled can cause hydrogen liberation and possibly hydrogen embrittlement.

In the splash zone, induced current cathodic protection is currently under development with a growing number of field applications. These require coating of the concrete surface with a conductive coat and provision of closely-spaced mesh or plates since the concrete above water, being partially dry, has a high resistivity. Impressed current systems require careful adjustment and monitoring.

The methods described above, from epoxy or zinc coatings to cathodic protection, are aimed at protecting the steel directly. Other methods which inhibit corrosion, acting to restrict chloride penetration, oxygen permeation, and to enhance resistivity of the concrete are treated in Section V, Durability.

2. Prestressing Systems

(a) *Strand*

The majority of prestressed concrete for marine structure is prestressed with seven-wire strands of high strength steel wire. Current industry standards are centered on steel having an ultimate strength of 270,000 psi. When such wires are prestressed to perhaps 60% of ultimate, they are subject to stress relaxation (long-term creep), which reduces the effective stress. Stabilized strand is now available that significantly reduces these long-term losses.

Higher strength steels, e.g., with an ultimate strength of 300,000 psi, are being manufactured in limited quantities by special processes. Indented strand is available where development lengths must be limited, especially where cyclic loading can lead to a decrease in bond transfer near the ends of the strands.

Strand is normally furnished black (uncoated). Galvanized strand is available on special order. Epoxy coated strand is available for use where cover is limited or exposure is severe, as in the decks of structures. Strand is also manufactured with polyethylene sheathing, the interstices between the wires and the casing being filled with an inert grease. Obviously, there can be no bond with concrete in the sheathed zones.

(b) *Bars*

High strength alloy bars are used for prestressing, especially for connecting precast elements and for attachment of steel embedments or segments to concrete structures. The available length of bars is generally limited to about 20 meters, so that mechanical splices (couplers) are used whenever longer lengths are required.

When making splices with a threaded coupler, care must be taken to ensure both bars are adequately engaged; because of friction, there is a tendency for the second bar to screw the coupler off the first bar. Couplers must be free to move under prestress.

Most prestressing systems employing bars have anchorages which are set by threaded nuts and, hence, have minimum slip, which makes them desirable for short tendons and the attachment of embedments.

(c) *Ducts*

For post-tensioning, ducts are required to form the hole. Spirally-wound, semi-rigid ducting is most commonly employed, although smooth pipe may also be used. The duct must be grout-tight. Black steel is most commonly employed, although both epoxy-coated and galvanized ducting have been employed.

Duct sheathing is light gage, hence subject to damage in handling or from vibration of concrete. It is also subject to corrosion in storage. For marine structures, the ducting should normally not be less than 0.8 mm thick. Where due to curvature requirements, thinner ducting is used, it should be placed and held during concreting with an internal mandrel.

Splices in ducts are especially critical. The ends should be sawn, not burnt, so as to be smooth and free from burrs. Sleeves should be used and, when placed, should be taped at both ends. Ducts should be delivered and installed with a plastic cap at the exposed end or ends, so as to prevent the entry of foreign material and water. Plastic ducts are generally unsatisfactory for use in marine structures due to cracking under stress, grooving as tendons are stressed, and lack of bond.

(d)

Grout for prestressing systems should have a maximum water-cement ratio of 0.45. Neat cement-water grout is used, to which an expanding admixture should be added in order to compensate for shrinkage. For tendons having substantial vertical projection, thixotropic admixtures may be added to prevent bleed water accumulation.

3. Fibers and Mesh

The use of layers of fine mesh reinforcement (ferrocement), or fibers, has been extensively applied in order to increase the apparent flexural and tension strength of concrete, especially in thin-walled structures such as boat hulls.

In order to overcome the problems of shear in such thin walls, deformed mesh, three-dimensional, has been developed. Fibers are randomly dispersed in the concrete and, hence, are inherently three-dimensional in behavior. They serve to bridge microcracks as they develop.

Most fibers have been made of steel, with either heads or deformed ends in order to develop bond. Plastic and glass fibers are also employed, the latter being made of alkali-resistant glass. Because of their low modulus, they have limited structural application in the marine field. Kevlar fibers have a moderately high modulus and hence are being used increasingly.

C. Composite Steel-concrete Construction

Steel structures in the marine environment offer advantages in structural efficiency and high strength-to-weight ratios. However, in most practical structural configurations, individual elements are subject to local failure through buckling in one of several modes.

Concrete structural configurations develop membrane (shell) action, which distributes local loadings and mobilizes a larger portion of the structure to resist local loads.

By combining steel plates with reinforced concrete, the advantages of both materials may often be combined. Either a single steel plate, backed by reinforced concrete, or a steel-concrete sandwich element emerges as an attractive structural system, especially where high compression is combined with bending or where concentrated local loads, e.g., ice impact, must be resisted by shell/slab members.

The technical challenge here, as in other steel-concrete composite design, is to transfer shear so that both materials function together at compatible strains. A number of systems have been evolved, including headed studs and connecting diaphragm plates. In one system the marine structure is first constructed in a shipyard using double steel plate construction. Headed stud connectors, overlapping in the intervening space, have been fused by friction welding. The structure is launched. At a deep water site, high strength concrete is placed.

This is an area of current intense research and development for both land and marine structures. It is considered to offer considerable promise for extending structural capabilities, not only in regions dominated by ice but also for regions of the world that are not blessed with deep water harbors and for deep sea applications.

D. Ferrocement

Ferrocement has been widely utilized for the hulls of boats and pontoons because it enables the construction of extremely thin shells. Layers of mesh can be placed over doubly-curved forms, and the fine concrete, actually a sand-cement mortar, is applied either by hand as plaster or by shotcrete. In the latter case, extreme care has to be taken to avoid trapping rebound. Ferrocement hulls have proven remarkably durable. Thin ferrocement shells are quite flexible.

One weakness of ferrocement lies in its low shear resistance: Here the use of three-dimensional mesh and fibers offer possibilities. Additionally, ferrocement has turned out to be quite labor intensive.

The technical and economical constraints described above limit its application to boats and pontoons generally less than 20 meters in length. However, ferrocement may have application in special regions of larger structures, for example, in zones which must exhibit flexibility in service, such as the corners of cryogenic storage tanks.

IV. Design

A. Special Considerations for Marine Structures

The application of concrete to marine structures usually requires more complex and extensive considerations than those associated with typical land structures due to the many different stages and load conditions which the structure must undergo, the dominance of the environment, the character of the loads (cyclic, dynamic, and reversing), and the need to consider accidental overloads. Forces during construction, for example pile driving or lifting, may impose different and more serious stresses than the structural element will see in service.

B. Design Loads for Marine Structures

Design loads for marine structures must include, in addition to their dead load and functional (live) loads, the following, as applicable:

Hydrostatic loads
Wind, waves, currents, and tsunamis
Ice and icebergs
Snow and ice accumulations and internal (bursting) ice forces
Earthquakes and seaquakes
Accidental loads such as vessel collision, explosion or fire, dropped heavy objects, and loss of design underpressure.

The design must address the stages or phases through which the structure passes, including:

Construction
Transporation
Installation
Operation
Extreme event, e.g., design storm
Removal or retrieval

The design philosophy usually employed is based on limit states. These include the following:

(a) Ultimate limit state
(b) Serviceability limit state, including considerations of cracking, displacements and dynamic motions, and leakage
(c) Fatigue under repeated loads
(d) Progressive collapse limit state

The system toughness is evaluated under the progressive collapse limit state, which states that, under an accidental load or an overload above the factored design value, the failure of one element shall not lead to catastrophic collapse. Local damage and repairable damage may be permitted. These requirements lead to the provision of redundant load paths and incorporation of ductility, e.g., confinement, in structural concrete members.

C. Bond, Splices, and Development Length

Because much of the primary reinforcing steel is subject to cyclic, reversing, or impact loading, bond may degrade. Therefore, insofar as practicable, bars should be anchored in a compressive zone. Development lengths should be increased 60% to 100%, as compared to those specified in typical building codes for quasi-static loads.

Lap splices are extensively employed in most marine structures. They are preferably located in zones where bond development requirements are a minimum. Lap lengths should be increased, as noted earlier, and tied at both ends. With closely spaced bars, lap slices may create a screen, preventing the proper flow of concrete during placement and seriously reducing the available tensile capacity across the section, sometimes leading to laminar cracks. In such cases, consideration is given to mechanical splices. Mechanical or welded splices are also needed for large bars (larger than 32 mm, no. 11), and for those carrying heavy compression forces, as in columns and shafts.

D. Punching Shear

To resist high punching shear forces and transverse beam shear, stirrups are usually employed, especially in the thick shell and slab elements used in off-shore structures. Spirals or stirrups are also used to provide confinement of members subjected to high compression loads, both increasing the ultimate strain of the core concrete and preventing the compression bars from buckling.

To be most effective under extreme or repeated loads, the tails of stirrups should be turned in and anchored within the core in the same manner as required for earthquake design. Stirrups can normally develop only about 70% of their yield strength due to crushing of the concrete under the short radius bends.

Mechanically headed bars locked behind the in-plane reinforcing on both faces not only enable higher percentages to be achieved but can develop their full yield strength. One type of these bars, the T-bar, can be readily inserted through the primary steel, then turned 90° to lock.

E. *Cracking*

Cracking of reinforced concrete members under load has special significance for marine structures, especially where water in the cracks can exert a resistance to subsequent closing. Under repeated loading, hydraulic fracturing may occur. Under freezing conditions, ice jacking may force the cracks more widely open.

Although often cited as a primary cause of corrosion, structural cracks which close after the load is removed, and even permanent structural cracks of small width, are not normally a serious cause of corrosion. Current practice limits crack widths in permanently submerged concrete to 0.3 mm and in the splash zone to 0.15 mm.

Even though the face of a concrete structural element may not be subject to design tension under loads, cracks may originate due to thermal strains, shrinkage, Poisson's effect from prestress, and fatigue.

It is therefore good practice to provide a minimum amount of reinforcement on each face such that it will not be stretched beyond yield when the tensile zone of the concrete is cracked. In other words, the required minimum steel percentage is related to the tensile strength of the concrete, with consideration of tension stiffening effects.

Spacing of reinforcing bars should be kept as small as reasonably practicable so as to distribute any cracks and minimize their widths. Typically, a maximum spacing of 300 mm (12 inches) is used.

F. *Membrane Shear*

Many walls and diaphragms in marine structures are subject to membrane shear as they transfer lateral loads from waves, ship berthing, soil, or ice down to the foundation. These shear forces result in inclined principal stresses, at an angle to the conventional orthogonal pattern of reinforcing steel. Therefore, not only are the bars less efficient but crack widths may be doubled.

This leads to a requirement for increased steel quantities. One practicable solution is to provide vertical prestressing, which turns the resultant principal tensile stress to the horizontal, where conventional reinforcement can efficiently resist it.

G. Prestressed Concrete Piles

Prestressed concrete piling are extensively used in marine structures, ranging in sizes from 10 inches square to 3.5 meters (12 feet) in diameter. Specified practice for prestressed foundation piles on land provides minimal values of prestress and spiral in order to reduce costs. In the marine environment, other factors must be considered.

Handling, placing, and driving stresses require adequate prestress. Since soft soils are often encountered, tension rebound stresses may be excessive during driving, resulting in horizontal cracks that may subsequently lead to pile breakage as a result of repeated hammering at the crack. This final failure is a case of high-amplitude low-cycle fatigue.

Not only should steps be taken in driving to reduce tensile stresses, but adequate prestress should be provided to close any cracks which do open. This typically requires a pre-compression of 800 to 1200 psi.

Spiral reinforcement for prestressed piles is required to counter circumferential tensile stresses which arise from a variety of causes. During manufacture, thermal stresses during curing and the lateral stresses due to prestress may initiate vertical cracks. During driving these lateral stresses (Poisson's effect) are intensified. Hydraulic ram and water hammer pressures may develop in hollow core piles. In service, thermal and moisture gradients may cause vertical cracking.

Bursting forces may develop in hollow-core piles due to differential hydraulic pressures and freezing.

Thus careful consideration has to be given to the required quantity of spiral reinforcement for piles in the marine environment, especially for hollow-core or cylinder piles. This is usually two to three times that contained in the typical pile standards for onshore piling.

H. Multi-Axial Stresses

Structural elements of concrete marine structures are frequently subject to multi-axial stress conditions due to concurrent effects such as external hydrostatic pressure combined with lateral wave loading and vertical dead loads. When all three principle axes are under compressive stress, the ultimate capacity will be slightly increased. The ductility, that is, the strain and energy capacity of the member, will be substantially enhanced, sometimes by a factor of three or more.

Prestress creates an active state of compression in one direction. Well-anchored reinforcement, such as stirrups, can create a passive enhancement (confinement) due to Poisson's effect. These techniques can be utilized by the designer to meet the requirements for ultimate strength and avoidance of progressive collapse.

Conversely, care has to be exercised when one or more of the axes is subjected to tensile stresses, as this of course reduces the ultimate strength and ductility under compression in the other direction.

I. Fatigue

Marine structures are inherently subject to cyclic loading such as waves, which may even reverse the stress conditions in a member, alternating from tension to compression. Repeated cracking, especially when submerged, can lead to hydraulic propagation of the crack, progressive washing out of fines, and eventual fatigue failure. This fatigue may cause loss of bond at the interface between coarse aggregate and the cement-sand matrix. It also leads to an amplified stress range in the reinforcing or prestressing steel in the cracked zone. Failure is usually by rupture of the steel.

This phenomena is also demonstrated in the driving of prestressed concrete piles, where after cracking in tension by handling or rebound driving stresses, the subsequent continued high compression cycles leads to failure. Prestressing is therefore widely employed in marine structures in order to prevent cracking and eliminate or minimize the net tensile stress peaks.

Many tests have shown that prestressed concrete members in air cycles from moderately high compression (e.g., 50% f'_c) into slight tension (e.g., less than 50% f_{ct}) will have a very high endurance against fatigue failure. However, recent investigations indicate that for the same member submerged, the endurance is substantially reduced, presumably due to water in microcracks undergoing cyclic pore pressure increase, which in turn propagates the cracks. This phenomenon occurs with normal sand and gravel concrete. For structural lightweight concrete, there is little if any change from dry to submerged conditions, presumed again to be due to the chemical bond and thermal compatibility between paste and aggregate. It is believed that the addition of condensed silica fumes in the mix may help avert the phenomenon due to secondary crystallization in the microcracks, thus eliminating them. The combination of lightweight aggregate and silica fume has been shown to significantly improve fatigue resistance in both air and water.

V. Durability

A. General

The marine environment places extremely severe demands upon the materials and structures. One of the principle attributes of concrete for marine structures is its inherent durability and long-term maintenance-free performance in service. This durability is not automatically achieved with concrete but rather requires great care in design and execution. Durability should receive equal attention to structural response to loads.

Concrete marine structures, largely developed in moderate temperature climates such as Europe and North America, are being extrapolated to such hostile environments as the Arabian-Persian Gulf and the sub-Arctic (Figure 4). These extreme environments constitute a field laboratory, where the aggressive phenomena are accelerated, so that successful measures and defects can rapidly be determined.

Fig. 4. Precast concrete piers for Saudi Arabia–Bahrain Bridge utilize concrete mixes especially designed for durability in extremely harsh environment (courtesy of Ballast Nedam).

B. Special Problems of the Marine Environment

The special environmental problems in the marine environment that lead to degradation of service performance include a number that have been treated elsewhere. Under Section III, Materials, the means of preventing attack on the individual material components were described. These included:

(a) Selection of proper cement type
(b) Use of pozzolans
(c) Use of air entrainment admixtures in the freeze-thaw environment
(d) Limitation on chlorides in the mix
(e) Selection of durable aggregates with special reference to soundness and alkali-reactivity
(f) Protective coatings on reinforcing steel, including epoxy coating and galvanizing

Under Section IV, Design, additional marine-related phenomena were addressed along with recommendations for prevention of degradation:

(g) Fatigue under cyclic loading
(h) Cracking and crack propagation

C. Microcracks

Fundamental to any understanding of concrete durability is the role of microcracks and their influence on durability.

Both aggregates and cement paste are highly impermeable. Concrete, however, has perhaps 1000 times more permeability than either. This is due to the microcracks at the interface between the paste and the aggregates and reinforcing steel. These microcracks are formed by a number of actions:

(a) Bleed water collecting under the coarse aggregate
(b) Shrinkage of the paste
(c) Thermal incompatibility between paste and aggregate and between paste and reinforcing steel.

Most degrading phenomena, including fatigue, abrasion, corrosion, and freeze-thaw attack (to a degree) are related to this matter of microcracks.

D. Zones of Attack

Within the marine environment, several zones have been indentified which have widely varying exposure to attack. The most vulnerable zone is the splash zone, where alternate wetting and drying, thermal changes, salt concentration due to evaporation, and readily available oxygen combine. Here we see the most severe corrosion of reinforcement as well as freeze-thaw attack, ice abrasion, and local impact damage. The upper limits of the splash zone are not well defined because spray, fog, and moisture laden air can extend up to 20 or 30 meters (60 to 100 feet) above water level. Decks of marine structures, both fixed and floating, are especially critical. From below, the humidity stays high, and the beams and slabs are frequently splashed. From above, salt water alternately wets and evaporates, leaving salt crystals, while temperature gradients are extreme.

The submerged zone, permanently under water, is relatively benign by comparison. Temperature and moisture conditions are relatively constant, and oxygen supply in water is only about 2% of that in air.

E. Prevention of Corrosion of Reinforcing Steel

Corrosion of the reinforcing steel is caused by chlorides and oxygen in the sea water and air, which penetrate the concrete, leading to an electrochemical action, with rusting of the steel and spalling of the concrete. The most effective prevention measures involve the use of several related procedures in combination:

1. Provision of an Impermeable Concrete

(a) Low water-cement ratio.
(b) C3A content of cement in range of 8%–10%.
(c) Addition of pozzolan or silica fumes.
(d) Selection of aggregates with low thermal coefficients, e.g., limestone or lightweight aggregate as compared with silica aggregate.
(e) Avoidance of high concentrations of reinforcing steel.
(f) Application of coatings or sealants on the surface.

2. Protection of Reinforcing Steel

(a) Maintenance of high pH by adequate cement content, so ferric hydroxide coating will form.
(b) Limitation of chlorides in the mix water and on the aggregates.
(c) Coating the steel with epoxy or zinc.
(d) Cathodic protection (usually used only as a remedial measure after severe corrosion has occurred.)
(e) Limitation of and prompt repair of cracking, especially multiple

closely intertwined cracks such as those due to impact. Sealing of cracks by coatings or epoxy injection.

(f) Adequate cover of concrete over the reinforcement. In most marine structures a nominal cover of 50 mm (2″) is used.

Many practitioners in the past have relied on cover alone to provide protection against corrosion. The protection provided by cover, however, is a complex matter, depending not only on its dimension, i.e., the thickness of concrete over the steel, but its quality (e.g., impermeability), the temperature, and its relation to the size of the coarse aggregates and the reinforcing bar diameter.

To prevent the linking of microcracks, cover should be at least $\frac{1}{2}$ to 2 times the size of the largest coarse aggregate particles and $1\frac{1}{2}$ to 2 times the diameter of the reinforcing bar. Adequate paste is required at the surface, since this is typically much more impermeable than the concrete inside. High temperatures, e.g., warm seawater in the Arabian Gulf or Red Sea, etc. can greatly increase the permeability. Too great a cover can be self-defeating by permitting excessive cracking.

F. Ensuring Durability of Concrete

Concrete itself is subject to degradation due to a number of phenomena.

1. Freezing and Thawing

Freeze-thaw attack is especially serious for concrete that is saturated. Thus special attention needs to be directed to the tidal and splash zone. If the air temperature is below freezing, as the tide rises it will thaw the concrete, then as the water falls, freezing will occur. Within one season, more than 100-cycles can occur. Salt water seems to be more damaging than fresh water, possibly due to secondary salt crystallization upon freezing and evaporation.

To minimize and prevent freeze-thaw degradation, several steps are employed.

(a) Air entrainment is mandatory. These finely divided air bubbles allow internal volume changes without leading to microcrack propagation. Air contents of 5% to 7% in the fresh concrete and proper air-void parameters, as determined by microscopic examination, are essential. See Section III A.5.

(b) Make concrete as impermeable as possible to prevent saturation. See Section V.D.1.

(c) Minimize water content in aggregates. Water absorption should be less than 6% in 24 hours to avoid taking in water from the fresh mix. Aggregates should not be saturated prior to use. Lightweight aggregates or other aggregates with water absorption over 2% should not be pumped, as pumping forces water into the aggregates.

(d) Use external sealants, such as silanes, which prevent water intrusion while allowing water vapor to escape.

(e) Avoid or limit impermeable external coatings such as epoxies, which can trap and condense water vapor that migrates from the warm to the cold side.

(f) Where water is stored inside, above the external waterline—as, for example, in ballast tanks—provide an impermeable membrane, for example, epoxy or steel lining, to prevent saturation from inside out.

2. Sulphate Attack

Aggregates and concrete are both susceptible to expansive reactions from concentrated sulphates. The concrete must, of course, be partially or wholly saturated for the chemical attack to occur.

Aggresive sulphate attack, originally recognized during the construction of fresh water canals and dams in the western United States, has become a major concern in the Mideast on land. In both areas, tricalcium aluminate (C3A) in the cement was identified as the major culprit, although unsound aggregates are also a factor. Thus Type V cement is specified for such an exposure.

Fortunately, this is not a major problem with marine structures in seawater, where the chloride content inhibits sulphate attack to a large degree. Especially where the concrete is relatively impermeable, due to the high content of cementitious material typical of reinforced and prestressed concrete structures, sulphate attack is almost not a problem. Pozzolans will give additional protection against sulphate attack. Low C3A cements such as Type V cement are not desirable for use in marine structures in salt water because of their several-fold increase in permeability to chloride penetration.

For marine structures in fresh water, where high sulphate concentrations exist, attention must be given to prevention of attack. The use of Type V cement and the incorporation of pozzolans appears essential.

Unsound aggregates are often found in desert areas where there has been inadequate weathering to naturally select durable stones. These can usually be identified in a sodium sulphate accelerated test.

3. Alkali-Aggregate Reaction

Many aggregates exhibit an expansive reaction due to the alkali reaction in the presence of an electrolyte. This is usually characterized by conical "pop-outs" at the surface, progressively extending inward. Alkali contents of the cement are therefore limited in many specifications. Aggregates, especially those from new sources, should be tested for reactivity. The test is fast and low cost.

Reactive aggregates occur usually in the form of flints and cherts. The flint and opaline nodules in limestone and chalk are usually reactive. Sands, with their greater surface area and smaller individual volume, should be checked as well as the coarse aggregate.

4. Biological and Chemical Attack

Sulphur-reducing bacteria can attack concrete near the waterline by generating a weak sulphurous acid. One of the anaerobic bacteria that is guilty of this behavior has the descriptive nomenclature "*Theobacillus concretivorous*." Dense impermeable concrete is essentially immune to this attack.

Marine Borers, rock-eating molluscs, were until recently believed to not be a problem for high quality concrete. Recently, however, in the Arabian Gulf especially, aggressive borers have attacked the limestone aggregate and bored holes all the way to the reinforcing steel with their acids. The steps adopted for the Saudi Arabia–Bahrain bridges were to use igneous aggregates and external epoxy coating.

Oils and petroleum products generally do not pose a hazard. Crude oil has been stored for many years in concrete tanks on the North Sea platforms without degradation of the concrete. Some refined products containing fatty acids can cause attack. For these uses, provision of a steel or epoxy lining is indicated.

G. Prevention of Abrasion and Erosion

Abrasion can occur from moving ice, sand, gravel, and rocks. Until recently most attention was directed to the use of hard, abrasion resistant aggregates. More recent investigations have shown the overriding importance of the mechanical and chemical bond with the coarse aggregate.

Crushed rock is more resistant than smooth gravel. Structural lightweight aggregate concrete can be more abrasion-resistant than normal sand-rock concrete. The use of silica fume appears to enhance the abrasion resistance. Coatings, such as dense epoxy and dense polyurethane give excellent abrasion resistance. Care must be taken to avoid creating a barrier that will trap water vapor, which can then freeze.

H. Protection of Prestressing Systems

The tendons of prestressing systems in concrete marine structures are basically protected by the concrete cover, the same as with normal reinforcing steel. Pretensioned tendons are usually given 50 mm cover. When the member is cast horizontally, with the tendons also being horizontal, then, with a highly workable mix and good vibration, the interstices in strands will be filled with cement paste.

With post-tensioning tendons, grouted tendons are normally employed, the tendon being within a steel duct. Tendons are typically protected by water-soluble oil until grouted. The best practice today is not to attempt to wash off the water soluble oil by injecting water, because water then tends to become trapped in the duct. Instead, the duct is kept dry until after the tendon is stressed and anchored. Then grout is pumped in from one end and out the other until the color and consistency are the same as at entry. Particular care has to be taken to force grout out through the anchorages, instead of trapping oil and water there. To do this, special details are required for each post-tensioning system.

Vertical tendons and tendons with a substantial vertical projection require special methods to prevent sedimentation and trapping of bleed water at the top. Thixotropic admixtures added to the grout cause it to gel once pumping stops. Use of a standpipe at the top anchorage and a secondary grout injection there can complete the filling. Both steps are often necessary, especially for long vertical tendons. Otherwise the trapped water at the top may freeze at early ages or may leave a void at later stages in which corrosion may initiate.

Anchorages for post-tensioning systems of marine structures are perhaps the most difficult to protect. The preferred means is to use a recessed anchorage. For structures destined for temperate or tropical waters, the anchorage and the pocket should be painted with a bonding epoxy. Then the recess is filled with a fine (small aggregate) grout in such a way as to avoid shrinkage or bleed gaps, e.g., using a "window-box" technique. The concrete fill should have small reinforcing bars tying the fill to the structure. These are installed in the structural concrete and later bent into the pocket just before concreting. Finally, the exterior is coated with epoxy. In sub-Arctic exposures the epoxies do not appear suitable. Latex modified mortar may be used instead to fill the pocket.

I. Interaction of Factors

Interaction of many factors is usually involved when concrete durability is degraded. It is important to recognize that many of the aggressive factors and phenomena often work synergistically to aggravate the disruption. For example, freeze-thaw attack may lead to extensive loss of cover and develop increased permeation of water and chlorides, leading to corrosion. Conversely, corrosion of the reinforcing may cause cracks and spalling, exposing more surfaces to freeze-thaw attack.

While it is often not practicable to apply all the measures for durability previously listed, the designer and constructor must give thorough consideration to the special environment in which their particular marine structure will serve and provide appropriate protection measures.

VI. Construction

Special construction methods become necessary for marine structures.

A. Tolerances

Because many marine structures must float at some stage in their life, geometric and weight tolerances become extremely important (Figure 5). For

Fig. 5. Shafts of Statfjord B Condeep offshore platform require extreme control of dimensional and weight tolerances (photo by author).

critical structures, continued as-built surveys are required. Weight control requires accounting for concrete unit weight, water absorption, ballast, actual reinforcing steel quantities, embedment plates and attachments, and temporary gear such as cranes, scaffolding, etc. The external geometry determines the displacement while floating and hence the freeboard, draft, and stability.

B. Underwater Concreting

1. Tremie Method

The most widely used method for placing concrete under water is the tremie method in which concrete flows down a tremie pipe under a gravity head. The tremie pipe is usually 12 times the diameter of the coarse aggregate (Figure 6).

Typical tremie concrete mixes are:

Coarse aggregate: gravel, 3/4″ maximum
Sand/coarse aggregate ratio, 45–50%
Cement 600 lbs/cy
Pozzolan 100 lbs/cy
Water-cement ratio, 0.42 maximum
Water-reducing admixture
Retarding admixture
Air entrainment, 6% or use of a plasticizing admixture

Note that some proprietary admixtures incorporate several functions. Superplasticizers are not recommended due to problems of control, slump loss, etc.

For starting a placement at moderate depths, use of a sealed plate and an empty tube is best since use of the "go-devil" or "ball" method of starting has caused numerous problems.

Tremie concrete can be satisfactorily placed around reinforcing steel, provided sufficient clear openings are left for concrete flow. To achieve this, bars are often bundled.

2. Grout-Intruded Aggregate

Grout-intruded aggregate concrete has been used very successfully for confined placements. All coarse aggregate, with no fines, is placed in the confined space. Grout is then injected upward through preinstalled tubes. The critical aspect of this method is to ensure that the passageways between

Fig. 6. Placing tremie concrete to form the underwater base for the main pier of Columbia River Bridge, Oregon (courtesy of Riedel International and State of Oregon Department of Transportation).

the coarse aggregate particles have not been clogged with aggregate fines, silt, or marine growth. Special admixtures to reduce surface tension are used to promote flow.

3. Thixotropic Admixtures

Thixotropic admixtures are now available which minimize segregation and washout of the cement when concrete is deposited through water. Spectacular results have been achieved in the laboratory. Experience in practice indicates that these have definite applicability for limited and confined placements, e.g., repairs or filling the interstices between rock. However, they should not be discharged through any depth of water, but rather at the placement level.

C. *Thick-Walled Structures and Mass Concrete*

Concrete marine structures frequently require thick-walled members or even mass concrete. Their strength and durability requirements often lead to high cement contents. Heat of hydration and the thermal distortions which occur upon subsequent cooling can lead to cracking. To prevent such cracking, a number of steps are employed:

(a) Use low-heat cement.
(b) Use coarse ground cement.
(c) Substitute pozzolan for a portion of the cement.
(d) Substitute coarse-ground blast furnace slag for most of the cement.
(e) Pre-cool the mix by using ice as mixing water, by cooling the aggregates, by protecting them from the sun and/or by evaporation, or by injecting liquid nitrogen into the mix.
(f) Insulate the forms to prevent severe gradients at the surface.
(g) Apply a membrane sealing compound as soon as forms are removed.
(h) Blanket the concrete as soon as forms are removed.
(i) Provide enough face reinforcement near the surface to ensure that the cracks will close when the temperatures have equalized.

VII. Structural Application

A. Characteristics

Reinforced and prestressed concrete members are being increasingly employed because of their favorable performance characteristics, especially:

High compression strength
Adequate shear and tensile strength
Rigidity
Durability in the marine environment
Endurance against fatigue
Moldability in any desired configuration, including forms of double curvatures
Ductility under overload
Impact and abrasion resistance
Economy

It will be recognized from the preceding sections that most of these are attainable only with attention to proper design and construction. Similarly, the negative aspects referred to in Section 1—high weight, cracking, and loss of durability—for example, must be dealt with properly on the basis of sound engineering and construction practice.

Representative examples of some of the more notable applications are given in the following sections with general remarks as to the reason for selection of concrete and the particular problem areas that had to be countered.

B. Prestressed Concrete Piling

Prestressed concrete piling represents one of the most notable applications. This use has had an almost explosive growth over the last 30 years, providing support for overwater bridges, piers and quays, and offshore terminals, as well as mooring for boats at marinas. Prestressed concrete fender piles are being utilized at an increasing number of naval and commercial piers.

The primary attractions have been their configuration, which can be tailored to enhance soil-pile interaction; their durability; their ability to be transported and driven to the required penetration; and their high structural capacity.

Problem areas center around two forms of cracking. During driving, especially in soft and stratified soils, rebound tensile stresses may cause horizontal cracking, which later leads to fatigue failure under continued hammering. This problem is successfully overcome by providing adequate prestress, proper cushioning at the head of the pile, and proper selection and control of the hammer characteristics.

The second form of cracking has been vertical cracking, primarily in service, due usually to a combination of thermal and hydraulic stresses. This problem is overcome by the provision of adequate spiral reinforcement, adequate curing, and the adoption of appropriate pile configurations.

C. Intake and Discharge Pipes, Outfalls

Precast reinforced concrete pipe has been extensively employed, in diameters up to 20 feet (6 m), for ocean outfalls and intakes. The primary reasons for its choice are its durability; its rigidity against deformation; its weight, giving it stability; and its economy.

Problem areas, which have been largely overcome, have primarily been associated with its joints—the need to provide both leak tightness and flexibility. A number of systems have been developed using the bell-spigot arrangement together with neoprene O-rings. Longitudinal bending strength is limited, as compared to steel pipes, and hence bedding rock must be properly placed.

D. Overwater Bridge Piers

Many such piers utilize mass concrete as well as prestressed concrete piling. The mass concrete may be the tremie concrete seal or the footing block and pedestal. Above the latter, a solid or thick-walled hollow shaft may extend up above the water level. Concrete has been chosen because of its ability to be placed in the mass configuration and because of its durability and low cost.

The principal current problems appear to be those caused by heat of hydration, which raises the temperature of the center of the mass leading to severe thermal gradients towards the outside water and cracking.

These problems are overcome by selection of constituents of the concrete mix that will minimize the heat of hydration, by pre-cooling the mix, and by insulation of the surface. Provision of reinforcing steel on all faces, while providing only minor resistance to cracking, can distribute the cracking. When the temperatures finally equalize, the steel will tend to pull the crack back together.

Many problems have also occurred with underwater concrete due to improper mix design, improper or non-use of proper admixtures, and improper placement procedures, especially when starting the pour.

E. Wharves and Piers

The structural concrete is either cast-in-place, precast, or a combination of both. Selected for its durability, trafficability, rigidity, and economy, it also is extremely efficient structurally in distributing lateral and vertical loads.

F. Intake Chambers and Seawater Pumping Plants

Concrete is almost universally selected because of its basic structural suitability, economy, and durability. Problems have arisen in some extremely severe environments due to corrosion of the reinforcement steel. The environmental attacks are aggravated by the flow of aerated, hence oxygenated, seawater over the concrete.

G. Offshore Platforms

One of the more spectacular applications has been the monstrous drilling, production, and storage platforms constructed for the offshore petroleum industry, primarily in the North Sea. Many of these structures require 100,000 cubic meters or more of highly-prestressed and heavily reinforced concrete, molded into complex forms, often of double curvature. These structures must withstand the dynamic cyclic forces imposed by the waves.

Concrete has been selected because of its high fatigue endurance, economy, and availability. It has proven to be remarkably durable and free from maintenance. Its restrictions are those of high weight, which increase its draft at critical stages in construction, and high requirements for reinforcement and prestressing, which of course translate into cost. Advances in structural lightweight aggregate production and in very high strength concrete give the promise of reducing the weight and draft.

H. Arctic Structures

Based on successful applications to offshore platforms, reinforced and prestressed concrete structure have been deployed in the Arctic and sub-Arctic seas, where they are subject to sea ice impact, concentrated forces, and abrasion (Figure 7). Their selection has been based on their ability to distribute concentrated loads in shell (membrane) action to absorb the

Fig. 7. The Super CIDS Arctic offshore drilling platform incorporates a mid-body of prestressed lightweight concrete (courtesy of Global Marine Development Co).

energy of impact through ductility if they are adequately reinforced, and to resist abrasion and other degrading environmental actions.

Their principle problem has been that of excessive weight, which increases the draft during transport and deployment. This has been partially offset by the development and application of high-strength lightweight concrete. The special requirements here of resisting local concentrated forces of the order of 1000 tons/square meter, such as those due to iceberg impact, have given added impetus to the development of composite steel-concrete structural elements.

I. Breakwater Armor

Until recently, most exposed breakwaters have been armored by unreinforced concrete units, configured in various complex configurations so as to provide a maximum of porosity and to interlock to prevent displacement. Being unreinforced, corrosion of reinforcing steel was not a problem, so seawater could be used in the mix, provided suitable retarding admixtures were added.

In deeper water, with more severe wave environments, these complicated shapes were dislodged and frequently impacted, breaking into fragments. The Sines Breakwater in Portugal, for example, was destroyed by storm, with 95% of the Dolos units breaking, the fragments of early failures contributing to the breakage of those which initially remained intact. Now, experimental projects are trying out the use of armor units reinforced either by bars or steel fibers. Initial evaluation of the performance of those with the steel fibers is positive.

J. Floating Structures

From Lambot's boat in the nineteenth century on through World War I and II, the use of reinforced concrete has been attempted for floating barges and the hulls of vessels. Prestressed ferrocement showed that the hulls of racing vessels could be made of concrete. In 1975, a quantum jump was made with the construction of the 62,000 DWT floating cryogenic storage vessel, the Seki Ardjuna, placed in service in the Java Sea. A floating phosphate beneficiation plant, built in Singapore and outfitted in Hawaii, now operates on the West Coast of Mexico. Floating ferry docks vie with long prestressed concrete floating bridges in Puget Sound, Washington and Canada. A concrete floating container terminal now serves the Port of Valdez, Alaska.

Huge floating airports and offshore LNG storage plants have been carried through preliminary design and found economical, even before recent advances in concrete technology. They offer not only savings in first cost, but reduction or elimination in maintenance, drydocking, and periodic shutdown.

VIII. Current Research and Development

Research and development in concrete has experienced a strong resurgence with the realization that significantly higher strengths and improved properties are now available, primarily through the addition of superplastizers and silica fume. Concrete foundations are now in the final engineering stage for underpinning an existing offshore platform at a depth of almost 1000 feet. If and when Arctic transport develops, the use of composite steel-concrete for the hull appears to open major opportunities for both economy and safety.

The use of multi-axial stresses, through three-dimensional reinforcement, both active (prestress) and passive, offers opportunities for compliant structures with minimum response to short-term phenomena such as waves. Heavily-reinforced concrete structures are now planned to resist iceberg impact; here not only shell action but ductility are of extreme importance (Figure 8). Considerable research is currently being conducted in this area in the U.S., Canada, and Japan.

Polymer concretes have been under research for many years, with early promising results in strength and durability being offset by technical problems of field application and cost. Recent French research reportedly has developed moderate strength concrete with a specific gravity of 1.0! The potential implications of this and similar developments for floating structures is enormous.

The cold water pipes of proposed Ocean Thermal Energy plants, suspended 3000 feet (1000 m) below a surface hull, can now become practicable in concrete. Recently completed research in the U.S. has shown that prestressed light-weight concrete has the ability to safely contain cryogenic materials, even under cyclic loading typical of prolonged storms at sea.

Concrete is rapidly becoming a highly sophisticated material, with new cementitious materials, manufactured ceramic aggregates, and a wide range of additives and admixtures to control properties. The future development of concrete seems only limited by the imagination of engineers.

Concrete semi-submersible hulls are currently under design for service as tension-leg platforms for oil drilling and production in deep water. They will be tethered to concrete gravity-base tanks on the seafloor, which will serve both as anchorages for the TLP and to store produced oil.

Fig. 8. Proposed prestressed reinforced concrete gravity-base platform to resist waves and iceberg impact, Hibernia project, Newfoundland (courtesy of Mobil Oil Canada).

IX. Conclusion

If concrete structures are to achieve their full potential in the marine environment, it will be essential that both the "carrot" (application needs) and the "stick" (research) be continuously interactive. Communication between researchers, designers, constructors, and users on a worldwide basis will be essential. Fortunately, the mechanisms currently are in place to attain this synergistic self-perpetuating activity.

References

1. A.C.I., Committee 357 (1985). *Recommendations for Concrete Sea Structures.* Publication ACI 357-84, American Concrete Institute, Detroit.
2. A.C.I., Committee 357 (1985). *State-of-Art Report on Concrete Structures for the Arctic.* Publication ACI 357-1R-85, American Concrete Institute, Detroit.

3. F.I.P. (1985). *Recommendations for the Design and Construction of Concrete Sea Structures.* Federation Internationale de la Precontrainte, Thomas Telford and Sons Ltd., London.
4. Gerwick, B. C. Jr. (1971). *Construction of Prestressed Concrete Structures.* Wiley-Interscience, New York.
5. Gerwick, B. C. Jr. (1986). *Construction of Offshore Structures.* Wiley-Interscience, New York.
6. Gerwick, B. C. Jr., and Holland, T. J. (1981). *Tremie Concrete for Bridge Piers and Other Massive Underwater Placements.* Federal Highway Administration Report FHWA/RD-81153, NTIS, Washington, D.C.

8

Materials for Mooring Systems

MAMDOUH M. SALAMA

Petroleum R & D Department
Conoco Inc.
Ponca City, Oklahoma

I. Introduction

The continuous expansion of offshore activities through deepwater mining and deepwater oil and gas developments has resulted in increasing attention being focused on materials selection and the design of mooring systems.

Mooring systems can be classified into two main categories, the catenary mooring system and the tension mooring system, as shown in Figure 1. The catenary system is composed of a number of free-hanging mooring lines and is used to moor floating vessels such as semi-submersibles and tankers. Typical mooring line compositions are single lengths of chain, rope, or a combination of both chain and rope. The tension mooring system is made of either tubulars or parallel strand ropes and is used to restrain the motion of Tension Leg Platforms (TLPs).

Mooring lines are generally manufactured of steel, however, the use of synthetic ropes is currently expanding. Synthetic ropes are made from nylon,

ISBN 0-12-341828-3

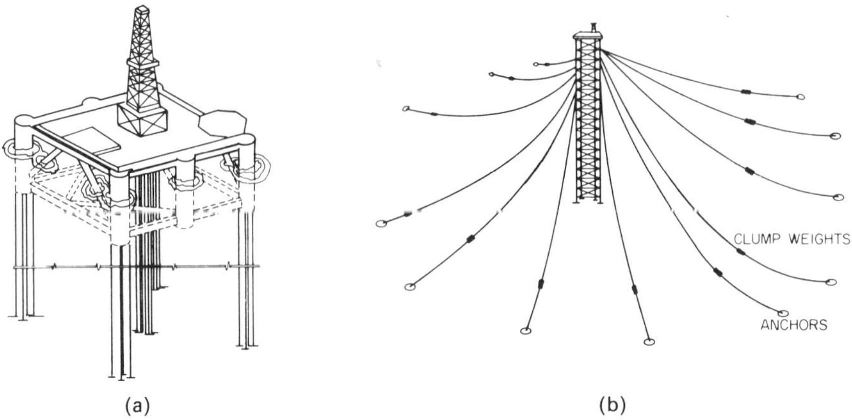

Fig. 1. Type of mooring systems: (a) tension system and (b) catenary system.

polyester, polypropylene, or aramid "KEVLAR" fibers. Table 1 gives a comparison of the physical characteristics of several of these mooring lines (*1*).

The design of mooring systems is generally governed by several objectives, which include the following (*2*):

(a) The mooring lines must function predictably and reliably in the hostile corrosive environment of the sea for at least their design lives.

(b) The mooring lines must with stand the severe fatigue loading conditions.

TABLE 1

COMPARISON OF SIZES AND WEIGHTS OF SEVERAL MOORING LINES WITH APPROXIMATELY 450,000 lbs. BREAKING LOAD

Description	Nominal diameter inches	Approx. weight in air lb/ft	Approx. weight in water lb/ft
Mooring chain	$1\frac{7}{8}$[a]	34.9	30.5
Galvanized spiral strand	$1\frac{7}{8}$	7.42	6.48
Galvanized 6 × 37 wire rope	$2\frac{1}{4}$	9.36	8.18
Stranded "KEVLAR" rope	$2\frac{1}{2}$	1.92	0.54
Nylon double braid rope	4–	4.23	0.52
Polyester double braid rope	$4\frac{1}{4}$	5.92	1.63

[a]Bar stock size

(c) The mooring lines must be designed to accommodate provision for corrosion protection, inspection, and replacement.

(d) The mooring lines must be designed such that the failure of one line will not cause failure in the remaining lines under the extreme storm conditions.

(e) The mooring lines' total length should be adjustable.

(f) The size and weight of the mooring lines must be minimized.

(g) The mooring line components must be fabricated from available materials at reasonable cost.

In addition to these general objectives, there are three specific parameters that have to be considered for each mooring system. These are strength, stiffness, and redundancy.

This chapter discusses some of the important considerations in the selection of a suitable mooring system and reviews the important material parameters for the different mooring lines. These parameters will iclude strength, fatigue, corrosion, wear, and toughness. Discussions will be limited to chains, ropes, and tubulars. Electromechanical cables will not be reviewed because, other than special communication and electro-mechanical characteristics, their design is similar to the rope design.

II. Chains

Chains are still one of the most widely used mooring lines. Chains are quite large and heavy in comparison to other mooring lines of similar break strength. However, they posses several desirable characteristics which make them suitable for a variety of applications. These are:

(a) Chains are relatively tolerant to corrosion and external abrasion.

(b) Chains can be easily stored and transported without the need for drums.

(c) The shape of the chain links provides positive attachment which makes handling easy.

(d) When used in short lengths, chains are quite inelastic and can be used to maintain the desired position of the vessel.

(e) When used in long lengths, the large weight allows the chains to hang in a large catenary, providing an increase in the apparent elasticity of the mooring line and allowing the accommodation of large motions without developing excessive tensile loads.

(f) The large weight can provide an increased holding power of the anchor system if part of the chain is allowed to lie on the ocean bottom.

(g) When a chain is broken or damaged it can be repaired easily using a special connecting link.

However, in cases where it is necessary to minimize the weight in order to achieve a proper buoyancy, the high weight of the chain system becomes a distinct disadvantage. This is particularly true for deepwater applications.

In spite of the fact that the operating loads on chains are less than one-third, and typically one-fifth to one-fourth, of the minimum, breaking strength failures have occurred at a relatively high rate (about 10%). These failures are attributed to local shock loading during handling, that can result in high dynamic amplification, and to local high stresses due to bending as would exist in the fairleads. Both unstable brittle failure and fatigue are the two critical failure mechanisms. Failures due to abrasion and wear have also been reported but to a lesser extent.

A. Brittle Fracture

Brittle fracture occurs in chain links when defects exist and the material has low notch toughness. Defects are, generally, either welding or casting flaws. Low toughness can be attributed to either improper chemistry or heat treatment. Most specifications for chains do not specify a fracture toughness requirement, however, they may specify a heat treatment requirement such as annealing. Verifying that such a heat treatment is properly performed requires either microstructure examination to ensure fine grain size or some mechanical test to assess the toughness of the materials. For critical applications charpy V-notch (CVN) or crack tip opening displacement (CTOD) criteria are specified.

1. Charpy V-Notch Criterion

The CVN criterion is sometimes called the transition temperature criterion and was adopted as a result of the National Bureau of Standards studies (*3*) on fracture of World War II ships. These studies define the transition temperature as the temperature at which the CVN energy equals 15 ft-lb (~20 Joules). The transition temperature requirement for a specific application is generally based on the minimum design service temperature (T) adjusted to take into account the effect of component thickness, type of loading, stress level, safety level, quality of design and fabrication, and level of residual stresses (*4*). The transition temperature T_r(°C) can be estimated by the following equation:

$$T_r = T + 30 + T_t + T_L + T_y + T_S + T_q + T_r,$$

where

T_t is the thickness temperature adjustment. It can be taken as 0°C for a thickness of 20 mm, −20°C for 30 mm, −40°C for 50 mm, and −50°C for 75 mm.

T_L is the type of loading temperature adjustment. It can be taken as 0°C for static loading, −10°C for cyclic loading, and −20°C for impact loading. Wave loading can be considered as quasi-static loading.

T_y is the applied stress level temperature adjustment. This can be taken as 0°C for S/S_y equals 0.20 and −30°C for S/S_y equals 0.50. Interpolation and extrapolation can be used for other ratios of applied stress S to yield stress S_y.

T_S is the safety level temperature adjustment. It would be desirable to use −10°C for critical applications and −20°C for most serious cases.

T_q is quality of design and fabrication temperature adjustment. A maximum adjustment of +10°C can be made if special care is exercised in design, fabrication, and inspection of the structure.

T_r is the residual stress adjustment. A maximum of +30°C is possible for fully effective thermal stress relief.

For a chain 76 mm diameter, the most commonly used on offshore mobile units, operating at 0°C, the transition temperature required can be estimated as:

$$T_r(°C) = 0 + 30 - 50 - 20 - 10 - 10 + 5 + 10 = -45°C.$$

This example illustrates some of the subjectivity introduced by using the transition temperature concept, which is influenced by the designer's judgement on the safety level and quality.

2. CTOD Criterion

The concept of Crack Tip Opening Displacement (CTOD) as an extension of linear elastic fracture mechanics beyond general yield has become widely used as a fracture mechanics parameter that characterizes the resistance of a material to crack extension (*5*, *6*). The basic advantage of using the CTOD approach is its ability to give the designer a mathematical tool to optimize design and operational requirements in terms of materials' toughness, applied stress, welding procedures, and defect acceptance. Lereim (*7*) established a simplified design curve for chains based on CTOD results:

$$\Phi = \frac{\delta}{2ae_y} = 0.9\left(\frac{e}{e_y}\right)^{1.5},$$

where δ is the CTOD value, a is the depth of a semi-elliptic surface crack, e is the maximum applied nominal strain, and e_y is the material's yield strain.

The above equation is different from those proposed in the British Standard Institute PD6493 (*8*) which were derived based on The Welding Institute (TWI) work. Lereim's simplified design equation appears to be more appropriate because a comparison between the PD6493 design curve and

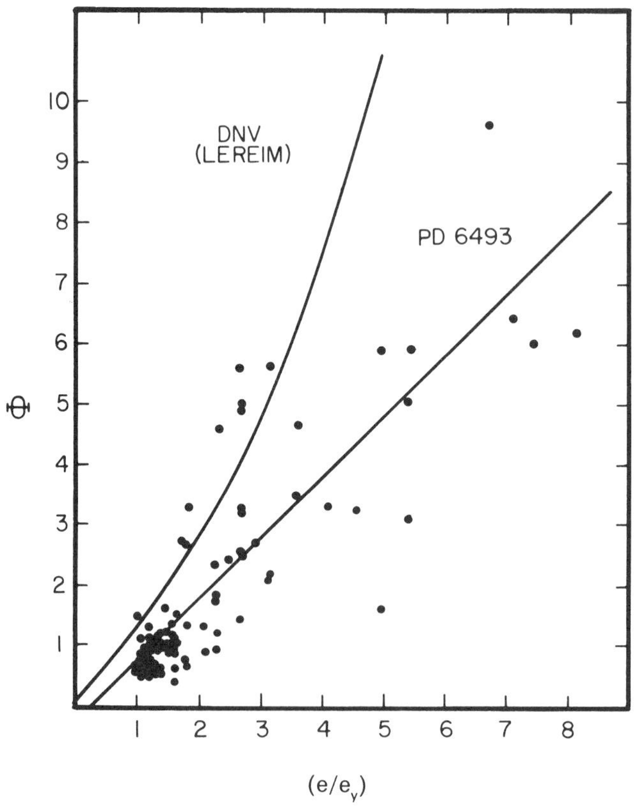

Fig. 2. Comparison between PD6493 and Lereim design curves and CTOD data.

Lereim's experimental results on chains yields poor correlation, as shown in Figure 2.

B. Fatigue Strength

Fatigue strength of chains is an important parameter for the design of permanently moored floating production systems. Experience with the mooring of ships has shown that the failure frequency of the chain connecting links is, in general, significantly higher than the failure frequency of common links. This conclusion is in agreement with the fatigue test results of anchor chains (*8*). The fatigue results on four types of connecting elements—common link, Kenter-type, Baldt-type, and D-shackle—(Figure 3) have shown a major difference in their fatigue strength (*9*) as shown in Table 2. Other fatigue data confirm this conclusion as shown in Figure 4, which presents a

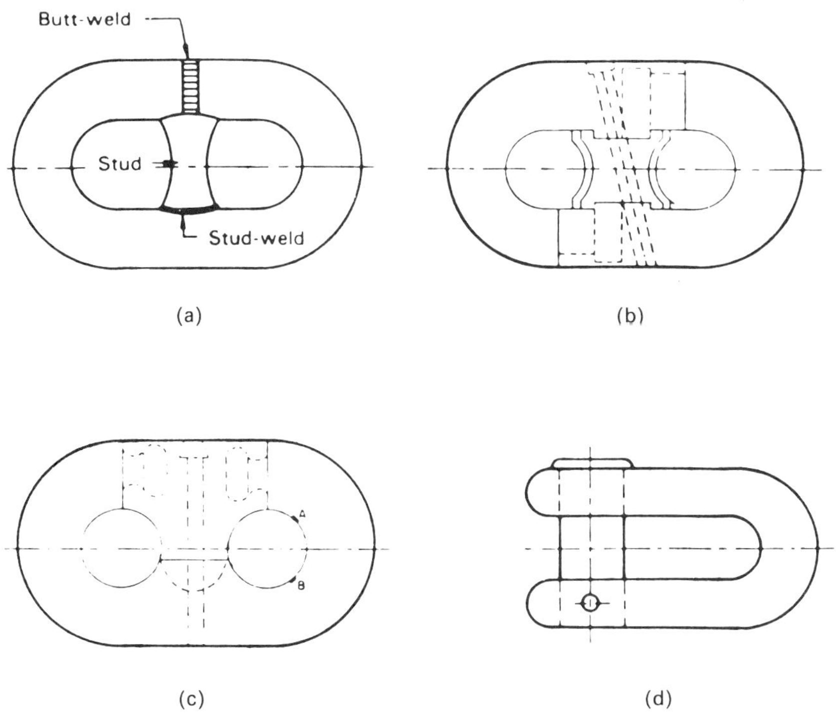

Fig. 3. Types of chain links: (a) common link, (b) kenter-type link, (c) Baldt-type connecting link, and (d) D-shackle.

TABLE 2

SUMMARY OF RESULTS OF FATIGUE TESTS

Type	Relative fatigue life
Common link	1
Kenter-type connecting link	0.2
Baldt-type connecting link	0.3–0.9
D-shackle	< 1

normalized S-N scatter band for both common and Kenter links (*9,10,11*). Van Helvoirt's (*10*) results suggest that seawater fatigue lives of common chain link can be derived from the AWS A-A curve (*12*) for structural steels, taking into account a stress concentration factor of four.

Using Paris crack growth equation and stress intensity factor solutions for different chain links, fatigue lives can be calculated. The derived numerical results have been shown to be in good agreement with experimental results (*7, 8*). The difficulty with this approach is in arbitrarily selecting the initial defect depth, which is generally selected between 0.05 to 3 mm to achieve the agreement.

C. *Wear Resistance*

Wear failure of chains is not very common. However, some wear failures have been reported in the literature. A failure at the junction between the fixed stopper link and the first free link in the catenary was reported by Shoup and Mueller (*13*). They concluded that the excessive wear was generated by large values of interlink force and motion induced by the dynamic behavior of the mooring pattern. The authors presented results of wear tests on two chain materials, U_3 and U_4. The grade of these materials is in accordance with API 2F specifications (*14*). The U_3 material is a silicon killed carbon-manganese steel with a typical yield strength of 75 ksi and ultimate strength of 116 ksi. In contrast, the U_4 material is an aluminum killed low carbon steel containing Mn-Ni-Cr and Mo with a typical yield strength of 95 ksi and ultimate strength of 130 ksi. These two steels appear to have similar wear resistance, as shown in Figure 5. In this case it appears that failure could have been avoided by selecting a different mooring system, placing large links below the chain stoppers, or using a stopper casting support which was free to rotate to eliminate wear generated by interlink motions.

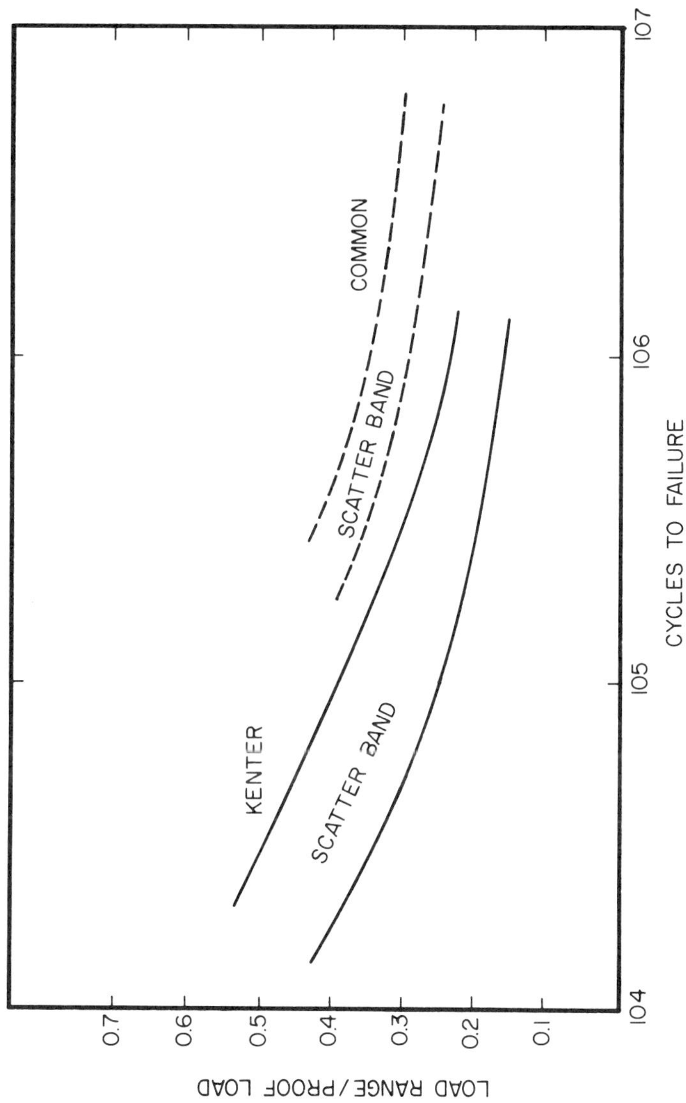

Fig. 4. Fatigue results of chain links.

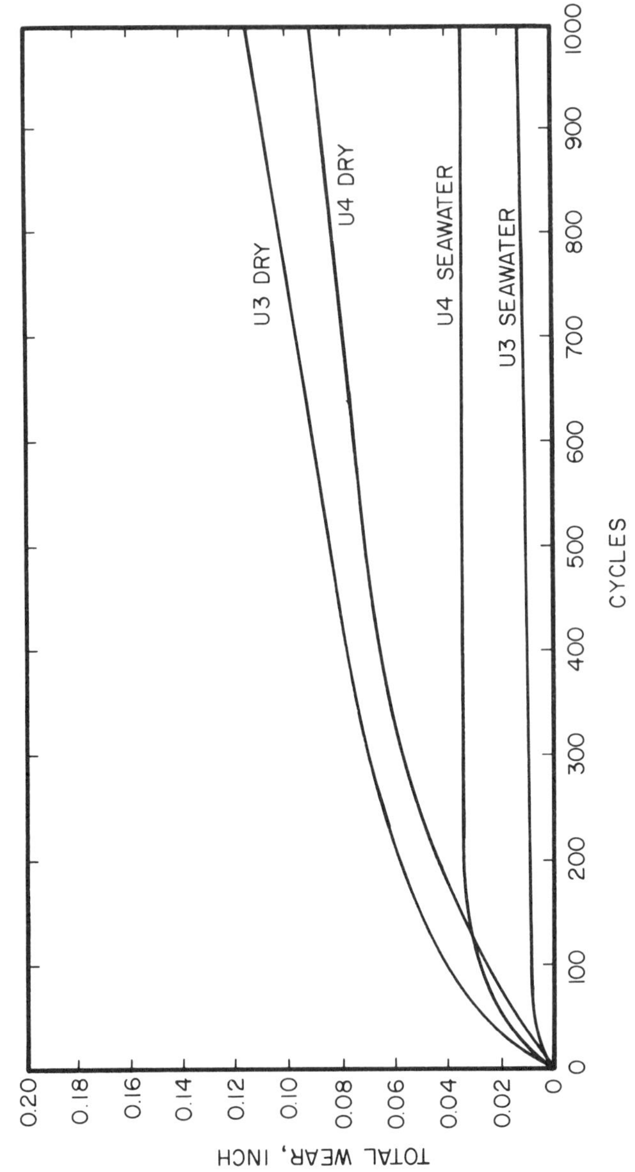

Fig. 5. Wear of U4 and U3 chain link steel at 1360 Lbs. (50 metric ton equivalent).

III. Ropes

Although there is about a half million rope types with different combinations of materials and construction, only few are appropriate for deepwater mooring applications. The rope material is either steel or synthetic fiber. Synthetic fibers include nylon, polyester, polypropylene, and aramid "KEVLAR." The three most favorable rope constructions, shown in Figure 6, are spiral strand, n-strand (where *n* is the number of strands which is typically 6), and parallel strand (this includes bridge strand and parallel yarn construction such as parafil).

The strength of a rope is not only affected by both its material and construction, but it is also greatly influenced by the type and workmanship of its termination. There are various methods for terminating ropes, however, the most common (Figure 7) are eye spliced, zinc or resin potted, and spike in socket terminations. For steel ropes zinc potted terminations are the most common, and for synthetic fiber ropes both eye spliced and resin potted terminations are used. The spike-in-socket termination is mainly used for parallel lay ropes.

A. Materials

1. STEEL

Wire ropes are generally made from high strength steel. However, for seawater application, corrosion resistance alloys such as stainless steel and copper alloys have been used. For steel wires corrosion resistance can be provided by either galvanizing (using zinc coating) or aluminizing (using aluminum coating). These techniques are closely controlled to give high quality, nonporous coating of even thickness with good adherence to the steel substrate. There are also approaches to improve corrosion resistance of steel ropes. These include blocking, sheathing, and cathodic protection (*15*). Blocking involves filling the interstices of the rope/strand with a heavy grease (petrolatum wax) during manufacture. The use of a continuous, extruded plastic sheathing for permanently arresting the ingress of corrosive media and preventing corrosion is considered an attractive form for corrosion protection. The use of high density polyethylene is recommended for this application. Cathodic protection involves incorporating continuous sacrificial elements, such as aluminum or zinc, into the rope or strand during manufacture.

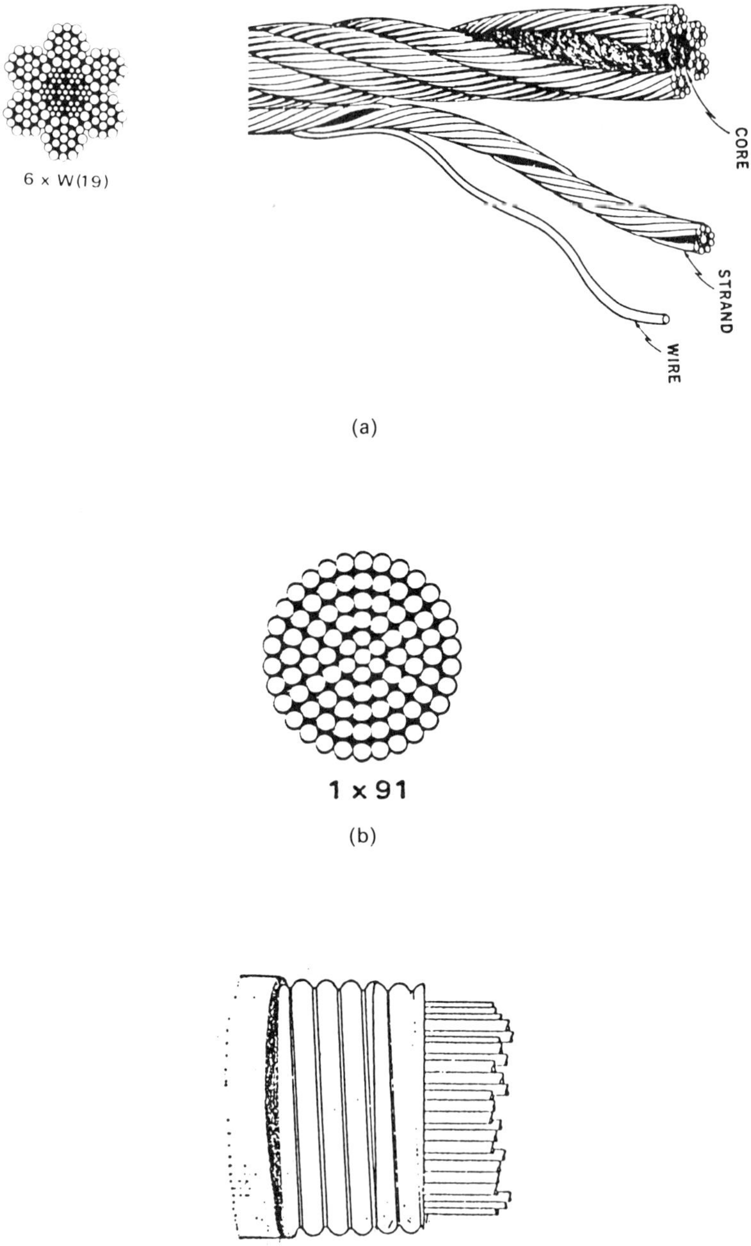

Fig. 6. Typical rope constructions: (a) spiral strand, (b) stranded, and (c) parallel strand.

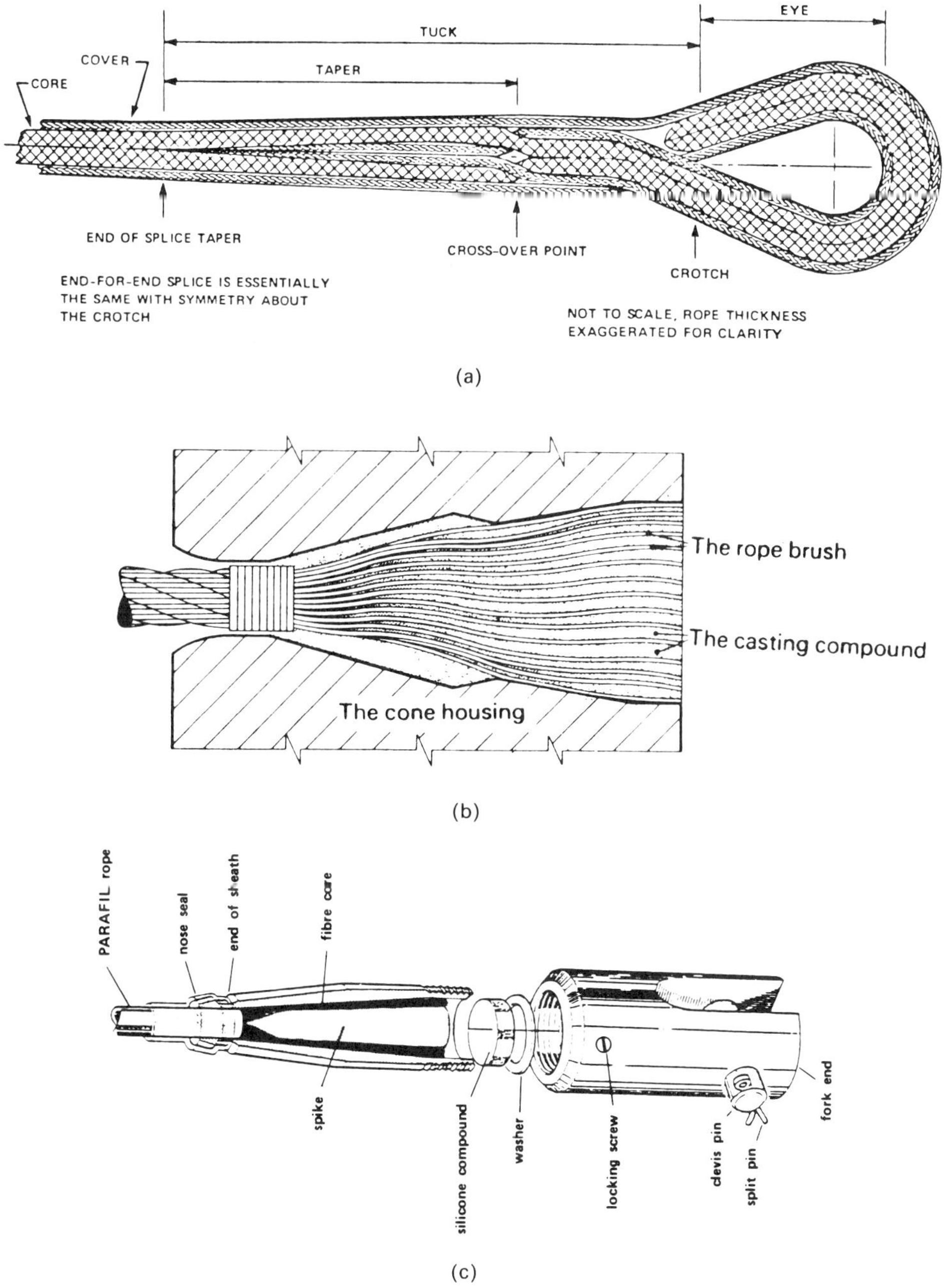

Fig. 7. Termination types: (a) eye spliced, (b) potted, and (c) spike in socket.

TABLE 3

SELECTION GUIDE FOR SYNTHETIC ROPES IN WET ENVIRONMENT

Material	Strength	Tensile fatigue	Bending fatigue	Abrasion resistance	Creep
Nylon	G	F	P	P	G
Polyester	VG	VG	G	E	VG
Polypropylene	F	F	P	F	F
"KEVLAR"	E	E	E	F	E

E = Excellent, VG = Very Good, G = Good, F = Fair

2. SYNTHETIC FIBERS

Synthetic fibers include nylon, polyester, polypropylene, and "KEVLAR." Table 3 gives general guidelines for the selection of synthetic ropes for seawater applications.

Nylon was the original high strength synthetic fiber for rope applications. Nylon ropes are characterized by exceptionally high elongation characteristics that can be utilized to advantage in cases where large amplitude motion must be accommodated. One recent application was reported by Smith, et al. (*16*), where nylon ropes were used as temporary mooring of the Hutton tension leg platform during the mating of the deck and the hull in shallow water. However, the great elasticity of these ropes can be a serious disadvantage because of the large stored energy which can be released explosively when a mooring line fails. Nylon has a good strength and good abrasion resistance when dry. It absorbs water, which causes swelling resulting in a 10% to 20% loss in strength and adversely affects its abrasion resistance (*17*).

Polyester is as strong as nylon and does not lose its strength when wet. Special coatings can be applied to the basic polyester filaments that improve wet abrasion resistance by a factor of eight over uncoated polyester and by a factor of 20 over wet nylon (*17*). Polyester has about one half the elongation of nylon.

Polypropylene fiber is about 30% weaker than nylon and polyester, and it has poor fatigue resistance. It degrades in sunlight and is subject to creep under sustained load. It is, however, a less expensive fiber than nylon or polyester, and it has a specific gravity less than water. Therefore, it is generally used if the rope needs to float.

Aramid "KEVLAR" ropes provide mooring lines with exceptionally high strength-to-weight ratios. The elongation of "KEVLAR" is typically at least twice that of a steel wire rope. While "KEVLAR" ropes are light, easy to handle, and unaffected by conventional corrosion in the ocean, they, similar to other synthetic ropes, are quite easily damaged unless special protection is provided. Protection by jacketing the entire lengths of rope with polyurethene, nylon, or other plastics has proven successful in greatly improving abrasion and cutting resistance while retaining flexibility. However, the jacket design must consider the known fact that aramid fibers have modest compressive properties. If "KEVLAR" rope is allowed to unwind, excess length is created which makes the strands want to bulge. The bulge can be prevented with a tight jacket that causes the strands to go into compression if the tension on the rope is not high enough. Kink bands can form when strands are subjected to compressive strains greater than 0.5%. If care to avoid the application of compressive strains is not exercised in the design and application of "KEVLAR" ropes, premature failures can result. Reiwald (*18*) reported on a field failure that was caused by this mechanism. Reiwald showed that the tensile strength of a rope can be reduced to less than 50% of its original value when subjected to a million compressive strain cycles of 0.8% strain. Some of his results are shown in Figure 8.

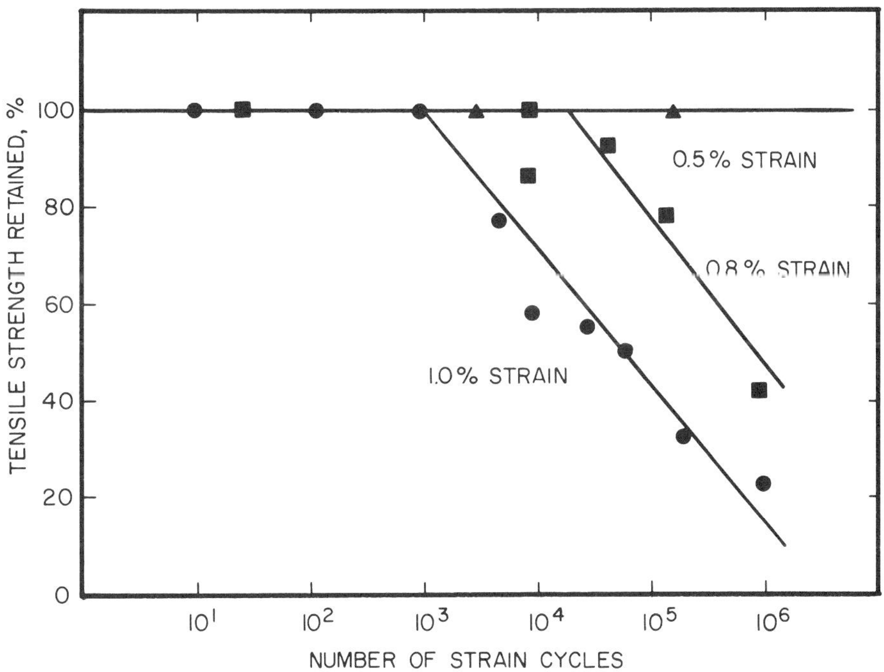

Fig. 8. Effect of compression cycling on strength of aramid fiber ropes.

B. Construction

Ropes are constructed from either individual wires or strands. The strand is fabricated by weaving individual wires into a small cord; these cords (or strands) are then wound into ropes. Spiral strand construction is made up of many layers of contrahelically wrapped steel wries. It offers a significant size and weight advantage when compared to chain, but it provides very little stretch and exhibits poor performance in bending. Spiral strands are not recommended for any application where a large number of bending cycles over sheaves or saddles is required, but straight sections of spiral ropes exhibit good cyclic tension fatigue strength.

A wire rope construction refers to stranded ropes and is described by the number of strands and the number of wires in each strand, i.e., 6×37 rope has 6 strands and each strand contains 37 wires. Wires do not have to be of the same size, e.g., for cases where wear is important the strand will have larger outer wires to provide a larger wear surface. For cases where flexibility is important the strands are made using small size wires.

The nomenclature for stranded ropes also describes the center core, the direction of winding, and whether the strands have small non-load-bearing filler wires to minimize movement between adjacent load-bearing wires. Ropes are manufactured with both fiber and steel cores to enable different properties to be obtained, i.e., elasticity (fiber core) and resistance to crushing (steel core). Wire ropes can be fabricated where the individual wires in the strand are either laid in the opposite direction to the lay of the strands (regular lay) or laid in the same direction as the lay of the strands (lang lay). Lang lay rope has more resistance to abrasion and bending fatigue than regular lay rope, but it has less resistance to deformation and kink. Also, lang lay ropes rotate easily when one end of the line is not fixed, because the rope tends to develop a very high torque value resulting in spinning and untwisting. These properties are not compatible with mooring lines and, therefore, lang lay ropes are not used.

Similar to steel ropes, synthetic ropes can be produced with stranded, braided, or parallel strand constructions. Stranded construction is similar to that of wire ropes. Braided construction has the advantage that the rope is non-rotating. Ropes with parallel strands or fibers are very efficient and provide the highest stiffness. For mooring lines, stranded and parallel lay ropes are the most attractive for catenary and tension systems, respectively.

C. Mechanical Characteristics

Ropes are designed to meet a broad spectrum of service requirements that include water depth, sea states, tensile strength, stiffness, and useful service life. Failure of ropes can be generally characterized as tensile overloading, fatigue, or a combination of both. Factors such as wear, corrosion, loss of lubrication, and deformation can result in premature failures by dramatically reducing both tensile and fatigue strengths. Therefore, it is prudent to establish the appropriate tensile strength and to assess the effect of various operating conditions on fatigue performance of wire ropes.

1. Tensile Strength

Tensile strength is the most commonly specified property of wire ropes. It is established by a straight pull of the rope until failure. This measured break strength may, however, not be achieved during the actual service of the mooring system if the end terminations provide less than 100% strength efficiency or if the mooring line is used while wrapped around a saddle or a sheave. Also, the strength of a rope can degrade during service life due to mechanical damage, wear, corrosion, kinking, and material fatigue.

Mechanical damage, corrosion, and wear can occur due to careless handling of the rope and the use of improper handling and installation equipment. Damage of the rope protective sheath, for example, can lead to accelerated corrosion, resulting in deterioration of strength. Degradation of the rope strength due to abrasion and wear occurs when the rope is dragged over rough surfaces and runs over improper or damaged pulleys, rollers, and fairleads.

Kinking can occur in wire ropes due to unwinding. Because of the geometric relationship of the load carrying wires and of the strands to the axis of the applied load, the rope will, in most cases, unwind in a direction opposite to its lay when a load is applied while one end of the rope is free to rotate. If the applied load is suddenly reduced, the rope will wind back to its unloaded state too fast, causing the rope to twist up on itself. This results in the formation of tight loops, which will be pulled, forming kinks when the load is reapplied. These kinks can reduce the rope strength by as much as 90% (*19*). This problem can be avoided by protecting the rope during deployment from the application of load until both ends are fixed or by using torque balanced constructions. Unfortunately, torque balanced ropes are not available in all sizes. Kinks can also form in synthetic fiber ropes when the fibers possess low compressive properties. This will be the case for aramid fiber ropes if subjected to compressive cyclic strains (*18*) as has been previously discussed.

Cyclic loading is known to result in a progressive deterioration of the rope strength due to fatigue. The level of reduction in strength depends on both the level of cyclic loading and the number of cycles. Additional discussions on this subject are made in the section on fatigue strength.

2. Fatigue Strength

Fatigue strength of ropes represents a very important parameter in the design of mooring systems for both deepwater and permanent installations. Unfortunately, only a limited amount of fatigue testing on ropes has been conducted under marine conditions. Also, most of the available data were developed using small ropes of less than 50 mm ($\sim$2 inch) diameter. Most of the data were obtained by bending tests, and few were obtained under direct tension. Since a rope with a good bending fatigue strength does not necessarily have a good tensile fatigue strength and vice versa, it is important to determine whether the long-term fatigue life of a mooring line is limited by the tensile or the bending fatigue characteristics. In general, the design of mooring lines that are wrapped around a saddle or a sheave will be governed by tensile fatigue if the wrap angle does not vary under normal operating conditions. If, however, the arc of contact between the mooring line and the saddle or sheave continues to change with more than one half the lay length, the service life of the rope will be governed by its bending fatigue strength (*1*).

In the past, these data were adequate because most applications, e.g., masts and bridges, required ropes with diameters between 30 mm ($1\frac{1}{4}$ inch) to 70 mm ($2\frac{3}{4}$ inch). However, current offshore mooring applications require larger diameter ropes, typically between 70 and 170 mm ($2\frac{3}{4}$ to $6\frac{3}{4}$ inch). The validity of extrapolating small rope fatigue data for use in the design of the large diameter ropes is questionable, because the internal compressive stresses and the relative motions that are developed within the rope increase rapidly with increasing diameter. Therefore, for cases where the failure mechanism is influenced by either the internal stresses or the relative motions, such as in bending fatigue, a change of the scale can affect the fatigue performance. Figure 9, which was developed by Waters et al. (*20*), based on the bending fatigue results of steel wire ropes by Rice (*21*), shows a reduction in fatigue strength as the diameter of the rope increases. In contrast to these results, Hanzawa et al. (*22*) have shown that under direct tensile fatigue tests the fatigue life of nine strand ropes increases as the rope diameter increases. Similar conclusions on three and six strand ropes can be reached from Lucht and Donecker's (*19*) results. This behavior, which is illustrated in Figure 10, suggests that the fatigue performance of large diameter ropes can be safely predicted from tests on small diameter ropes. This approach is not, however, unanimously accepted by rope experts, particularly when the failure occurs in the termination. An example is recent work by the author on aramid parallel lay ropes where the fatigue lives of 30 ton ropes were more than three orders of magnitude higher than lives for 500 ton ropes when subjected to the same cyclic stresses. Failures for these ropes were occurring inside the terminations.

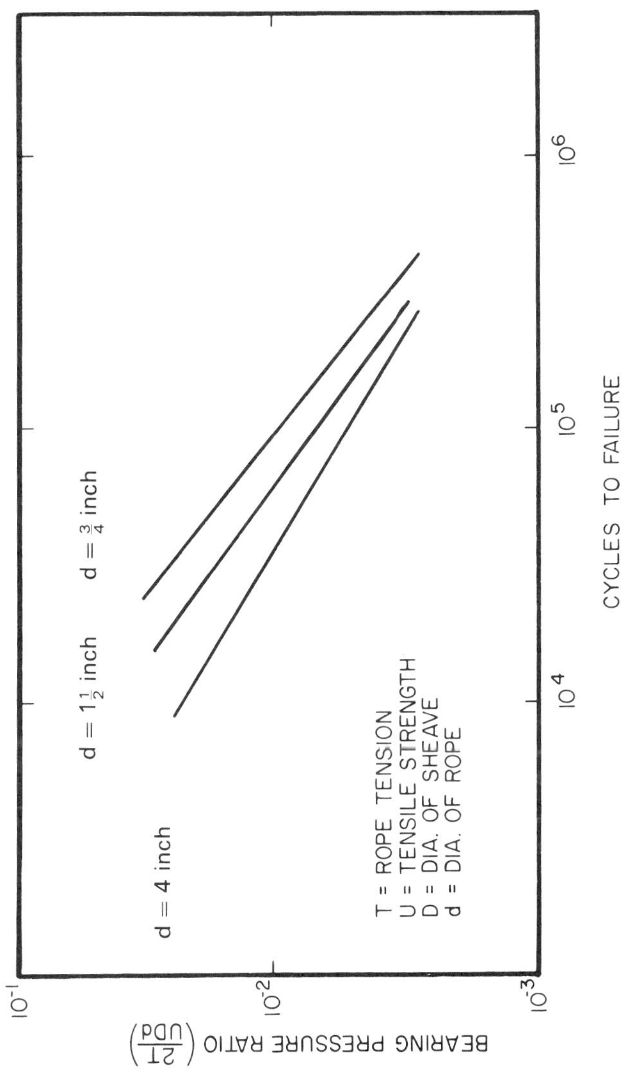

Fig. 9. Bend over sheave fatigue results.

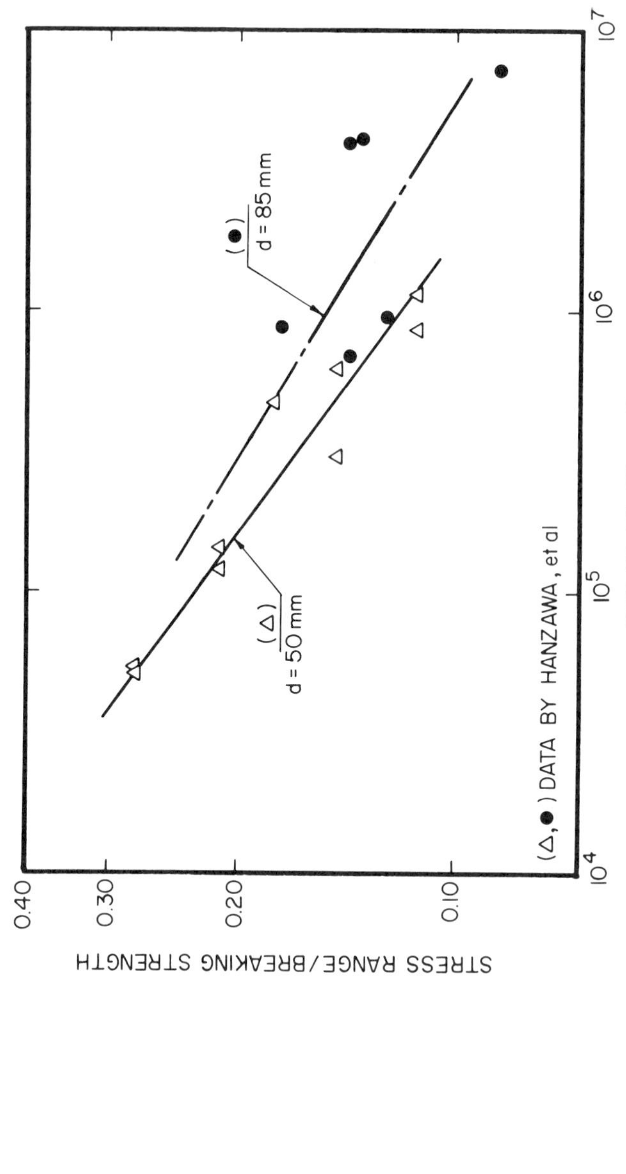

Fig. 10. Axial fatigue results.

IV. Tubulars

Tubulars are used primarily for the mooring of Tension Leg Platforms (TLP). Both thin wall (about 1 inch thick) and thick wall tubulars are being considered. Conoco's Hutton TLP in the North Sea utilizes thick wall tubulars as tendons. The tubulars are forged from high strength steel with 92.5 mm (3.64 inch) wall thickness (*23*). Thin wall tubular tendons are being considered by several oil companies for mooring TLPs in the Gulf of Mexico and offshore California.

Thin wall tubulars are a well-known commodity in the oil industry, hence they have attractions for use as mooring lines because of this familiarity. Facilities for the production of high quality steel tubulars with yield strength of 80 ksi (552 MPa) or less in various diameters and wall thickness, which would be suitable for this application, exist throughout the world. The junction of the tubular lengths will be made either by welding tubular lengths together or by screwed connections. If welding is used, the mooring line will be either field welded offshore using advanced methods such as electron beam, or the welding will be done on shore and the welded mooring line will be towed to location in one piece. For screwed connections it is most likely that two forged connectors, forming the pin and the box ends, will be butt welded to a steel pipe rather than being machined directly on forged ends. Materials proposed for this type of mooring lines include API 5LU X-80 (1Ni-0.5Cr-0.4Mo) and HY-80 for the pipe and ASTM A541 cl.6 ($2\frac{1}{4}$Cr-1Mo), A508 cl.2a (1.5Ni-0.5Cr-0.4Mo) or A372 cl.6 (3.0Ni-1Cr) for the connector (*24*, *25*). These steels have a yield strength around 600 MPa (87 ksi) and CVN toughness value that exceeds 200 J at $-20°C$.

Since TLP mooring lines are pretensioned to a high tensile load and are subjected to high cyclic tension load in sea-water, corrosion fatigue and stress corrosion cracking (SCC) are important design and materials selection factors. For forged tubulars with integral connections similar to those used in the Hutton TLP, the fatigue life is generally controlled by the threaded joint. Thus, fatigue design curves need to be developed for the specific joint (*23*). For tubulars with welded connectors, the fatigue life is controlled by both the threaded and the welded joint, and, therefore, fatigue information is needed for both. Fatigue design of the welded joint can be based on the ASME Section VIII S-N curve (*26*). Although this curve is intended to be used for components subjected to air environment, it can be used for properly protected components in seawater.

The use of high strength steels as mooring lines raises questions concerning the possibility of SCC. The SCC is a type of spontaneous cracking which can develop under the combined action of sustained tensile load and the exposure to a corrosive environment such as seawater. Detailed evaluation of candidate steels for TLP mooring lines showed that SCC under purely static loading is not likely to be a significant factor in their structural application (*25*). As an example, fatigue precracked specimens from Hutton steel did not show any indication of crack growth after 4500 hours of loading at a K_1 level of 120 ksi-in$^{1/2}$. However, preliminary results by Hauser and Crooker of NRL have shown that when a ripple-loading of 3% ($\Delta K = 3.8$ ksi-in$^{1/2}$) at 0.1 Hz was added, while maintaining the above K value, the specimens failed after 3000 hours (@1,000,000 cycles). Further work is required in this area before drawing broad conclusions from that single test.

V. Inspection

The inspection of mooring lines both before and during service is of great importance. It is unfortunate that, even with the present state of technology, no reliable nondestructive evaluation techniques are available for establishing the retirement before failure criteria for mooring lines. This is an area where further developments are currently being pursued.

Mooring lines are subject to both short-term and long-term degradations during their service life. Short-term degradation can result from accidents, mishandling, and operational problems. Long-term degradation can result from wear, corrosion, fatigue, and any combination of these mechanisms.

For ropes, the discovery of short-term damage can be relatively easy because it takes highly visible forms such as kinks, bends, bird cages, crushing, cork screwing, and broken strands. The long-term damage is the one that causes the greatest problem for inspection. The most frequently used inspection approaches are electromagnetic, radiographic, ultrasonic, acoustic emission, and infrared techniques.

Electromagnetic methods determine degradation by locally magnetizing the rope and observing any variations in the resulting field. Radiographic techniques pass a beam of high-energy, subatomic particles through the rope and onto a film that shows the variations in metallic density. Ultrasonic techniques introduce high-frequency sound waves that reflect from discontinuities. Acoustic emission approaches listen for the noises made by the microscopic and macroscopic changes occurring during the degradation process. Electromagnetic, ultrasonic, and acoustic emission techniques are by far the most widely used (*27*).

For TLP mooring lines, in-service inspection can be achieved by acoustic emission, ultrasonic, acoustic holography, or eddy current. An ultrasonic system is being used for the inspection of Hutton lines (*28*). Acoustic holography appears to be more attractive, particularly if testing can demonstrate its capability for external inspection of tubulars without the need to remove marine growth.

VI. Recent Advances

In several applications, including both marine and bridge structures, greater emphasis is being placed on axial stiffness of mooring lines. This led to the evolution of a long-laid parallel wire construction and large diameter strands. Strands of up to $5\frac{1}{8}$ inch (131 mm) diameter and 3.8 million pounds breaking strength have been produced for suspension bridges. Larger sizes up to 10 inches (254 mm) may be required for TLP applications. This increase in strand sizes has a major drawback of increasing unit weight and the associated handling and transportation problems. To address these problems a new concept utilizing pultruded 5 mm diameter carbon fiber or KEVLAR 49 composite rods instead of the high strength steel wires is being developed (*29,30*). Walton and Young (*29*) conducted fatigue tests on 300,000 pounds breaking strength strands and showed that the fatigue life of KEVLAR and carbon strands are five and ten times longer than steel strands, respectively. These composite strands were constructed from 61 wires (rods) laid up in a close packed hexagonal formations as shown in Fig. 11. The strands have an overall diameter of $1\frac{3}{4}$ in (45 mm) and a nominal strand pitch of 11 ft (3.3 m), which corresponds to a lay to diameter ratio of 75 to 1

The 61-wire strands were formed using a track based manufacturing system. The wire rods are supplied in the form of 6 ft diameter coils. The helical bundle is formed together without inducing any torsional stresses into the rods. Intermittent tape wrappings were applied to assure strand integrity during coiling and handling. To protect the strand and improve handling a thermoplastic jacket is fitted to the strand.

So far composite rods are pultruded with a vinyl ester thermoset polymer, but research is currently underway to evaluate the feasibility of fabricating rods using high performance thermoplastic resins such as polyphenylene sulfide (PPS), polyethersulfone (PES), polysulfone (PS) and polyetherether-ketone (PEEK). These resins offer advantages over thermoset resins in increased toughness and impact resistance, excellent moisture and corrosion resistance, and improved repairability.

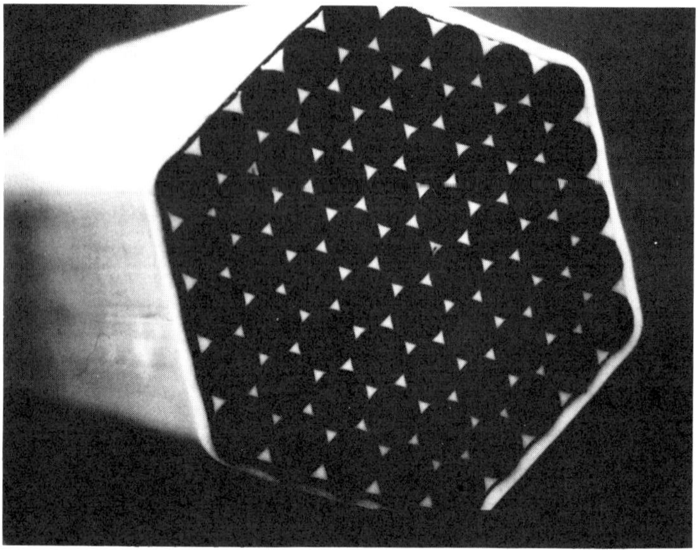

Fig. 11. Cross-section of experimental 1 × 61 composite strand.

References

1. Gibson, P. T. (1983). "Performance of Mooring Tension Members." Proc-Oceans '83, *Effective Use of the Sea: An Update*, IEEE, pp. 538–542.
2. Salama, M. M. (1984). "Lightweight Materials for Mooring Lines of Deepwater Tension Leg Platform." *Marine Technology* **21**, (3), 234–241.
3. Williams, M. L. (1958). "Correlation of Metallurgical Properties and Service Performance of Steel Plates from Fractured Ships." *Welding in the World* **37**, 445.
4. Salama, M. M., and Mehdizadeh, P. (1980). "Toughness Requirements for Offshore Structural Steels." *Energy Technology Conference and Exhibition, New Orleans 3-7, 1980*, Paper 80-Pet. 71. The American Society of Mechanical Engineers.
5. Anon (1979). *Method for Crack Opening Displacement (COD) Testing.* BS 5762: 1979, British Standards Institution.
6. Anon (1980). *Guidance on Some Methods for the Derivation of Acceptance Levels for Defects in Fusion Welded Joints.* PD 6493, 1980, British Standards Institution.
7. Lereim, J. (1985). "Summary of the 4-Year Research Project: Anchor Chain Cables Offshore." *Proc. 17th Annual Offshore Technology Conference, Houston, Texas*, Paper OTC 5060.
8. Lereim, J. (1985). "Anchor Chain Cables Offshore Strength and Reliability." In Proc. of Boss' 85: *Behavior of Offshore Structures.* Elsevier Science Publishers, pp. 807–812.
9. Stern, I. L., and Weatcroft, M. F. (1978). "Towards Improving the Reliability of Anchor Chain and Accessories." *Proc. 10th Annual Offshore Technology Conference*, Paper OTC 3206.

10. Van Helvoirt, L. C. (1982). "Static and Fatigue Tests on Chain Links and Chain Connecting Links." *Proc. 14th Annual Offshore Technology Conference*, Paper OTC 4179.
11. de Laval, G. (1971). "Fatigue Tests of Anchor Chain-Cable." *Proc. 3rd Annual Offshore Technology Conference*, Paper OTC 1503.
12. Anon (1976). *Structural Welding Code*. American Welding Society, AWS D1-1-76.
13. Shoup, G. J., and Mueller, R. A. (1984). "Failure Analysis of a Calm Buoy Anchor Chain System." *Proc. 16th Annual Offshore Technology Conference*, Paper OTC 4764.
14. Anon (1981). *Specification for Mooring Chain*. The American Petroleum Institute, API 2F.
15. Ronson, K. T. (1980). "Ropes for Deepwater Mooring." *Proc. 12th Annual Offshore Technology Conference*, Paper OTC 3850.
16. Smith, I., White, G. J., and Collings, C. S. (1985). "Synthetic Rope in the Mooring System of the Hutton TLP." *Proc. 17th Annual Offshore Technology Conference*, Paper OTC 5063.
17. McKenna, H. A. (1983). "Developments in Fiber Rope Technology." In: Proc. OCEANS' 83, *Effective Use of the Sea—An Update*, V.1, pp 527–533, published by Marine Technology Society.
18. Riewald, P. G. (1986). "Performance Analysis of an Aramid Mooring Line." *Proc. 18th Annual Offshore Technology Conference*, Paper OTC 5187.
19. Lucht, W. A., and Donecker F. W. (1977). "Factors Affecting Wire Rope Life in a Marine Environment." *Proc. 9th Annual Offshore Technology Conference*, Paper OTC 2924.
20. Waters, D., Eggar, D., and Plant, H. (1985). "Development of Fatigue Assessment of Large Diameter Ropes Used in Offshore Mooring." *Proc. 17th Annual Offshore Technology Conference*, Paper OTC 5000.
21. Rice, R. C. (1980). "The Correlation of Large and Small Diameter Wire Rope Bending Fatigue Behavior." *Proc. First Annual Wire Rope Symposium*, Published by *Washington State University, Pullman, Washington.*
22. Hanzawa, M., Yokota, H., Toda, Y., and Yokoyama, K. (1981). "Fatigue Behaviour of Large-Diameter Wire Ropes." *Proc. 13th Annual Offshore Technology Conference*, Paper OTC 3999.
23. Salama, M. M., and Tetlow, J. H. (1984). "Selection and Evaluations of High Strength Steel for Hutton TLP Tension Leg Elements." *J. Energy Resources Technology, Trans. ASME* V. **106** (1), 32–37.
24. Akahide, K., Hashimoto, O., and Narumoto, A. (1985). "Materials for Tension Leg Platform Tethers," In: *Proc. 4th Int Offshore Mechanics and Arctic Engineering Symposium* (J. S. Chung, et al., eds). ASME, V. 2, pp. 351–356.
25. Hauser, J. A., and Crooker, T. W. (1986). *An Investigation of Stress-Corrosion Cracking in Candidate Steels for Tension Leg Platform Tendons*. NRL Memorandum Report 5773, Naval Research Laboratory, Washington, D.C.
26. ASME Section VIII-Division 2, Appendix 5, Fig. 5-110-1, Am. Soc. of Mech. Engineering, New York.
27. Egan, R. A. (1977). "Nondestructive Testing of Wire Rope." *Proc. 9th Annual Offshore Technology Conference*, Paper OTC 2926.
28. Hein, N. W., Skilbeck, F., Herd, V. P., and DeVries, A. J. (1984). "Requirements for a Nondestructive Inspection System for the Hutton Tension Leg Platform Mooring System." *J. Energy Resources Technology, Trans. ASME* 106(2), pp. 260–265.
29. Walton, J. J., and Yeung, Y. C. T. (1987). "Flexible Tension Members From Composite Material." *Proc. of 6th International Offshore Mechanics and Arctic Engineering (OMAE) Symposium* (M. M. Salama, et al., eds). ASME, V.3, pp. 7–14.
30. Kim, K. S., Hann, T. H., and Williams, J. G. (1988). "Application of Composites in TLP Tethers." *Proc. of 7th OMAE Symposium* (M. M. Salama, et al., eds). ASME, V.3., pp. 1–7.

9

Fracture Control for Marine Structures

JERRELL M. THOMAS*

Failure Analysis Associates
2225 E. Bayshore Road
Palo Alto, California 94303

* The author gratefully acknowledges the contributions of his coworkers, Philip Besuner, Keith Ortiz, and Sheryl Adams, to the research that led to this chapter.

ISBN 0-12-341828-3

I. Introduction

Fracture control is the rigorous application of those branches of engineering, management, manufacturing, and operations technology dealing with the understanding and prevention of crack initiation and propagation leading to catastrophic failure. Until recently, fracture was controlled implicitly by low working stress levels and evolving design procedures based on trial-and-error experience. When fracture did occur, it often was not catastrophic due to the high degree of redundancy built into the structures. If a failure were catastrophic (often in prototype or early production structures), subsequent designs would often use large factors of safety and thus lower working stresses.

Fracture control has recently become an important design consideration. Modern high strength materials allow the designer to use higher working stresses, but often at the expense of lower ductility and less "forgiveness" in the material due to decreases in resistance to aggressive environments and/or crack-like defects. At the same time, better analytic techniques and understanding of structural behavior (e.g., dynamics) have led to reduced redundancy and smaller factors of safety. Thus, some of the controls implicit in past design methods have been removed. When these steps led to failures, they were followed by the introduction of explicit facture control methods.

In the 1940s, attention was drawn to the fracture problem by the cracking of a large number of World War II ships, in particular by the brittle fracture and sinking of Liberty ships. Study of this problem led to design rules which minimized stress concentrations. These rules, along with further research in the 1950s, led to the use of improved notch- and crack-toughness material by some designers. Thus, the engineering application of fracture mechanics to marine structures was born largely to prevent brittle fracture in ships. Today fracture mechanics is used to predict initiation and arrest of brittle (and several types of ductile) fracture, fatigue and other subcritical crack propagation rates, and critical crack sizes leading to final fracture (*1*, *2*, *3*, *4*).

While it is still rare to find formal documentation and procedures which emphasize integration of the subspecialties of fracture control, attention to each subspecialty has been increasing, and some integration is guaranteed by the negotiations and trade-offs needed to satisfy the subspecialists. Thus, fracture control, whether explicit or implicit, governs design stress levels, stress concentrations, welding procedures, toughness, and crack growth resistance. Fracture control may also provide for redundance or "fail-safety" to maintain the safety of a structure in the event of the fracture of a part. The philosophy may be simply described as:

(a) Prevent cracks when possible.
(b) Contain or tolerate growth of those cracks not prevented.

(c) Contain a fracture within a part or tolerate the loss of the part if a crack should grow critically.

Fracture control practices can be implemented as part of four major activities in the life cycle of a marine structure. First, material selection and quality control are aimed at the prevention of fracture and fatigue due to substandard material properties. Second, design provides the structure with resistance to crack growth and tolerance of damage (*5*). Third, construction is controlled and inspected to minimize substandard fabricated details (especially welds), initial defect sizes, and detrimental residual stresses (*6*). And fourth, operation and inspection* are carried out to maintain the integrity of the structure. Each of these activities as related to fracture control are examined in the sections to follow. Although the examples and illustrations are oriented toward offshore tubular-type structures, the principles presented apply to all steel marine structures (*7*).

II. The Scope of the Fracture Problem

A. Sources of Crack Initiation

Cracks or crack-like defects may initiate during the construction or operation of a marine structure. The first source of crack initiation encountered in the life of any structure is a defect in the original material. In steel plate such a defect might be a porous region, or a non-metal inclusion or lamination. In good practice the largest and most significant of these defects are detected and rejected before the plate is used.

There are several opportunities for cracks to initiate from the welding process. Poor weldability of the materials or poor welding technique can leave large crack-like defects in the weld. Certain joint configurations lead to heavy restraint of the welds, which results in high residual stresses and, sometimes, cracking as the welds cool and shrink. The worst examples of this type of cracking are normally caught in inspections during fabrication and can be prevented by using the proper preheat and other welding procedures.

Improved and special welding procedures can also alleviate problems with material embrittled by welding. Such material is susceptible to brittle fracture when loaded. Also, the use of material with special through-thickness ductility in critical locations can reduce the chance of fracture by lamellar tearing and of brittle crack extension in the rolling-plane of structural details subjected to significant out-of-plane loads (*8*).

* Inspection may be treated as a separate activity but, because most inspection occurs before and during the operation phase, is considered here as a part of that activity.

Finally, during construction, there are many opportunities to overload a joint. For instance, assembly may require the coordinated effort of several cranes. Mispositioning a crane, not balancing the loads correctly between cranes, or sudden impact loading could lead to joint overload and, hence, crack initiation. Cracks may also initiate in several ways during the operation of a vessel or structure. Impact damage is probably the most common cause. A collision or a dropped object are examples of this. In these cases, the operator usually knows when (and perhaps where) to check for crack initiation.

More subtle sources of crack initiation are corrosion and fatigue. Since these processes occur slowly and their cracks evolve over a period of time, continued or periodic surveillance is required to find these cracks.

Cracks due to overloads may be suspected after a severe environmental loading such as a storm or an earthquake. However, cracks may initiate but not be anticipated if the ordinary loading is not properly considered in design, i.e., if the structure is underdesigned. In the first case, overload, the operator usually knows where cracks may initiate and should find them easily. In the second case, underdesign, the operator probably doesn't expect cracks to initiate, so they might not be discovered until they become large enough to be detected by routine inspection or make themselves known by causing problems.

B. *Typical Examples*

With good welding practice and proper controls, cracks and crack-like defects large enough to degrade the structure are usually prevented or caught and rejected during the welding process. However, there are classes of defects that are known to exist in a welded joint but are allowed to remain. A small defect in the weld root is one example. When found, the defect and the joint are considered for their "fitness-for-purpose;" that is, they are evaluated to see whether, even with the defect, the joint will still accomplish its intended purpose. Thus, these cracks are not considered to be problems.

Lamellar tearing was a serious problem about ten years ago. The extra-thick plate used in North Sea platforms was particularly sensitive to this problem. Today, when conditions of heavy weld restraint and through-thickness loading occur, a special plate material is used, with high through-thickness (or z-direction) ductility, which resists lamellar tearing and in-plane brittle crack extension.

Crack indications are sometimes picked up with underwater ultrasonic testing only a few years after the structure begins operation. In the cases where the indications correspond to real cracks, many questions are asked.

The first always is, when did the crack occur? Sometimes the cracks initiated during fabrication but were not found; possibly those welds were not inspected thoroughly at that time. In other cases, it is suspected that they initiated from an overload or from fatigue. The operator's concerns are more acute in the case where no significant crack existed at the time of construction and the crack has suddenly appeared in a few years. For this serious situation of structural degradation with time, the operator must determine whether the structure is underdesigned or whether the crack is due to another problem and then must estimate how much time is available before remedial measures (inspection, repair, or replacement) are needed.

Corrosion can be a subtle problem; however, it doesn't always occur slowly. An improper or inadequate ground for an offshore welding operation can turn the underwater welds into electrolytic anodes, resulting in rapid, highly detectable, corrosion of the weld metal. The damage may be corrosion pitting or more severe knife-edge slices (crevice corrosion) into the weld.

Finally, it is known that some of the first offshore platforms installed in the North Sea have had problems with fatigue. In early North Sea designs fatigue was not explicitly analyzed, and the severe environment (wave load spectrum) for fatigue led to cracking. Another problem for later platforms has occurred with the horizontal framing supporting the well conductors at the first level below the surface. In those designs, vertical wave forces probably were underestimated and repeated joint overloads also led to fatigue cracking, even though fatigue was considered.

C. *Perspective*

It appears that there are two basic types of fracture problems. One type occurs because of poor workmanship or direct human error. Examples of this are significant crack initiations due to poor welding technique, construction overloads, and falling objects. Fracture control is achieved in these cases by preventing the errors from occurring.

The other type of problem relates to new technology or new frontier areas. In these cases, as experience is gained the problem diminishes in later designs. Examples are the weldability of new materials, lamellar tearing in thick plates, and fatigue in the North Sea. Experience has shown that the first generation of platforms to be installed in a frontier area has the most problems, the second has fewer, and by the third generation most of the problems have been worked out. Fracture control is best achieved in these cases by first being aware of the unique characteristics and demands of a technology or area and then by rapidly gaining experience and applying appropriate measures.

III. Material Selection

A. Philosophy

A popular philosophy in material selection is to assure that the material will behave at least as well as assumed in design calculations. For example, in tubular joints of offshore structures this means that the material must be able to accommodate large amounts of plastic deformation without fracture.

There is an important trade-off to be considered in material selection. To decrease the weight of the structure, a material with a higher yield stress, i.e., a stronger material, will often be chosen. There is, however, often an inverse correlation between the strength and the fracture resistance, or toughness, of a material. That is, a stronger steel is usually less tough, especially in proportion to its strength. Therefore, the trade-off between strength and toughness should be carefully considered. One way to avoid the strength-toughness dilemma is to bear the expense of more costly steels for which both properties are adequate. This option creates a more complex trade-off among strength, toughness, and cost.

B. Implementation

1. Desired Properties

General properties such as yield strength, ultimate strength, and ductility are routinely considered. In special cases where tensile stresses are expected through the plate thickness, through-thickness ductility may be needed to prevent lamellar tearing.

To limit defects which might initiate fracture, tolerance levels for porosity, inclusions, and laminations in rolled plate are needed. Since the presence of material defects is controlled by the steel manufacturing process, the process itself may need to be specified. The chemical composition of the steel is also routinely controlled to assure the material's key mechanical properties and weldability. Carbon-equivalent is the most important relevant measure for weldability and some other key properties.

Brittle fracture, initiated by a crack or crack-like defect, is a frequently catastrophic failure mode in structural steels, initiated by a crack or crack-like defect that occurs suddenly with little or no warning (*9*). High toughness, which is closely related to ductility, is the key property needed to prevent brittle fracture. For example, material selection is typically based on the fracture analysis diagram (FAD), of which the most important element is the nil-ductility transition temperature (NDT). The NDT represents the temperature below which fracture is almost entirely brittle and the probability of ductile failure is negligible. The FAD plots the nominal dynamic stress required to propagate a given flaw size to failure in a Naval Research Lab (NRL) drop-weight test plate, as a function of the test temperature, which is calibrated against the NDT (see ASTM Standard E208-69) (*10*). This test, which dynamically bends a sharply-notched plate, closely simulates the strains and strain rates of a dynamic fracture initiation at the highest-stressed locations (hot spots) of welds in tubular joints. Cracks are always initiated dynamically during this test; a specimen passes the test only if the crack is arrested before it can break the plate. A family of S-shaped, stress-versus-temperature curves is plotted for different initial flaw sizes. Based on these data, if the material is to withstand moderate flaws at stresses well above yield, the NDT of the material must be at least 45°F below the design temperature. The NDTs for materials in other applications can be similarly determined. Thus, to avoid brittle fracture, according to this philosophy, the NDT of the materials should be below temperatures less than, and defined in terms of, minimum operational temperatures.

As the state of knowledge in fracture mechanics increases, other toughness measures are increasingly used. These include fracture toughness (K_{IC}) or its nonlinear counterpart (J_{IC}), Charpy V-notch (CVN), and crack tip opening displacement (COD) (*11*).

2. Specifications

Standard mill tests are used to determine material defect levels, carbon-equivalent, strength, and ductility. The specification of brittle fracture properties such as toughness is less standardized. Traditional linear elastic fracture mechanics (LEFM or K_{IC}) fracture toughness testing methods, while of interest in establishing conservative bounds, are typically of little or no value as specifications. This is because most current practices dictate that, in critical structures, if K_{IC} can be determined using thicknesses less than or equal to structural details, brittle fracture resistance is already too low (*12*). Given the usually unsatisfactory ability of K_{IC} tests to assure desired toughness levels, the typical practice is to specify notch toughness criteria for either NRL drop-weight tests or Charpy V-notch absorbed energy impact tests. Charpy energies thought to be slightly above the lower shelf of the impact energy-versus-temperature curves (i.e., the energy corresponding to the NDT) are selected. Thus, the Charpy tests and drop-weight tests are intended to assure that the NDT of the material is below the necessary temperature.

Another way of specifying toughness is through the crack tip opening displacement, or COD (also CTOD), of the material (*13*). This is a fracture mechanics-based measure of the amount of plastic strain absorbed at a crack tip in a ductile material before fracture occurs under static loading. Properly related to local stress and strain, the COD is suitable for design calculations. This test is more useful than the Charpy test for establishing quantitative relationships among loads, geometry, material properties, and crack size. It is, however, more expensive and employs larger specimens than the Charpy test so that its use is normally restricted to material qualification, structural certification, and defect evaluation rather than quality control. COD is the accepted measure of fracture toughness for British practice in fitness-for-purpose evaluation of weld defects in ductile steels (*14*). Its use is currently making its way into American practice, especially for fitness-for-purpose specifications and structural and defect evaluations.

C. *Toughness Testing*

Quality control and assurance for material selection is accomplished by testing samples of the material to be used in fabrication (*15*). There are standard tests, in particular those specified by the American Society for Testing and Materials (ASTM), that cover composition, toughness, and strength (*11*). Toughness testing is most closely related to fracture control.

Toughness values are used for quality assurance, as part of fracture control plans, and for detailed assessment calculations. However, there is considerable debate as to the best test to use for these varying purposes. There are four basic types of toughness tests used for marine structures:

(a) The Charpy V-notch (CVN) impact test
(b) The drop-weight test (DWT), or the closely related dynamic tear test
(c) The crack tip opening displacement (CTOD or COD) test
(d) Fracture mechanics tests to measure critical stress intensity factors (K_C or K_{IC}) or critical values of the J-integral (J_C or J_{IC}). Since COD results can be used (usually) more effectively in fracture mechanics analyses, and since current trends dictate that marine materials should be tough enough to invalidate K_{IC} tests of specimens taken from the offshore structure, little description is provided below for K_{IC} and J_{IC} testing.

As described below, these tests vary in a number of ways: cost and difficulty, whether they measure crack initiation toughness or crack propagation toughness or both, and whether they are static or dynamic.

1. Charpy Test

This test is the simplest, most familiar, and most widely used toughness test. It is covered by an ASTM standard, A370-77. A small (10 mm square in

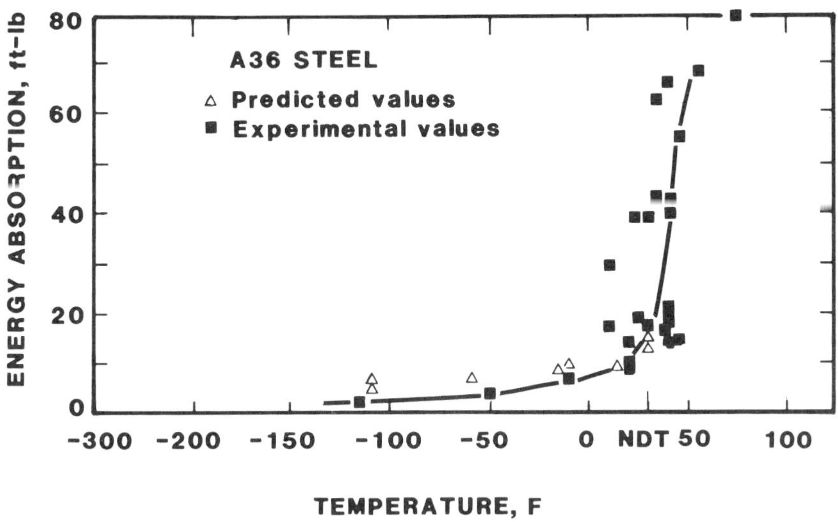

Fig. 1. Experimental data and predicted values of Charpy V-notch energy absorption for impact loading of standard CVN specimens.

cross section) notched specimen is broken dynamically by a swinging pendulum. The energy to break the specimen is recorded. This CVN energy (CVE) is the toughness parameter most commonly extracted from the test results, although the fracture appearance may also be used as a parameter. The CVN energy increases with test temperature from the lower shelf, through the transition region, up to the upper shelf at higher temperatures. A plot of energy versus temperature is shown in Figure 1.

The CVN test is essentially qualitative and the results cannot be directly related quantitatively to allowable stresses in the structure without extensive data demonstrating the correlation between CVE and such parameters as K_{IC}. Its use is based on satisfactory experience with materials that meet specified minimum standards for energy absorbed in the test. Fractures were never observed in Liberty ship plates during World War II when the CVN energy of the steel was 15 ft-lb or greater. This 15 ft-lb criterion is still commonly, but not exclusively, used. However, it may provide an inadequate guarantee against fracture for some steels or for higher design stress levels.

The advantages of the CVN test are that the specimen is cheap and simple to fabricate and test, and that it is universally familiar. It is widely used for quality control purposes. The disadvantages of the CVN test include the fact that the rate of loading and the thickness of the specimen are usually not very similar to those experienced by the structure. The results of the test cannot be directly related to allowable stresses in the structure. Further, the energy absorbed includes both initiation energy and energy to propagate a crack through the specimen. It has been argued that this is a drawback on the grounds that the most important toughness property for hull steel is resistance to and arrest of dynamic crack propagation, since cracks will in any event initiate at weld flaws. In this argument (which conservatively assumes that such crack initiation can occur dynamically), the fracture control function of the base plate is to arrest dynamic crack propagation, not to prevent dynamic crack initiation. Hence, according to this argument, the test used should measure resistance to a dynamically propagating crack. This is the "fracture-safe" philosophy. The DWT test described below is such a test.

2. Drop-Weight Test

The DWT was developed at the Naval Research Laboratory by Pellini and coworkers (see, for example, NRL report 6957, ref. 12). Crack initiation energy is reduced to low levels by using a brittle starter weld (or, in the closely related dynamic tear test, by using a very sharp pressed notch). Thus the test measures, predominantly, propagation and arrest energies. The specimen is considerably larger than the Charpy specimen (ranging from $5/8 \times 2 \times 5$ inches up to $1 \times 3.5 \times 14$ inches). This is important since apparent material toughness generally decreases as specimen size increases (all other factors being equal). The DWT test uses material thicknesses more representative of those used in offshore structures than does the Charpy test. The test is covered by ASTM Standard E208-69 (*10*).

The main parameter extracted from the DWT test is the nil-ductility transition (NDT) temperature. This is the maximum temperature where the initial flaw propagates to at least one edge of the plate, at the plate surface, when the nominal stress in the surface is at yield. For temperatures higher than NDT, the crack will be arrested even at yield stress levels. The material is chosen so that the service temperature will always be above NDT. Any crack in a similar, monotonically-decreasing bending stress gradient should therefore be arrested. Those in the offshore industry who use this test consider that it is more representative of tubular joints—in terms of flexural loading, presence of a notched weld, limited yielding, realistic plate thickness, and lower strain rate—than is the Charpy test. Furthermore, they feel that it is a more realistic representation of the most dangerous failure mode in the structure, where a crack may be initiated in a weld and not arrested in the plate.

The DWT has the disadvantages that the specimen is more costly to fabricate and test than the Charpy specimen. Correlations have been formulated between Charpy test results and NDT which allow the CVN test to be used instead of the DWT to measure NDT, although at some cost in accuracy. DWT test results can be approximately related to allowable stress levels in the structure using the Fracture Analysis Diagram. The correlation is both qualitative and approximate, however.

3. Crack Tip Opening Displacement

A further type of test which has been proposed to replace the CVN test is the crack tip opening displacement (CTOD or COD) test (see British Standard 5762, ref. 13). COD can be measured when the crack initiates in the specimen or when the specimen reaches maximum load. Measurements made at or close to crack initiation are used in the design philosophy where reliance is placed on the static initiation barrier. COD measurements are popular in Europe, particularly Britain, and are rapidly gaining popularity in the American offshore industry. The COD test is normally a "static" test, i.e., it usually measures resistance to initiation or propagation under slow loading conditions. This is in marked contrast to the CVN and DWT tests, which measure values only under dynamic, impact loading conditions. COD values in this context are used as "specification tests," i.e., a minimum allowable value is specified. However, as noted below, in a different type of fracture control approach, COD values can be linked quantitatively and directly to allowable stresses and crack sizes in the structure (although with some degree of uncertainty), unlike the results of CVN and DWT tests.

The British standard for this test calls for a notched three-point bending specimen to be slowly loaded until it fractures. The measured opening of the notch can be translated into the opening at the tip of a crack when fracture occurs. American standards for this test are being developed. Among the difficulties with interpreting the test results are: Where is the crack tip? Does "fracture" mean "fracture initiation" or "final fracture?" How are differences in stress gradient between specimen and structure best accounted for?

4. Failure Assessment Diagram (FASD) Procedure

The Failure Assessment Diagram Procedure (for which the published acronym is usually FAD, which unfortunately matches that of the Fracture Analysis Diagram Method) is rapidly gaining acceptance as a practical competitor or supplement to such quantitative elastic-plastic fracture methods as the COD tests. The FASD procedure gains much of its practicality and generality from the fact that it addresses empirically two failure modes simultaneously and can be used in conjunction with almost any elastic-plastic fracture test ranging from ordinary unnotched ultimate-strength tensile specimens to COD specimens.

The FASD procedure, like the COD procedure, is considered to be a state-of-the-art technique for conservative evaluation of weld defects and has been formally validated in Supplement I to a Central Electricity Generating Board (CEGB) report by Harrison et al. (*16*). Figures 2 and 3 have been reproduced from Supplement I. These figures document the key FASD format and the validation results generated during the development of the FASD procedure.

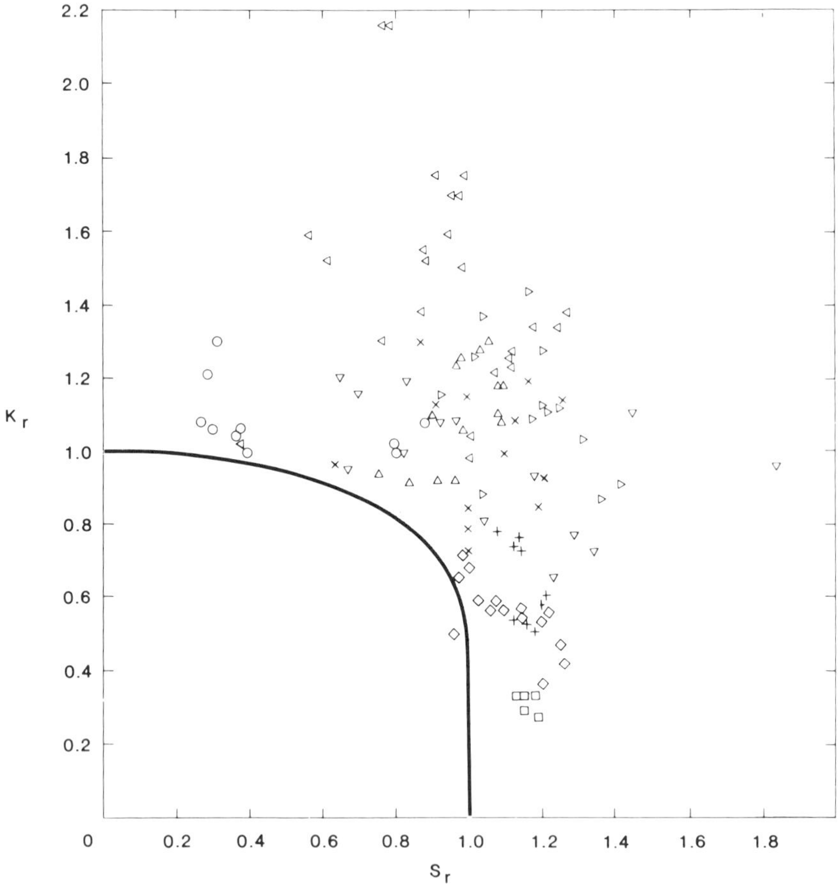

Fig. 2. Failure assessment diagram compared to test specimen failures (symbols distinguish nine various test groups). [Figure A from CEGB *Assessment of the Integrity of Structures Containing Defects*, Supplement 1.]

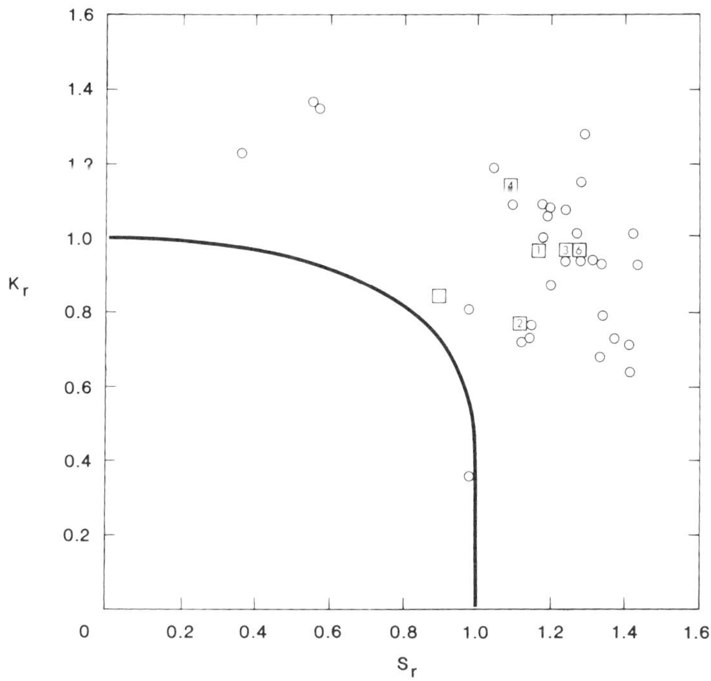

Fig. 3. Failure assessment diagram compared to vessel failures (symbols distinguish various test groups). [Figure B from CEGB *Assessment of the Integrity of Structures Containing Defects*, Supplement 1.]

The procedure consists of evaluating the ratio S_r of the applied (σ) to plastic-collapse ($\bar{\sigma}$) stresses and the effective ratio K_r of applied (K) to critical (K_Q) crack toughness under the elastic-plastic conditions being evaluated. The inherent practicality, generality, and safety factors in the FASD procedure come from: (1) the use of conservative characterizations of both applied loadings and material properties and (2) the conservative techniques used by Harrison et al. to envelop their large experimental data base, which includes a wide variety of different types of fracture specimens and full-scale (pressure vessel) structural simulators.

Note that one data point in each figure falls slightly within the assessment line. According to the supplement authors, "These points are considered unreliable for reasons discussed in the test. It is concluded that these figures confirm that the Failure Assessment Diagram Procedure... provides an appropriate limit line for the avoidance of failure in ferritic structures."

IV. Design

A. Philosophy

Traditionally, the designer is responsible for guarding against a structure's possible failure modes as well as producing a near-optimum functional and economic design. Before the occurrence of the previously mentioned Liberty ship failures, only two general types of failure modes were considered in typical design efforts: yielding (excessive plastic deformation) or ultimate failure under generally tensile or bending loads, and elastic or plastic buckling or other instabilities under compression loads. With the advent of higher strength steels under higher operational stresses, other failure modes have become increasingly important (*17*). Fracture, defined as unstable rapid crack extension—either elastic or plastic—leading to partial or complete failure of the member, is often considered. Subcritical crack growth under fatigue, environmentally accelerated fatigue, stress corrosion leading to loss of section and fracture, and bulk corrosion are all important failure modes in marine structures.

Fracture control is considered on two levels in the design process: local and global. One major concern at the local level is the design of tubular joints and other welded structures (Figure 4). Their detailed design for strength and sufficient toughness implicitly controls fracture by preventing fracture initiation. The most explicit consideration of fracture control in design is usually in the fatigue design of the joints (*18*, *19*, *20*).

On the global level, the concern is with the structure's tolerance to fracture, either due to fatigue or overloading. Redundancy and design approaches for limiting the extent of crack propagation are the primary global techniques for avoiding catastrophic failure in the presence of a fracture.

There are two main concepts regarding inspection of the structure during its operation. First, if a tubular joint can tolerate a large, easy-to-find crack without significantly degrading its performance, then the inspection procedures do not have to be designed to locate small, harder-to-find cracks. Second, if the structure can tolerate a seriously damaged joint, then, again, the inspection procedures can be more relaxed. These points will be discussed both in this section and in the section on operation and inspection.

B. Implementation

Structural design concerns two aspects: the loadings placed on the structure by the environment and by its operation, and the resistance of the structure to these loadings. On the loading side of the equation, fracture

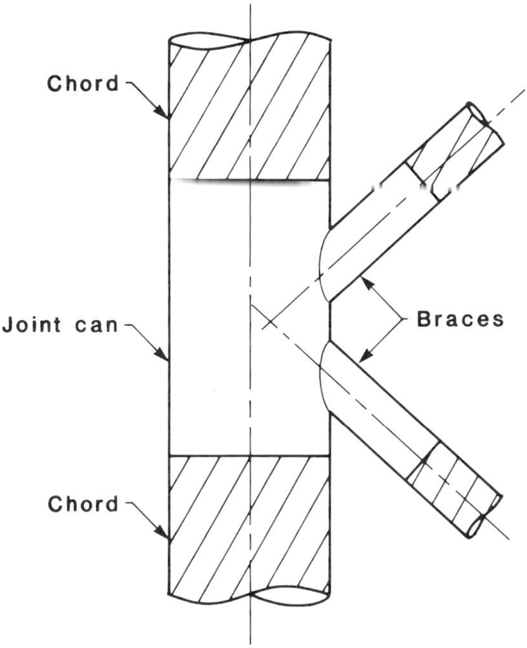

Fig. 4. Simple tubular joint. Note: The unshaded area is a node.

control is implicit in the design for normal operating conditions, extreme environmental conditions (e.g., storm, earthquake, and ice), and forces due to installation (e.g., load out, transport, and launch). Fracture control is explicitly considered in the design for fatigue loadings, which are often dominated by the many repetitions of wave loadings.

On the resistance side of the equation, fracture control is implicit in tubular joint design for strength, and explicit in tubular joint design for fatigue. For the most part, when considering the structure's failure modes, fracture control is the main goal; however, many designers do not think of it as such in that they are simply meeting design codes (*1*), requirements, and constraints. These design rules are depended upon to have enough implicit safety factors and "forgiveness" to preclude serious failures.

1. Allowable Stress Fatigue Design

A very simple fatigue design procedure that is an extension of the implicit fatigue control by safety factors on static stress is sometimes used for some of the simpler offshore applications. The procedure is to analyze the design extreme environmental loading (wind, wave, and current) and limit the maximum stress range (maximum stress ecountered as the wave passes through the structure minus the minimum stress) to predetermined allowable values (*21*, *22*). The allowable stress range can be specified in a region of nominal stress or, alternatively, as a larger value in regions of stress concentration (hot spots).

Therefore, using this procedure, only the effects of the design wave need be analyzed. The entire spectrum of different wave heights occurring throughout the life of the structure need not be analyzed. The justification for this is that a certain long-term distribution of wave heights is implied by the analysis.

The allowable stress range is based on prior generic analyses of typical structures in the same environmental conditions and on a particular S-N curve (*23*). That is, complete fatigue analyses have been made, developing the full distributions of stress cycle occurrences. The fatigue damage due to these stress distributions and a particular S-N curve has been calculated in each case. The design allowable stress range that gives the desired fatigue life (service life times a factor of safety) has been back-calculated and specified for design (*24*).

2. Fatigue Analysis for Wave Loadings

For structures that do not have an experience base to justify the allowable stress approach, especially if the dynamic response is thought to be an important factor, a full fatigue analysis is needed (*25*). Fatigue analysis techniques are typically based on the same approaches, S-N curves and Miner's rule.

The S-N curves (stress versus number of stress cycles to failure) are usually based on constant amplitude load cycle tests in the laboratory. The S-N curve is plotted for tests covering a wide range of stresses. For design purposes, an S-N curve which either bounds the test data or is the statistical mean curve minus two standard deviations is typically used. The statistical scatter in tubular joint fatigue test data is wide.

Fatigue tests can be performed on standard specimens or on details representative of the application. Great care must be exercised in applying the data in a manner consistent with how the data was acquired. For example, to use data from standard (smooth) laboratory specimens in design, the stress concentrations caused by discontinuities in the design must be accounted for in calculating the applied stress. Practice is evolving toward using data from tests that contain details representative of the application. Even so, since no measurement of the precise peak stress can be made, the data must be calibrated to hot spot stresses or nominal stresses.

Since constant amplitude cyclic stresses rarely occur in applications, a methodology is required for using data from constant amplitude tests for variable amplitude applications. Applying the often-used Miner's rule to such accumulated damage, the damage due to n cycles of stress, for which N cycles would cause failure, is defined as n/N. The damage due to cycles of various stress amplitudes is the sum of the damage due to each amplitude independently, e.g., $n_i/N_i = n_1/N_1 + n_2/N_2 + n_3/N_3 + \ldots$. Thus it is assumed, when applying Miner's rule, that fatigue failure occurs when the summation reaches 1, which implies that there is no interaction between the cycles of stress of different amplitudes.

Loading History

Wave loading is the fundamental concern in fatigue calculations (*26*). In rare cases, the effects of current and wind are also considered. There are two basic ways to describe the long-term wave environment over the service life of the structure.

The first description is simply to count the expected number of waves of various heights (an assumed wave period is associated with each wave height). For example, of 1000 total waves, 600 are of heights from 0 to 3 feet, 300 from 3 to 6 feet, 99 from 6 to 9 feet, and 1 greater than 9 feet. The loading history is thus divided into wave-height groups or blocks.

The second description counts the number of occurrences of different sea states, rather than individual waves. A sea state is typically described by two parameters, the significant wave height and the mean zero crossing period. The significant wave height is the average height of the highest one third of the waves in a sea state. The mean zero crossing period is a measure of the average wave period; more precisely, it is the average time between the occurrences of the water surface rising above mean sea level (zero) at a point. If certain assumptions are made concerning the statistical distribution of wave heights, the significant wave height specifies the entire spectrum. Typically, the environment is assumed to remain in the same sea state for about three hours. The number of occurrences of each sea state is then marked on a scatter diagram, which has significant height and mean period as its x- and y-axes. For example, there may be ten occurrences, i.e., 30 hours, of the sea state having a significant height of 4 feet and a mean period of 10 seconds. This is a more detailed description of the wave environment than the wave height count. In fact, the wave height count can be derived from the scatter diagram of sea states, but the reverse is impossible.

Refinements on the above are possible. With either description the directionality or heading of the waves can be considered by using a different wave height count or scatter diagram for each wave heading. More sophisticated descriptions, including the spreading of wave directions (in a sea) about an average general heading, or including the combination of waves arriving from a distant storm with locally generated waves, are possible with the sea state scatter diagram.

Structural Response

Even though wave loading is dynamic, in many cases it occurs slowly enough, compared to the structure's first dynamic mode period, that the dynamic effects of structural deformation may be neglected in an analysis.

For fixed platforms, the structure's response is calculated by static analysis at several points in time as the wave is stepped through, assuming that the load applied at that instant is the only load acting. That is, the inertial effects are neglected. For floating structures the dynamic response of the vessel as a rigid body is often calculated and combined with the static response.

If structural dynamic response is important, there are several methods of performing dynamic analysis. The most basic is to pass a series of identical waves through the structure and integrate the structure's response in time until a steady state is reached. Another possibility is to decompose the wave load into its Fourier series components, again assuming identical waves, and to directly solve for the steady state response. The two methods should give theoretically identical results. The response of structures under random, non-identical waves is possible to compute; however, it has been only rarely used in design. Another refinement is to consider the effect of the relative velocity between the water and the moving structure on the loading.

Stress Analysis

Using a static or a dynamic analysis, the calculation of stresses is fairly standard. The structure is analyzed using finite element methods. From this analysis the nominal stresses at numerous locations in the structure are calculated and searched for locations of highest stress.

The next level of refinement is to compute hot spot stresses, especially at welds. The hot spot stress is defined as the stress that would be measured by a strain gage in a laboratory situation. Since strain gages have finite size, it is impossible to place them and measure surface stresses at the precise point where fatigue cracks initiate in tubular joints, i.e., at the weld toe. Typically, they are placed within 0.25 inch of the weld toe. Note, however, that the stresses vary extremely rapidly in this region due to the notch effects of the weld. So variability in the experimental measurements would be expected. Also note that the stress being measured may not be the maximum principal stress, but the (weld-crack-driving) stress perpendicular to the weld toe.

The reason this stress is used is because S-N curves for design-specific details are typically derived this way. That is, the stress (or strain) used in the S-N curve is not the maximum principal stress at the crack, but the stress perpendicular to the crack a finite distance away. Thus, (if the in-service structure/geometry is similar enough to that used to produce the S-N curve) the next refinement in stress analysis, finding the stress at the weld toe or crack itself, is not necessary. The notch effects of the weld are already included in the S-N curve.

Finite element techniques are fairly well established for calculating hot spot stresses beginning with nominal stresses. The primary difficulty with using them is generating the element mesh for the rather complex geometries at discontinuities such as joints in tubular structures. Generally, thin shell or plate elements are used. Stresses from thin shell element models need to be corrected to get the "true" stress that would be measured on the surface of the plate at the weld toe or other discontinuity. For this reason, isoparametric thick shell and solid elements (with full 3-D elasticity capabilities) are sometimes used.

Another accepted technique for hot spot stress analysis of tubular joints is to use parametric stress concentration factor (SCF) equations. The SCF so calculated, times the nominal brace stress, give the hot spot stress. There are numerous sets of such equations. One of the most popular was derived by Kuang et al. (*21*). These equations are based on regression (curve fitting) of the results of numerous thin shell finite element analyses on various joint geometries and loadings (other analytical or experimental results could be fit as well). These equations are often used for comparing SCFs computed by other parametric equations or from finite element analyses. So Kuang's equations are, in a sense, an industry benchmark.

Another popular set of parametric equations was derived by Wordsworth and Smedley (*22*). These investigators measured hot spot strains in acrylic models of tubular joints. They have had success with the cheap acrylic material and were even able to model the weld reinforcement. Large discrepancies have been found to exist between the Kuang and Wordsworth and Smedley equations for certain conditions. Apparently the same is currently true for any pair of the numerous sets of parametric equations that have evolved over the years. The computation of hot spot stresses, which ideally should be a straightforward exercise in stress analysis and presentation of results, is still viewed by many as a black art.

Stress Cycle Counting

With either a static or a dynamic analysis there are two methods of counting stress cycles. The simple method is known as the discrete wave or the deterministic method. This method is used when the wave climate is described by wave height blocks. The stresses due to a representative wave from each block are analyzed. The maximum stress range calculated is then used to calculate damage in the S-N curve. The number of cycles of that stress range is assumed to be the same as the number of waves in the block.

When using the discrete wave method, it is important to use enough wave blocks to adequately define the peaks and valleys of the structure's response, which occur due to both geometric effects on the loading and dynamic effects on the response. The wave picked to represent each block should also be chosen carefully so that is is truly representative, not only of the range of wave heights in the block, but also the range of wave periods.

The above method is also known as the deterministic method because the number of stress cycles is known in advance, only the stresses need to be calculated from predetermined waves. The second method is known as the probabilistic method or the spectral method. In this method the stress ranges and the number of cycles are estimated using probability-theory-based equations. The method relies heavily on linear systems analysis techniques in the frequency domain.

A sea state description of the wave climate, such as significant height and mean period, is required. Given these parameters the sea state spectrum can be estimated using the Pierson-Moskowitz formula or some other empirical formula. This spectrum is actually a power spectral density function, which is one way of representing a random process such as wave amplitude or wave-induced stress. The sea state spectrum is thus the input to a linear system, the structure. For fatigue analysis a transfer function relates wave heights or forces to hot spot stresses; thus, the output is the hot spot stress spectrum. From the stress spectrum, the number of stress cycles and their distribution can be calculated probabilistically for each sea state. This calculation is repeated for each sea state, and the fatigue damage as measured by the S-N curve is accumulated for all sea states. Note that since more than one stress cycle per wave can occur, the number of stress cycles calculated by this method is generally greater than the number of waves, as assumed in the discrete wave method.

Most of the computational effort in the spectral method is to comput the transfer functions. A transfer function describes the response of the system to a unit harmonic loading over a range of frequencies. In this case, the response is the hot spot stress amplitude due to waves of unit amplitude over the relevant range of frequencies. Determination of the transfer function thus theoretically requires running many waves of unit amplitude through the structure at different frequencies (and, if under consideration, directions) and calculating the hot spot stress for each wave. About 12 frequencies are usually enough for each direction. Thus, if eight directions are considered, then actually 96 wave analyses are needed for the fatigue analysis, versus the only eight or 16 wave analyses used for strength design. If the wave analysis must be dynamic, the cost of the fatigue analysis skyrockets.

The wave force, and thus the hot spot stress response, is nonlinear with respect to wave height. So the wave analysis is not performed with unit amplitude waves, but with waves of larger amplitude chosen to give the correct average response. The stresses due to the larger waves are then normalized back to unit amplitude wave responses. This technique is known as equivalent linearization. Approximations that are introduced by linear analysis techniques are not always insignificant.

Fatigue Lives

For other than smooth specimens, there are two basic types of S-N curves. One type is based on hot spot stresses. Since the hot spot is away from the weld, the S-N curve must account for the stress riser at the weld.

The other type of S-N curve is the generic curve. To use these curves, nominal brace stress range is used instead of hot spot stresses. As with the allowable stress fatigue design method, this type S-N curve assumes certain values for the stress concentration factors. European research indicates a degrading fatigue performance in thick plates which can be represented by special S-N curves. Similarly, most S-N curves are based on tubular joint tests in air for a structure. Left unprotected in a seawater environment, the fatigue performance would be expected to degrade and special S-N curves would be required. However, some continuing research in this area suggests that with proper cathodic protection the fatigue lives of submerged joints approach those of joints in air.

All the variations of fatigue analysis have as their goal the calculation of fatigue damage using an S-N curve, and the interpretation of that damage in terms of fatigue life (*27*, *28*, *29*). Recall that the total fatigue damage is the sum of the damages due to each stress cycle range occurring during the life of the structure, $D_T = n_1/N_1 + n_2/N_2 + n_3/N_3 + \ldots$ Nominally, the total damage must be less than 1.0.

If D_1 is the average damage occurring during one year, its reciprocal, $1/D_1$, is the expected fatigue life. Obviously, the expected fatigue life of the structures must be at least the intended service life. Due to the many uncertainties involved, it is desirable to add a safety factor (SF) to obtain the design fatigue life. For example, with a service life of 15 years and SF equal to 2.0, the design life should be 30 years.

3. Global Fracture Control

The last line of defense a structure has against cracks is its ability to withstand their effects without a catastrophic failure. There are two components to this defense: the reserve strength of the structure against such a failure and the redundancy of the system.

The reserve strength of the structure is, in turn, due to two related reasons. The first is that individual members and connections normally do not fail brittlely once their design strength is reached. Due to the safety factors built into design formulas, the failure load of an average part is well above the design load. When the "failure" load is reached, failure of tubular members and connections is normally ductile. That is, plastic deformation redistributes loads throughout the cross section rapidly enough to forestall catastrophic rupture and, therefore, some load carrying capacity remains.

The second mechanism for structural reserve strength is that a structural detail is rarely stressed uniformly. Usually, before the first part or cross section reaches its ultimate load and fails, it redistributes some load to adjacent parts or cross sections, again through plastic deformation or displacement control.

The second line of defense is the fail-safe redundancy of the structural system. Here, redundancy is used as a synonym for fail-safety, rather than as the mere existence of multiple load paths as in a statically indeterminate structure. That is, a redundant structure in the present context will be able to survive the failure of a major structural member.

IV. Construction

A. Philosophy

In terms of fracture control, the main goal in construction is to limit built-in stresses, the size of initial defects, and large variations from specified material properties and structural dimensions. To this end, assembly and welding procedures are designed to minimize restraint and residual stresses, and inspections are performed to eliminate critical defects. Care is taken during construction of the structure to avoid accidental overloads that might cause cracking. Additionally, of course, great care is taken to assemble the parts of the structure correctly, so that it behaves as it was designed to behave.

B. Implementation

In construction, fracture control is primarily concerned with quality assurance of incoming material and the welding of parts together. There is general agreement that it is difficult to overemphasize the importance of well-designed and fabricated welds to fracture control. Welding involves controls on how welds are made, the inspection of these welds, the rejection or acceptance of weld defects, and the repair of defective welds (*30*, *31*, *32*).

1. Welding

Preparations are made for the weld to assure weldability and fusion. The base metal is preheated if necessary, and welding rods may be kept dry in special ovens (especially important for low hydrogen rods in humid fabrication yards). The fit-up of the parts is watched closely, especially to control the weld root opening.

The weld profile should merge smoothly with the adjoining base material, without excessive undercut, etc. The weld has the effect of a notch on the stress distribution through the plate thickness, possibly causing a large stress riser at the weld toe. In fatigue sensitive joints the weld profile may be "improved" to decrease this notch effect. Butter passes and grinding of the weld to control the profile are two methods of improvement. Given a nonagressive environment, grinding can substantially improve the fatigue performance of weld toes.

Once the weld is complete, it may be left in the as-welded condition or subjected to treatments to relieve residual stresses. The need for such treatment is influenced by material selection, welding practices, and part thicknesses. Post-weld heat treatment is a common technique that is difficult for large structures. Some work has also been done with grinding or peening the weld to reduce residual stresses.

2. Fabrication Defects

In terms of fracture control there are two primary types of defects of concern: welding discontinuities and gross errors. The gross errors are macroscopic errors such as member misalignment or insufficient gaps between members in a joint. Such errors can lead to congested welds, undue residual stresses, and cracking. The structural engineer is usually consulted in these cases to assess the severity of the problem.

The principal welding discontinuities found in welded joints are: inadequate root penetration, incomplete fusion, undercut, slag inclusions, porosity, and cracking. These crack-like defects and cracks could compromise the strength of the weld, lead to brittle fracture, or initiate a fatigue crack. All of these discontinuities exist in welds to some extent. Because of inspection system limitations, not all such defects are found by inspections, and those that are found may not be accurately characterized (e.g., sized). Of equal concern is the problem of defining the significance of cracks that are found and determining the disposition of a weld containing such a crack.

The decision to accept a weld with discontinuities or to reject it is based on the severity of the defect. In the past, strict limits have been placed on the size of acceptable defects, such as no larger than 1/8 inch depth × 1 inch surface length. There is currently a trend away from such absolute standards and toward acceptance criteria based on the defect's impact on the fitness of the joint to serve the purpose for which it was intended; this trend is popularly known as "fitness for purpose," "fitness for service," or "engineering critical assessment" (*33*).

Current fitness-for-purpose evaluations are based on fracture mechanics models of the behavior of crack-like defects in the welded joint. The defect is checked against brittle fracture initiation and other possible modes of failure, under the design and other critical loadings. If the modeled weld can withstand tests for failure under single extreme loads, the growth of the flaw in fatigue (multiple loading) is also considered to see if the critical crack size could be reached in the design life of the structure. If the flaw does not adversely affect the joint's fitness-for-purpose, it is acceptable. Otherwise, the weld is rejected and repaired.

The propagation of the crack is considered using simple crack growth laws and the expected stress cycle history, which may be available from the design fatigue analysis. A two-dimensional model is likely to be used, possibly with conservative assumptions to simplify the crack growth model or analysis. For instance, it is known that the displacement-limited, three-dimensional stress redistribution occurring around the crack is beneficial in slowing down the growth rate. However, experiments show that, in many cases, once the crack goes through-thickness the growth rate rapidly increases, leaving only a little additional fatigue life of a joint. Assuming the two-dimensional solution calculates faster crack growth rates, the modeled crack would conservatively reach the critical crack size sooner. Computer programs have been implemented to solve these problems automatically.

In addition to brittle fracture initiation, other possible modes of failure may be considered to determine the critical crack size. In tubular joints, yielding due to overloading of the remaining cross section, plastic instability, or leakage from a through-thickness crack might be considered. Even arbitrary limits might be chosen, such as a certain fraction of the part thickness, since a through-thickness crack is often far more costly to repair than a part-through crack.

Should the weld be judged unfit, either by a fitness-for-purpose evaluation or by an absolute standard, it must be repaired (*34*). A weld repair consists of removing the defect from the weld, rewelding the joint, and reinspecting the work.

A defect is usually ground out under the close scrutiny of a team of inspectors. Their purpose is to assure that the defect "indication" found in the inspection corresponds to the actual size of the defect in the weld. Thus, defect removal is a very slow tedious process; only a thin layer of material is removed at one time, and then the exposed surface is inspected for the predicted defect. This serves as an important calibration of the inspection process, especially if embedded flaws are indicated by ultrasonic testing.

After the weld defect has been removed, rewelding proceeds according to the steps described for the original weld. However, more attention to following the correct procedures, such as preheating, etc., can be expected. The resulting weld should be as good as an original weld and, because of the extra attention, might even be above average. Inspection of this weld and other welds is described below.

C. *Inspection*

In the construction phase, the most effective inspection scheme is to prevent the introduction of defects into a weld, rather than to find them after they occur. To this end extensive qualification tests are required for both welders and welding procedures (*34*). Inspection of welds in complex geometries is difficult and special care is needed. Also, when defective welds are found, it is important to investigate the possibility of systematic error.

1. Welding Qualifications

Welding qualifications are designed to prevent large weld defects from occurring in the first place. Since special skill is required to make certain welds, welders with the necessary skill must be identified. The AWS D1.1 describes different skill levels to which a welder may be qualified (*30*). These skill levels are set by the type of weld (full or partial penetration, groove, butt, fillet, etc.), the position of welding (flat, horizontal, vertical, overhead, etc.), and the welding process necessary to the job. As in pipeline construction, special welders may be designated to do the root welds.

The welding procedures themselves must also be qualified (*30*). For example, the fabricator might propose to make the root weld using hand-held welding rods and fill in the body of the weld using a semiautomatic procedure. Since different procedures result in different heat inputs, cooling rates, chemistries, and microstructures, the toughness of the weld should be tested. The test pieces should be representative of the actual plate thicknesses and diameters to be welded in the structure.

2. Inspection of Welds

The inspection of production welds is vitally important (*35*). Field inspections are of three kinds: observation of the welding process, visual examination of the weld, and nondestructive examination (NDE) of the weld.

The welding process is observed to assure that qualified welders and procedures are used. Sometimes, different colored hard hats are issued to welders of different qualification levels to make it easy to see that a qualified welder is working. Critical welds are watched carefully.

Upon completion, the weld is inspected visually to check its profile and see that it merges smoothly with the adjoining base material without excessive undercut. Pocket gages and magnifying glasses may be useful for this purpose. A visual examination may detect small surface cracks, such as in the weld or at its toe, or larger defects such as cratering. For fillet welds, a visual examination is likely to be the only inspection. For other welds, some sort of NDE is likely.

Nondestructive examination of some tubular joints is very difficult. This may be due to complex geometry or to the fact that the weld is only accessible from the outside of the joint. Under these conditions, the radiographic inspections (gamma ray or X-ray) that are often used for simpler more accessible joints are impossible. In these circumstances the burden of NDE inspection falls upon two techniques, ultrasonic and magnetic particle. A third technique using dye penetrants is sometimes used when one of the above cannot be performed.

For example, ultrasonic testing has become the most important inspection technique for tubular joints. In this test, the ultrasonic echoes of flaws are measured electronically by skilled technicians. Since the interpretation of the test results is difficult, especially for tubular joints, the API has published the *Recommended Practice for Ultrasonic Examination of Offshore Structural Fabrication and Guidelines for Qualification of Ultrasonic Technicans*, API RP-2X (*34*). The test results are best for flat plates, and least conclusive for the acute-angle brace-to-chord weld areas in K- and Y-joints. Because no single test is conclusive by itself, ultrasonic testing is often used in conjunction with another NDE technique.

For tubular joints the second major technique is magnetic particle inspection. Here an electric current is passed through the metal, creating a magnetic field. Flaws in the metal alter the current flow and distort the magnetic field; this distortion can be seen in magnetic particles dusted on the surface. One drawback to this technique is that it becomes less effective for detecting flaws the deeper the flaws are found in the plate. Typical inspection requirements call for 100% of tubular joint welds to be both ultrasonically tested and magnetic particle inspected.

In some instances, dye penetrants may be used in lieu of one of the above methods: for example, when the weld is inaccessible to the testing equipment. In this method, a dye is washed over the metal and penetrates surface cracks. Dye left in the cracks can be detected visually. However, dye penetrants cannot detect subsurface defects.

Besides tubular joint welds, pipe seam and girth welds must also be inspected. Here radiographic methods can be used (usually gamma radiation because of the equipment's portability). Typical inspection requirements call for these welds to be 100% radiographed and/or ultrasonically tested. Miscellaneous welds are inspected by a variety of methods.

Treatment of Defective Welds

In common practice, when a defective weld is found and rejected, the possibility of systematic error would be checked first. If not already done, the entire weld around the joint would be inspected to see if the problem is inherent in the joint. For example, the fit up may have been poor or the welding process may be at fault. The welder himself may be at fault, so his work on other joints would be examined as well. The quality of a welder's work remains nearly constant, so if a defect is found in one of his welds, other defects are likely to exist. If too many defects are found in a welder's work, he would be disqualified (at least until retraining) for that type of welding.

VI. Operation

A. Philosophy

The primary goal of fracture control during operation is to minimize the risk of operating the structure in the marine environment. Two conditions of the structure must be considered: routine conditions, when the structure exists as it was designed to exist, and damaged conditions, when cracks (or other damage) may be present. Under design conditions, fracture control is concerned with protecting the structure from damage, either due to accidents (e.g., boat collisions) or due to normal operating conditions (e.g., corrosion). In damaged conditions of the structure, fracture control is concerned with preserving the integrity of the structure: Cracks must be found and their significance evaluated before they are dangerous and, if necessary, repairs must be made.

To these ends there are three important trade-offs to consider. First, the inspection level (type and amount) chosen for periodic inspections may be high or low. The higher the inspection level, the more it will cost and the more likely cracks will be found while still small. But for redundant "fail-safe" structures, small cracks may not be dangerous, and low level inspections aimed at finding large cracks or even ruptured members may be more cost effective. Second, there are trade-offs between (1) doing a repair on the crack and (2) doing no repair on the crack, but changing the routine operation of the structure. Remembering that the goal is to minimize the risk of operation, a repair, in certain circumstances, may prove to be the riskiest option an operator has, and the choice of avoiding the risk by changing the routine operating mode may be best. The third trade-off is between the several types of repair that might be done. Besides the different repair procedures available, the operator can often choose between short-term and long-term fixes depending on his needs (*36*).

B. *Implementation*

Fracture control practices during operation can be divided into several tasks: Minimize the formation of cracks, find and monitor any cracks that may start, assess the crack's impact on the integrity of the structure, and take the necessary remedial steps.

1. Routine Operation

The fracture control aspects of routine operation are mostly concerned with minimizing high stresses, cracks, and other damage. Cracks can initiate during operation due to accidents or improper maintenance. Collisions and falling pieces of equipment are common examples of accidents initiating cracks. These can be avoided by closely observing the operating procedures written for the structure or vessel. Improper maintenance of the cathodic protection system can lead to corrosion damage, including crack initiations

(*37*). Improper or inadequate maintenance of appurtenances, such as cathodic protection sleeves, can lead to these objects detaching from their supports, falling through the structure, and causing impact damage. Again such damage can be avoided by closely following the operating procedures.

2. Evaluation of Cracks

Suppose that a crack has been found in the structure. The concern is now with the evaluation of the crack's impact on structural integrity and with remedial measures (*35*). There are four steps in the evaluation of a crack's impact on structural integrity. The size and location of the crack must be determined. The effect of the crack on the joint or structural member must be studied, as must the effect of the damaged element on the structure. Finally, a course of remedial action must be decided upon. A fifth step, which is often included in such evaluations, is to determine the cause of cracking, if not already known.

The size and location of the crack are of primary importance (*36*). The length of a surface crack can be determined with a photographic survey; however, crack depth cannot be measured from photographs. Nondestructive evaluation (NDE) techniques have been applied underwater (*32*). As is the case in construction inspections, ultrasonic testing has gained in popularity. With ultrasonic methods the crack depth can also be estimated. A recent innovation in NDE is the use of plastic molds to get an impression of the crack. Under certain circumstances, especially if tensile stress helps "open" the crack, both surface length and depth are measurable from the impression.

The assessment of the crack's effect on the joint or the structural member in which it has initiated is also known as a fitness-for-purpose analysis. The analysis steps are the same as for fabrication defects. The potential of the crack for brittle fracture may be evaluated by a fracture mechanics analysis (*38*). The fatigue life may also be estimated by a fracture mechanics analysis. However, note that the critical crack size might be determined by criteria other than brittle fracture; for example, a through-thickness crack may be judged very costly to repair and, therefore, critical. Also, the effect of a crack on the ultimate strength of some complex structures such as tubular joints is not yet understood. So the critical crack size for a fatigue life analysis may be merely a definition or conjecture for the purpose of performing the analysis. However, if the growth rate has accelerated rapidly at the defined size of a critical crack, the fatigue life—comprising mainly crack initiation and slow growth—while small, may be a very weak function of the critical crack size definition.

For redundant structures the analysis of the effect on the structure usually begins with the assumption that the joint or the member has separated. A relevant measure of the diminished strength of the redundant structure is often the ratio of the diminished ultimate strength to the as-designed ultimate strength or some other relative measure. In some circumstances a new fatigue life analysis for the "damaged" structure is in order.

Once the effect of the crack on the joint or structural member and the effect of the damaged element on the structure have been assessed, a decision must be made. Should the crack be repaired? If yes, then how should it be repaired? It is obvious that leaving the crack in the structure without repairs may add risk to the operation of the structure. That risk should be understood from the above analyses. It is less obvious that repairs may also add risk, which is more difficult to quantify. The decision to repair and the choice of repairs must therefore consider not only the cost of each choice, but also the risk of each. As commonly practiced, however, this decision is often more intutitive than analytical—"engineering judgement" being the vehicle for the decision analysis.

A failure analysis is sometimes included as a fifth step in the evaluation of cracks; that is, a determination is made of the cause of the crack, if not already known. Such an analysis could identify generic problems with the structure and thus predict the locations of other possible cracks. This step is especially important if the crack appears in an unexpected spot. In this context, the appearance of a crack serves as an indicator of the structure's performance and calibrates the fracture potential for other spots in the structure.

3. Repair of Cracks

The repair of a crack underwater is extremely expensive, potentially one or two orders of magnitude more expensive than the comparable repair done on dry land. There is great incentive for the operator to avoid such repairs when possible, either by preventing the cracks from starting (i.e., through design, material selection, etc.) or by justifying the operation of the structure without repairing the cracks (*39*). This incentive is greater for fixed platform operators than for the operators of mobile vessels, because the mobile vessels can be regularly drydocked, or at least elevated in calm shallow water for repairs. In such cases repairs of cracks that occur during operation are often similar to repairs of cracks discovered during construction.

When a repair must be made, the repair procedure chosen depends on many factors. First, is the size of the crack small or large? If cracks are small enough, they are often repaired by simply grinding them out with a large radius, without rewelding. A functional definition of a small crack might thus be "one that can be repaired by grinding." A large crack is generally considered to be anything bigger than a through-thickness crack. Large through-thickness cracks are significantly more costly to repair than small, part-through cracks. Other factors include the cost of the repair, the time required to do the work, the diving conditions, the weather window, and the confidence that the repair can be made properly.

Fixed offshore structures represent special problems in weld repair, since virtually all repairs must be made at the site of operations. Underwater rewelding is not unusual, but special precautions are required. Given other options, this is a less favorable repair because of the increased difficulty. Underwater rewelding is difficult to do, in part, because of the possibility of rapid quenching by the surrounding water. Special precautions are required to prevent low quality and brittleness. Among the special precautions available is the opportunity to apply the underwater wet or wet-backed weld in such a way as to produce less geometric stress concentration and residual stress.

A more reliable repair requires that the entire problem joint (or node) be removed from the structure and raised out of the water. (This is not always feasible, especially if a leg joint is involved, and is not a normal operation.) The crack itself is then rewelded and inspected while dry. The repaired node is placed back into the structure to complete the repair. The underwater welds to reconnect the node are circumferential welds, which are easier to do than the tubular joint weld, and are far enough away from the highly stressed region so as not to be a problem themselves. However, the brace material toughness is often lower than that of the node material, necessitating perhaps more control of nodal rewelds than would otherwise be required. Of course, a high price is paid for the higher degree of confidence, not only in terms of money and time, but also in risk. While the node is out, the structure is left in a more precarious position than with the cracked joint in. Also, due in part to weather variability the risk of off-shore repair operations is proportional to their duration: The longer the operation, the riskier it is.

A recent development in tubular joint repair is the use of grouted clamps (*40*). These clamps are prefabricated onshore to fit around the brace and chord of a cracked joint with good clearance between the inside of the clamp and the outside of the tubulars. The clamp is taken underwater in two pieces, which are bolted together around the cracked joint. Then the annulus between the clamp and tubulars is filled with high strength grout to complete the operation. This procedure was originally developed for North Sea repairs, where the diving conditions are poor and the weather window is short, making the speed of the repair operation most important. The immediate risk of this repair of a joint is consequently lower, and the confidence higher than node removal, repair, and replacement.

The grouted clamp repair is an advancement over the use of friction clamps, which are similar, but fabricated to snugly fit the tubulars without an annulus. These clamps are held in place by steel-on-steel friction. The advantage of grouted clamps over friction clamps is that they can be fabricated with much looser tolerances with respect to the geometry of the problem joint. Friction clamps are difficult to fit up due to out-of-roundness of the tubulars, etc. The use of clamps is still relatively new and research continues on long-term performance of clamped joints, how a clamp affects a joint's stiffness, and how the clamped joint can be inspected.

"Repair" does not always mean that the crack itself is fixed. It may be that the structure can live with a small crack as long as it doesn't grow larger in fatigue. Then it is appropriate to "repair" the crack by removing the cause of the cracking or of crack extension. Typically, this means removing sources of high stress concentration or high loads. For example, gusset plates may have been used in the original design and may now be seen to do relatively more harm than good. Another example, a redundant brace which may bring too much load into a joint might be removed without any problem, allowing the load to flow down another fail-safe load path.

Finally, some distinction should be made between temporary and permanent repairs. Should the failure analysis reveal a severe fatigue problem, rewelding the crack can only be a temporary solution. A permanent solution must also mitigate the cause of the cracking. Thus, the permanent repair may require, for example, that braces be added to reduce the cyclic stresses. An immediate temporary repair is often adopted to get the structure through a season of rough weather, to be followed a year or more later by the permanent solution.

4. Non-routine Operation

Crack repairs are not the only remedial measures available to the operator. Non-routine operation, that is, changes to routine operations, may be chosen instead of repair or to supplement repair. In light of the high risks associated with offshore work and underwater repair (e.g., to divers), such options may occasionally be the best way to minimize risk. Even doing nothing at all might be the best course of action in some cases.

Removing the hazard associated with the operation of the cracked structure is a good way to reduce the risk. For example, oil storage might be removed from a cracked platform, thereby reducing the pollution hazard associated with platform failure. If danger to the operating personnel is an increased threat during storms, the platform might be de-manned more frequently (i.e., for smaller storms) than would be the case during routine operations. The hazard posed to any nearby platforms by the cracked platform must be a part of these considerations.

When the crack is small or otherwise tolerable, another remedial measure is to inspect the structure more closely and monitor the growth of the crack, rather than take any of the above actions. This approach can be especially valuable if the failure analysis is inconclusive. Examining the crack metallurgically and fractographically can also help to determine whether or not it is growing as the result of fatigue or whether it is a stable defect introduced sometime during construction. In this way, the crack may serve as another measure of the performance of the rest of the structure.

C. *Inspection and Monitoring*

The emphasis of this discussion is not on the "how to" aspects of inspection, but rather on the integration of inspection with fracture control plans. The backbone of an inspection program is the periodic inspection of the structure specifically to find cracks. Inspections to check the condition of the structure may be conducted on special occasions when warranted. Also, the structure might be continuously monitored to check its general health. Given the uncertainty connected with many inspections, it is good policy to combine several independent inspection types which can be used to check each other.

1. Periodic Inspections

Periodic inspections are considered the most important means for the operator to review the condition of a structure.

There are two different fracture control philosophies for these inspections. One philosophy approaches the inspections with a "fine-toothed comb," while the other uses a "broad brush." Both have independent merits and may provide beneficial synergism if combined.

The "fine" approach is designed to find small cracks in critical joints before they can become dangerous. That is, the attempt is made to catch the cracks early in their development. These cracks can then be repaired by grinding or other minor repair methods.

In order to find "small" cracks, the surface of the joint being inspected must first be cleaned of marine growth (down to bright metal), usually with a water jet and wire brush. Then, according to some experts, cracks at least 100 mm ($\sim 4''$) long on the surface can be detected visually, and cracks at least 30 mm ($\sim 1''$) long can be detected reliably by nondestructive examination (NDE) (magnetic particle inspection is used for this in the North Sea).

Other practitioners are more optimistic or pessimistic on the crack size that is detected with high probability. Such examinations are extremely time consuming; therefore, only a few (20–30) joints can be examined during a typical annual inspection. The particular joints examined during each inspection may be rotated each time so that all critical joints are eventually examined, for example, every five years. The operator may decide which joints to inspect each time based on the importance of the joints, their fatigue lives, the convenience of their inspection, or other criteria.

The "fine" approach to inspection has been adopted by North Sea operators and regulators. It has the advantage of early warning, which makes the repairs less expensive. But, it also assumes that the operator knows which joints to inspect and when. This is a drawback.

The "broad" approach is designed to find relatively big cracks which are easily detected by visual examination alone. These cracks must generally be through-thickness cracks and are generally more costly to repair. The approach assumes that the joint has a great tolerance for cracks and that the structure is safe with a severely cracked joint. Since a visual examination can be made fairly quickly, as compared to other NDE techniques, the entire structure can be checked during every inspection period.

2. Special Inspections

Special inspections, in addition to the usual periodic inspections, may be warranted when there are indications that damage might have occurred. For example, an accident such as a collision may have occurred; or the structure may have experienced an earthquake, a mudslide, a particularly severe storm, or some other overloading; or a potential for cracking may be indicated by the detection of cracks somewhere else in the structure (by the periodic inspection) or by cracks found on another "similar" structure. When a special inspection is warranted, the inspection method and procedure used should be whatever is appropriate to find the kind of cracking suspected.

3. Monitoring

A number of structural monitoring techniques are currently being developed with the goal of providing a continuous review of the "health" of a structure, versus the periodic "snapshots" available today. Monitoring techniques also have the potential to reduce costly and risky diver activities.

Of the various techniques being studied, the most work has been done on ambient vibration methods. In these methods, the natural vibration modes are measured to detect any changes over time which might be due to a loss of structural stiffness resulting from cracks. One problem with detecting cracks in this way is that, until the joint separates, small changes in the general structure's stiffness must be measured as a crack grows. This difficulty is due to the same properties that protect the structure against cracks, i.e., tolerance of the joint to the cracks and structural redundancy. The small signal to be detected may be masked by equipment operating noise and vibration. So while ambient vibration monitoring is potentially a useful tool for checking the general health of a structure, the application of this technique to small crack detection is still in development.

A related technique is to measure the local vibration modes of the individual members, such as braces in offshore platforms. Each brace has its own characteristic "ring" when struck, and a change in its ringing will result when cracks degrade the stiffness of its end restraints, the joints. This is a relatively new technique which has not been studied as much as the more global ambient vibration method (some early tests indicate that equipment reliability is still a limiting factor of this technique). Because it checks one brace at a time, it is less sensitive to structural redundancy; also, it is still uncertain how large a crack would have to be to cause a confidently detectable change in the ringing.

A third technique is to measure acoustic emissions. When a crack grows, it makes noises (acoustic emissions) which are detectable with sensitive listening devices. The quality of the signal created—the rate and level of emissions—depends on many factors, including the type of material and the rate of crack growth. At the moment, the low signal-to-noise ratio is a major obstacle, even after filtering. As with the ambient and local vibration techniques, this one also struggles to overcome equipment reliability problems as well as interference from normal production and drilling noise.

An innovative monitoring technique which could reduce, if not eliminate, diver activities is the use of pre-cracked specimens directly welded to platform structures near critical hot spots (*41*). These specimen coupons may be constructed of the same material, including welds, as the monitored region and ideally should be attached to the structure so as to pick up the full stress and corrosive environment spectrum without affecting the fatigue performance of the structure to which it is attached.

The major difference then between the coupon and the structure is that the pre-crack is purposely set so as to be larger and much more likely to grow than the maximum weld defects expected in the structure. The reliability of monitoring equipment is another problem which must be overcome. In theory, through use of coupons of varying crack size, a very accurate reading of damage accumulation could be achieved. Practically, the technique has not yet been proven to be an effective means of monitoring.

Most other monitoring techniques being studied emphasize the limitation or removal of diver activities; for instance, divers might be replaced by remotely operated submersibles or remote sensing devices. However, from the fracture control perspective, all monitoring techniques share the same important aspects: the continuous (or almost continuous) scrutiny of the structure and the capability to survey the entire structure.

VII. Comparison of Current Practices

Although there are certain sound practices that are applicable to any marine structure, fracture control for each application must be tailored to some degree. In this section, the similarities and differences of two major practices are examined to demonstrate how the peculiarities of a particular situation should influence fracture control practices.

The development of offshore petroleum resources in the North Sea became an important influence on worldwide offshore practices during the 1970s. Since fatigue was such a major problem during the early development, there has been considerable interest in fracture control among North Sea operators. Much of the full-scale fracture-related research in the world is conducted by European laboratories. Through documentation, codification, and research publications, the North Sea standard of practice has become established as one of two main approaches to fracture control of fixed steel offshore structures, the other being the Gulf of Mexico standard of practice.

International offshore practices tend to fall then into two types: a Gulf of Mexico type and a North Sea type. The type of practice followed in a particular location is affected by the type of offshore environment (Gulf type versus North Sea type), who the operators are (American versus European), and the time at which offshore development began (before versus after the 1970s). For instance, in the Persian Gulf, the Gulf of Mexico type standard can be expected: The environment is mild, the operators have American ties, and development began before the 1970s. In the Tasman Sea (New Zealand), the North Sea type of standard can be expected: The environment is severe, the major operators are European, and development began in the 1970s. Local ties to classifying agencies (e.g., DNV, Lloyds, ABS) also influence the type of practice.

This section compares the Gulf of Mexico type of standard practice with the North Sea type for those aspects that affect fracture control. Other relevant aspects of worldwide practice not typical of the Gulf of Mexico or the North Sea will be mentioned where appropriate. The section is divided into three major parts. The first deals with differences in the offshore and operating environments. The second treats the current practices in the four major fracture control activities, and the third discusses the selection of fracture control plans for frontier areas.

A. Offshore and Operating Environments

As summarized in Figure 5, the environments—both the physical offshore environment and the financial and regulatory (philosophical) operating environment—have played major roles in establishing the different fracture control practices. The severe fatigue loading of the North Sea has already been mentioned. Other contributing factors in the North Sea offshore environments include the great water depths, the cold temperatures, the large distances from land, and the short weather window for offshore construction and inspection activities. In other offshore environments other hazards, such as earthquakes, mudslides, and ice loadings, may affect the fracture control plans.

The physical environment aside, there are two aspects to the operating environment that have affected North Sea fracture control practices. The first aspect is related to the risk of operations and the second to regulations.

First, the risk aspects: Compared to the Gulf of Mexico, where most of the thousands of offshore structures are in relatively shallow water (less than 300 feet), many North Sea structures are major deepwater installations. These structures are expensive to build and to replace. Thus, the economic consequences of a single platform failure, or even of a major repair, could be devastating to a North Sea operator. Also, compared to the Gulf of Mexico, the length of experience operating in the North Sea is much shorter. The uncertainties are therefore higher, increasing the probability of failure at a given level of loading. Thus, the total risk (cost of failure times probability of failure) could be much greater were not the North Sea operating companies extra conservative in their designs.

Second, the regulations: The conservatism of the operators is compounded by the conservatism of the North Sea regulators. In American waters, the Federal government requires verification of a platform's design, fabrication, and installation through the OCS Platform Verification Program (*35*). This program requires the platform to be designed and constructed to acceptable standards of practice; for instance, tubular joints are to be designed by a procedure comparable to that set forth in the API RP-2A recommendations (*1*). A third party reviews the design, verifying the design process. The operator is required to submit general information to the government, such as a summary of the fatigue analysis, and the government, through the

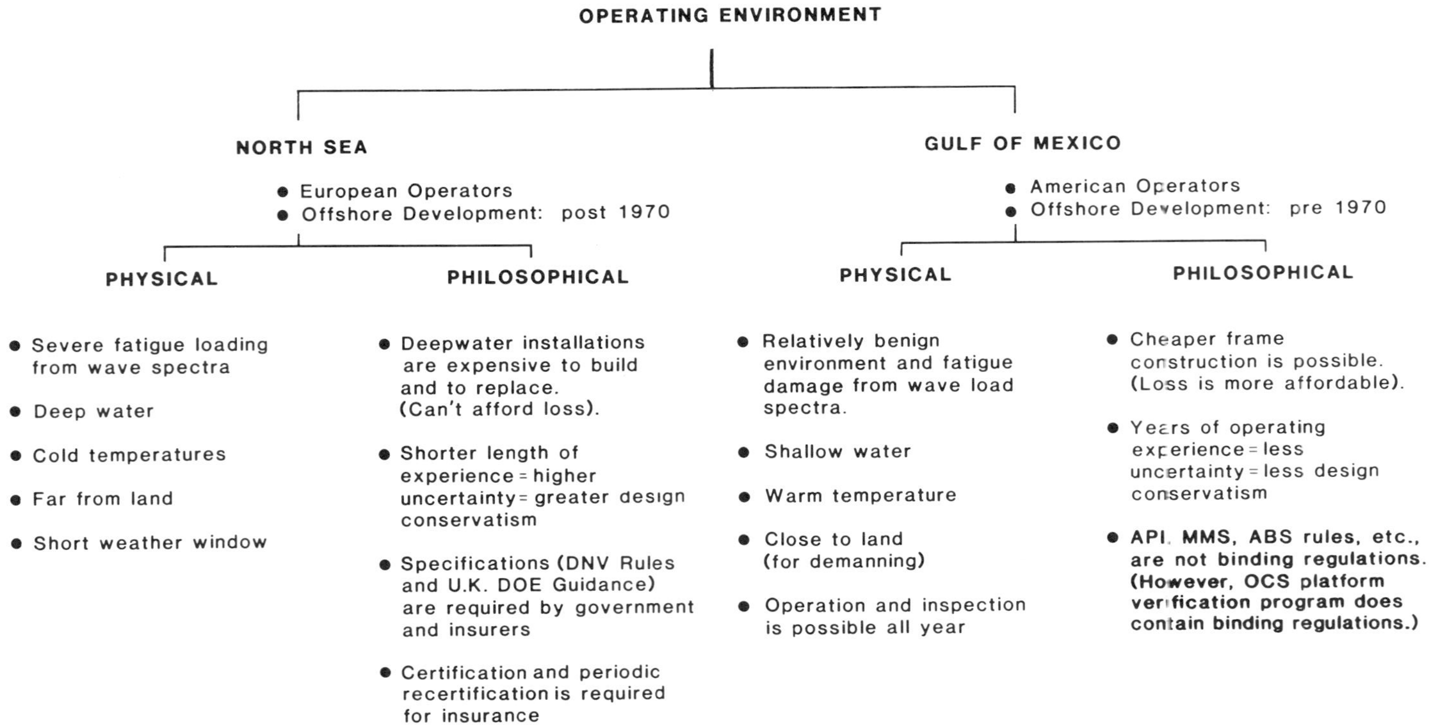

Fig. 5. Differences between the operating environments of fixed offshore platforms in the North Sea and the Gulf of Mexico.

Minerals Management Service, issues a drilling permit. The MMS requires periodic inspection of the structure during its life (*42*).

In comparison, in the North Sea adherence to the much stricter Det Norske Veritas's "Rules" (*2*) and/or the UK Department of Energy's "Guidance" (*3*) is required. Acceptable design practices for these waters are specified in much greater detail by these documents than by their American counterparts. Further, a certificate of fitness is issued by the government (i.e., either the UK or Norway) after the platform has been installed, which must be renewed periodically (every five years or sooner). Rigorous inspection requirements are at present made for the recertification, although this practice may be changing in favor of one more like that required by the MMS. Thus, government regulators and their agents play a much more active role in the building and operation of North Sea platforms than in the Gulf of Mexico platforms.

The extreme conditions of the offshore environment, the higher potential risk, and the strict regulatory environment have combined to help develop a more cautious approach to fracture control in the North Sea. The discussion continues next with the major differences between current Gulf of Mexico practices and current North Sea practices and describes how the environment has helped create these differences.

B. Major Differences in Fracture Control Practices

Figure 6 summarizes the differences in practices between the North Sea and the Gulf of Mexico platforms.

1. Material Selection

It was recognized early in North Sea development that the toughness of the materials used is important in controlling fracture. A low nil-ductility transition temperature (NDT) is especially important because of the low service temperatures experienced there. It is significant that the most severe storms occur in the North Sea when the water temperatures are the coldest, but in the Gulf of Mexico the hurricane season occurs during warm months. Thus, the most demand for toughness (for the North Sea) is placed on a material when its supply is lowest, while the reverse is true in the Gulf of Mexico.

European research and development of fracture toughness measures in ductile materials has resulted in the selection of materials for crack tip opening displacement (COD) properties. Recent platform designs for the

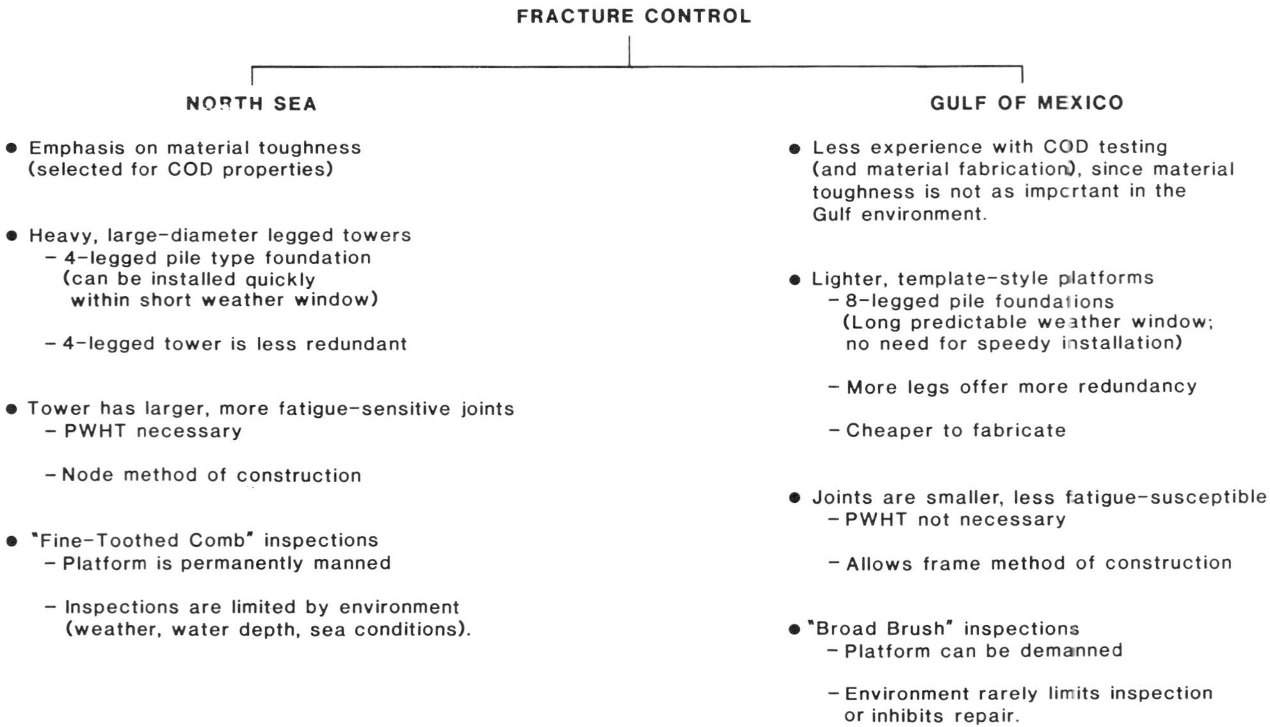

Fig. 6. Differences between current fracture control practices in the North Sea and the Gulf of Mexico.

North Sea have seen COD testing explicitly required by specifications and regulations.

To obtain these material properties on a regular basis, the close cooperation of the steel making industry has been required. European steel makers supplying North Sea fabricators have the necessary experience with COD testing to satisfy this requirement, as have the Japanese. American steel makers have had less experience in this, but are gaining. In contrast, some U.S. operators have reported difficulty in getting even simple Charpy impact tests when contracting with steel makers in some developing countries; assurance of toughness then must be attempted through other means—for example, through correlations with strength testing.

The widespread use of COD testing is one of the most explicit fracture control measures used in North Sea practice.

2. Design

Fixed steel structures designed for the North Sea tend to be heavier than their Gulf of Mexico counterparts. For platform designs, North Sea operating companies seem to favor large-diameter legged towers over the template designs found in the Gulf region. These differences affect fracture control through the redundancy of the structure and the size of the joints.

Foundation design is one of the main reasons towers are favored for North Sea environments. Due to the short weather window, installation time is limited. Pile groups clustered around the four corner legs can be installed in less time, because the driving equipment does not have to be relocated for each pile. This leads to a four-legged tower-type design and, in some cases, to self-floaters. Other considerations, such as fabrication costs, are also important.

The framing of a four-legged tower tends to be less redundant, because there are fewer braces, than the framing of an eight-legged Gulf of Mexico template. Thus, the loss of a single brace in the structure has a more severe effect on the tower's structural integrity. The tower is therefore inherently less tolerant of cracks.

The large diameters of a tower's legs necessitate large tubular joints. These joints are thickened further by the effects of dynamics and fatigue. The resulting joints are both large in diameter and in thickness. Detrimental size effects then come into play: The fatigue strength (measured in units of stress) of tubular joints decreases as their size increases. They are also more difficult to fabricate.

Fracture control is therefore affected by the typical design of North Sea platforms. Tower-type designs result in lower structural redundancy and, in combination with more fatigue content in wave load spectra, larger, more fatigue-sensitive joints. These structures are less tolerant of cracks than are the template type structures typical of the Gulf region.

3. Construction

The large joint sizes used in North Sea structures require post-weld heat treatment (PWHT). This requirement led to the development of the node method of fabrication, which involves attachment of the bracings to separately-fabricated nodes. From the fracture control viewpoint, this has several advantages. First, the node can be fabricated in a more controlled environment, assuring a higher quality of fabrication for the critical joints. Second, access to both sides of a tubular joint weld for welding and for inspection is possible, allowing better control over weld defects at the root. Third, special material can easily be used for the brace-end stub when extra fracture resistance is necessary. Fourth, the node method shifts the weld stress away from the node intersections.

Thus, the predominant use of the node method of fabrication, especially for critical joints, has several advantages in limiting initial defect size.

4. Operation

The inspection philosophy for the periodic inspections of North Sea structures is to find small cracks before they can become dangerous. The "fine-toothed comb" approach to inspections is followed, and, in fact, required by regulation.

There are two major operational reasons for this inspection philosophy. One is that the platforms are permanently manned and have much larger contingents than those in the Gulf of Mexico. Evacuation of the platforms for large storms is often impossible because of the short notice before a storm occurs and the large distances to land bases, so the consequences of structural failure are much higher than in the Gulf of Mexico. The second reason is that inspections are limited by the environment. The water depth limits the accessibility of critical joints to diver inspections, and the short weather window for diver activities limits the duration of those inspections.

Under these circumstances, it is logical that an inspection philosophy emphasizing a careful inspection of the joints most likely to develop cracks is used in the North Sea.

C. *Discussion*

North Sea practices have evolved in an environment that places severe demands on materials (crack resistance), limits inspection opportunities and precision, and necessitates the use of structural designs that are not particularly tolerant of cracks. In the face of devastating failure consequences, a very cautious attitude towards fracture control prevails. Gulf of Mexico practices, on the other hand, have evolved in a more benign environment, enabling the design of structures that are very tolerant of cracks in a region that produces little stress and lower failure consequences. Both practices have advantages and disadvantages, but it is important to remember that each is a product of its environment, physical and philosophical.

Perhaps the most exciting question in fracture control today is, what is the best practice for new frontiers? As mentioned before, the current international practice is to adopt one of the two main approaches, either in whole or with small modifications. This can clearly be inappropriate if the choice is not made wisely and could in fact be dangerous, especially if fracture control provisions are taken out of an important context.

For example, consider a region where earthquakes are an important consideration. If the fatigue environment is relatively benign, it may be tempting to follow the Gulf of Mexico practice. But earthquakes, unlike hurricanes, occur without warning, so evacuations are not possible. Therefore, a risk analysis would show the risk of platform failure to be more like that of the North Sea situation. The effect of earthquakes on fracture control plans depends on the individual platform's location and response.

The choice of which fracture control practice to follow or the design of a comprehensive fracture control plan must address all aspects of fracture control. Starting with the basics of fracture and fatigue, the fracture control plan should integrate the practices in material selection, design, construction, and operation. Ultimately, fracture control must be considered as a part of the entire risk of an offshore project; the risk of fracture must be weighed against all other risks.

References

1. American Petroleum Institute (1982). *API Recommended Practice for Planning, Designing, and Constructing Fixed Offshore Platforms.* API RP-2A, Thirteenth Edition, Dallas, Texas.

2. Det Norske Veritas (1974). *Rules for the Design, Construction and Inspection of Fixed Offshore Structures.* Oslo, Norway.

3. United Kingdom Department of Energy, Petroleum Engineering Division (1977). *Offshore Installations: Guidance on Design and Construction.* Her Majesty's Stationery Office, London.
4. Fisher, P. J. (1981). "Summary of Current Design and Fatigue Correlation." In: *Fatigue in Offshore Structural Steels*, Conference Proceedings, London, February 24–25, 1981.
5. McClelland, B. editor (1982). *The Design of Fixed Offshore Structures.* Van Nostrand Reinhold, New York.
6. American Bureau of Shipping (1982). *Rules for Building and Classing Offshore Installations.* ABS Special Committee on Offshore Installations, New York.
7. Marshall, P. W. (1979). "Fixed-Bottom, Pile-Supported, Steel Offshore Platforms." ASCE Convention, Atlanta, Georgia, October 1979.
8. Rolfe, S. T., and Barsom J. M. (1977). *Fracture and Fatigue Control in Structures.* Prentice Hall, Inc., New Jersey.
9. Dieter, G. E. (1976). "Brittle Fracture and Impact Testing." In: *Mechanical Metallurgy*, McGraw-Hill, New York, pp. 490–527.
10. American Society for Testing (1981). *Standard Method for Conducting Drop-Weight Test to Determine Nil-Ductility Transition Temperature of Ferritic Steels.* ASTM Standard E208-69, ASTM, Philadelphia, Pa.
11. American Society for Testing (1981). *Standard Methods and Definitions for Mechanical Testing of Steel Products.* ASTM Standards A370-77, ASTM, Philadelphia, Pa.
12. Pellini, W. S. (1969). *Evolution of Engineering Principles for Fracture-Safe Design of Steel Structures.* Naval Research Laboratory Report 6957.
13. British Standards Institution (1979). *Methods for Crack Opening Displacement (COD) Testing.* British Standard BS 5762, London, England.
14. British Standards Institution (1980). *Guidance on Some Methods for the Derivation of Acceptance Levels for Defects in Fusion Welded Joints.* BSI PD 6493:1980, London.
15. Carter, R. M., Marshall, P. W., Swanson, T. M., and Thomas, P. D. (1969). "Material Problems in Offshore Platforms." In: *Proceedings of the Offshore Technology Conference*, OTC 1043, Houston, Texas.
16. Harrison, R. P., Loosemore, K., and Milnay, I. (1977). *Assessment of the Integrity of Structures Containing Defects.* CEGB Report R/H/R6-Revision 1 and Supplement I to this report.
17. Marshall, P. W. (1977). *A Review of American Criteria for Tubular Structures—and Proposed Revisions.* IIW Doc. XV-405-77.
18. Marshall, P. W. (1974). *Basic Considerations for Tubular Joint Design in Offshore Construction.* WRC Bulletin 193.
19. Marshall, P. W., and Toprac, A. A. (1974). "Basis for Tubular Joint Design." *Welding Journal, Research Supplement*, May 1974.
20. Marshall, P. W. (1982). "Tubular Joint Design." In: *The Design of Fixed Offshore Structures* (B. McClelland, ed). Van Nostrand Reinhold, New York, Chapter 18.
21. Kuang, J. A., et al. (1977). "Stress Concentration in Tubular Joints." *SPE Journal*, August 1977.
22. Wordsworth, A. C., and Smedley, G. P. (1978). "Stress Concentrations at Unstiffened Tubular Joints." *Proceedings of the European Offshore Steels Research Seminar*, Welding Institute, Abington, Cambridge.
23. Rodabaugh, E. C. (1980). *Review of Data Relevant to the Design of Tubular Joints for Use in Fixed Offshore Platforms.* WRC Bulletin 256.
24. Vugts, J. H., and Kinra, R. K., "Probabilistic Fatigue Analysis of Fixed Offshore Structures." *Proceedings of the Offshore Technology Conference*, OTC 2608, Houston, Texas.
25. Kinra, R. K., and Marshall, P. W. (1979). "Fatigue Analysis of the Cognac Platform." *Proceedings of the Offshore Technology Conference*, OTC 3378, Houston, Texas.

26. Kan, D. K., and Petrauskas, C. (1981). "Hybrid Time-Frequency Domain Fatigue Analysis for Deepwater Platforms." *Proceedings of the Offshore Technology Conference*, OTC 3965, Houston, Texas.
27. Marshall, P. W. (1976). "Failure Modes for Offshore Structures Part II—Fatigue." *Methods of Structural Analysis*, ASCE Conference, Madison, Wisconsin.
28. Tomkins, B. (1982). "Fatigue Design Rules for Steel Welded Joints in Offshore Structures." *Proceedings of the Offshore Technology Conference*, OTC 4403, Houston, Texas.
29. Dover, W. D., and Dharmavasan, S. (1982). "Fatigue Fracture Mechanics Analysis of T and Y Joints." *Proceedings of the Offshore Technology Conference*, OTC 4404, Houston, Texas.
30. American Welding Society (1979). *Structural Welding Code—Steel.* AWS D1.1-79, Third Edition, Miami, Florida.
31. American Welding Society (1976). *Guide for Steel Hull Welding* AWS D3.5-76, Miami, Florida.
32. Marshall, P. W., Crick, H. F., and Marks, P. T. (1982). "Welding and Inspection." In: *The Design of Fixed Offshore Structures* (B. McClelland, ed). Van Nostrand Reinhold, New York, Chapter 27.
33. Marshall, P. W. (1982). "Experience-Based, Fitness-for-Purpose Ultrasonic Reject Criteria for Tubular Structures." AWS-WI Conference, Atlanta, Georgia, May 1982.
34. American Petroleum Institute (1980). *API Recommended Practice for Ultrasonic Examination of Offshore Structural Fabrication and Guidelines for Qualification of Ultrasonic Technicians.* API RP-2X, Dallas, Texas.
35. U.S. Geological Survey (1979). *Requirements for Verifying the Structural Integrity of OCS Platforms* (especially Section 9). U.S. Geological Survey Conservation Division—OCS Platform Verification Program—Order No. 8.
36. Marshall, P. W. (1979). "Strategy for Monitoring, Inspection, and Repair for Fixed Offshore Platforms." *Proceedings of the Conference on the Behavior of Offshore Structures* (BOSS '79), London.
37. Burnside, O. H. et al (1981). "Long-Term Corrosion Fatigue of Welded Marine Steels—First Year Assessment." U.S. Coast Guard Interim Report by Southwest Research Institute.
38. Institution of Civil Engineers (1981). *Fatigue in Offshore Structural Steels.* (J. R. Smith, H. O. Crisp, N. A. Randall, and F. Sterry, editors). Thomas Telford Ltd., London.
39. Marshall, P. W. (1982). "Fracture Control and Reliability for Welded Tubular Structures in the Ocean." Presented at the 9th NATCAM, Ithaca, New York.
40. Tebbett, I. E., and Robertson, D. A. (1981). "Novel Underwater Strengthening System for Tubular Joints." *Proceedings of the Offshore Technology Conference*, OTC 4110, Houston, Texas.
41. Wade, B. G., Grover, J. L., and Egan, G. R. (1979). "Development, Design, and Installation of In Situ Corrosion Fatigue Gage." Failure Analysis Associates for Shell Oil Company.
42. National Research Council, (1979). *Inspection of Oil and Gas Platforms and Risers.* Marine Board, Washington, D.C.

Index

V

W

Y

Z

Contents of Previous Volumes

VOLUME 5

VOLUME 6: PLASTIC DEFORMATION OF MATERIALS

VOLUME 7: MICROSTRUCTURES OF IRRADIATED MATERIALS

VOLUME 8

VOLUME 9 : CERAMIC FABRICATION PROCESSES

VOLUME 10: PROPERTIES OF SOLID POLYMERIC MATERIALS

Part A

Part B

VOLUME 14: METALLURGY OF SUPERCONDUCTING MATERIALS

VOLUME 15: NEUTRON SCATTERING

VOLUME 16: EROSION

VOLUME 17: GLASS II

VOLUME 18: ION IMPLANTATION

VOLUME 25: EMBRITTLEMENT OF ENGINEERING ALLOYS

VOLUME 26: GLASS IV

VOLUME 27: ANALYTICAL TECHNIQUES FOR THIN FILMS

VOLUME 28: MATERIALS FOR MARINE SYSTEMS AND STRUCTURES